Methodological Approaches to STEM Education Research Volume 3

Methodological Approaches to STEM Education Research Volume 3

Edited by

Peta J. White, Russell Tytler, Joseph Paul Ferguson and John Cripps Clark

Cambridge Scholars Publishing

Methodological Approaches to STEM Education Research Volume 3

Edited by Peta J. White, Russell Tytler, Joseph Paul Ferguson
and John Cripps Clark

Series: Contemporary Approaches to Research in STEM Education

This book first published 2022

Cambridge Scholars Publishing

Lady Stephenson Library, Newcastle upon Tyne, NE6 2PA, UK

British Library Cataloguing in Publication Data
A catalogue record for this book is available from the British Library

ISBN (10): 1-5275-8844-0
ISBN (13): 978-1-5275-8844-8

CONTENTS

CONTRIBUTORS

Teresa Beck
Teacher Training Center, Technical University Chemnitz, Germany, teresa.beck@zlb.tu-chemnitz.de

Robin Bellingham
School of Education, Deakin University, Australia, r.bellingham@deakin.edu.au

Amanda Berry
Faculty of Education, Monash University, Australia, amanda.berry@monash.edu

Jill P. Brown
School of Education, Deakin University, Australia, jill.brown@deakin.edu.au

Susan Caldis
School of Education, Macquarie University, Australia, susan.caldis@mq.edu.au

Coral Campbell
School of Education, Deakin University, Australia, coral.campbell@deakin.edu.au

Jared Carpendale
Institute of Education, Massey University, New Zealand, j.carpendale@massey.ac.nz

John Cripps Clark
School of Education, Deakin University, Australia, john.crippsclark@deakin.edu.au

Seamus Delaney
School of Education, Deakin University, Australia, s.delaney@deakin.edu.au

Urban Eriksson
Lund University, Sweden,
urban.eriksson@fysik.lu.se

Joseph Paul Ferguson
School of Education, Deakin University, Australia,
joe.ferguson@deakin.edu.au

Michael Fitzgerald
Las Cumbres Observatory, California, USA & School of Education,
Deakin University, Australia, psyfitz@gmail.com

Shelley Hannigan
School of Education, Deakin University, Australia,
shelley.hannigan@deakin.edu.au

Linda Hobbs
School of Education, Deakin University, Australia,
l.hobbs@deakin.edu.au

Paul Hart
University of Regina, Canada,
Paul.hart@uregina.ca

Kennedy Kam Ho Chan
Faculty of Education, The University of Hong Kong, Hong Kong,
kennedyckh@hku.hk

Leissa Kelly
School of Education, Deakin University, Australia,
leissa.kelly@gmail.com

Gahyoung Kim
Faculty of Education, Monash University, Australia,
Jinny.kim@monash.edu.au

Melinda Kirk
School of Education, Deakin University, Australia,
melinda.kirk@deakin.edu.au

Lucinda McKnight
School of Education, Deakin University, Australia,
l.mcknight@deakin.edu.au

Jo Raphael
School of Education, Deakin University, Australia,
jo.raphael@deakin.edu.au

Emily Rochette
Melbourne Graduate School of Education, The University of Melbourne, Australia, emily.rochette@unimelb.edu.au

Saeed Salimpour
Max Planck Institute for Astronomy, Germany & School of Education, Deakin University, Australia, astrophysics@saeedsalimpour.com

Carly Sawatzki
School of Education, Deakin University, Australia,
carly.sawatzki@deakin.edu.au

Madeleine Schultz
School of Life and Environmental Sciences, Deakin University, Australia,
madeleine.schultz@deakin.edu.au

Chris Speldewinde
School of Education, Deakin University, Australia,
christopher.speldewinde@deakin.edu.au

Bronwyn Sutton
School of Education, Deakin University, Australia,
bsutton@deakin.edu.au

Cassandra Tytler
Independent Scholar, cass@cassandratytler.com

Russell Tytler
School of Education, Deakin University, Australia,
russell.tytler@deakin.edu.au

Colleen Vale
Faculty of Education, Monash University, Australia,
colleen.vale@monash.edu

Jan van Driel
Melbourne Graduate School of Education, The University of Melbourne,
Australia, j.vandriel@unimelb.edu.au

R. Scott Webster
Independent Scholar, scott.webster63@gmail.com

Peta J. White
School of Education, Deakin University, Australia,
peta.white@deakin.edu.au

Wanty Widjaja
School of Education, Deakin University, Australia,
w.widjaja@deakin.edu.au

Lihua Xu
School of Education, Deakin University, Australia,
lihua.xu@deakin.edu.au

FOREWORD

URBAN ERIKSSON

LUND UNIVERSITY PHYSICS EDUCATION RESEARCH (LUPER) GROUP, SWEDEN

I started my academic career studying astronomy, physics, and mathematics at Lund University in the late 80s. However, and due to a strong interest in understanding how to learn science, I turned to become an upper-secondary teacher in these subjects. I loved challenging the students with the physics of everyday phenomena and the wonders of the universe! After some years teaching at adult education schools, I got a junior lecturer position at Kristianstad University in Sweden, teaching physics and astronomy to student teachers at all the different programs. Here, I first met science education research and I started to get an understanding as to why I felt successful in my own teaching; I seemed to use teaching techniques that were found effective by education research. I had fantastic colleagues that tried to drag me into science education research and I really tried at first. But, even though supported by such great people as Prof. Gustav Helldén, my efforts failed as I could not understand the language of science education; the methods and ways of working with data seemed incomprehensible to me at the time and I went to study astronomy instead. After half a PhD (a PhD in Sweden is four years full time) I was still so incredibly fascinated by the students' hurdles in understanding the universe. My colleague, Prof. Andreas Redfors, then turned my attention to a PER group at Uppsala University and I continued my studies there, now focusing on Astronomy Education Research (AER) under the supervision of Prof. Cedric Linder. This was the turning point in my career, and I was now ready to fully embrace the methods and techniques used in science education research. I did struggle at first but thanks to Cedric's efforts and infinite patience I came to understand the ways to work with different methods to collect and analyse data to be able to answer my research question. After a few years I earned my PhD, which is part astronomy and part education research. Eight years later I am now head of the LUPER group at Lund University, supervising PhD students, a postdoc, and Master's students, when I am not head of the Department for Educational Sciences at the same university.

One of my biggest problems at the time of my PhD was what methods to use. Now, supervising PhD students, I see them struggling with the same problems as I had. We have had so many good discussions on the methodological issues related to framing a research plan, including collecting and analysing data; this is understandable since our research field is young and diverse. There are so many different methods in the literature; in Sweden we have a saying: 'You do not see the forest for all the trees'. This is similar to what many novices in our field experience when starting to do research in Discipline-Based Education Research (DBER) in STEM or STEAM; one risks drowning in the sea of methods available. Moreover, when reading research papers, the methodology sections are often remarkably short and do not really give enough details on the challenges the authors have had while working on their projects and reporting on them in a paper. Many novice PhD students in PER, AER, or DBER, come from science disciplines and are thus familiar with the methodologies used there–*quantitative methods*–and have great difficulties accepting that one can do research using *qualitative methods*. I feel a strong empathy with my PhD students here as the transition is not easy.

During my career I have not really found any conference that takes a particular focus on methodological issues in DBER, until I came across the *Contemporary Approaches to Research in Mathematics, Science, Health & Environmental Education* symposia (CAR), organised by Deakin University since 1993. Unfortunately, it was not until 2020 that I became aware of this symposium, thanks to Prof. Russell Tytler, and I was so relieved that I finally found one! Here, I have found rare opportunities to discuss and refine methodological understandings from a broad spectrum of related science disciplines. However, for such a symposium to be successful, the organisers must create a safe and trusting environment, so that participants are willing to share their thoughts and be open to constructive scholarly critique. I have found that CAR symposia offer such a unique environment and allow presenters to open up for discussions on possible weaknesses and strengths in their presented projects, and participants engage in and critique the work in a very professional, yet friendly, manner, something I find unusual at many other conferences and symposia.

This year's symposium focuses on practical and theoretical aspects of research methodology and was designed to promote substantive discussions of methodological issues. This year also saw a broadening of the methodological focus to include new materialists and post qualitative inquiry.

The presentations for this year's symposia, and hence chapters in this book, held a very broad range of topics in the designated focus area. They stretch from methodological issues and considerations in thesis writing in cosmology education research (Salimpour et al. - chapter 15) to netnography (Rochette - chapter 9), and even the political aspects of methodology (Webster - chapter 12). Several of the chapters deal with aspects of environmental and climate change education (Sutton - chapter 4; Tytler, C. - chapter 3; Raphael et al. - chapter 5), and related subjects (Bellingham - chapter 2), and are thus very relevant today. Some chapters focus very much on the methodological aspects of DBER, such as Xu et al. - chapter 16, where they raise four methodological issues faced in their project and present thoughts on these in a clear and concise way; and Beck - chapter 10, who describes the 'Documentary Method'. These contrast with most other chapters, where the authors have chosen to clearly flesh out the methodologies used in a subject context to highlight its advantages and possible disadvantages. Finally, there are chapters dealing with various aspects of teaching, both from student perspectives and teacher perspectives (Hobbs et al. - chapter 8; Brown et al. - chapter 13; Clark & Ferguson - chapter 14; Delaney & Schultz - chapter 7; Tytler, R., & White - chapter 6). Some chapters are more focused on certain methods, such as Kirk & Ferguson's chapter 17 on how to methodically navigate the research literature using bibliometric visualisation tools.

Finally, Paul Hart's - chapter 1 exploration of post-qualitative inquiry gives a comprehensive overview of qualitative research and includes the 'posthumanist new materialist philosophy'.

In total, the contributing authors and chapters give a broad overview of the field and can function as a very good starting point for anyone interested in learning more about the current status of research methodology in STEM education research.

Thanks to the organisers for arranging and continuing the CAR symposium and thanks for the opportunity to share my thoughts with you on the very interesting chapters in this book.

Enjoy reading it.

INTRODUCTION

A 'POST' PERSPECTIVE ON METHODOLOGY

Methodological Approaches to STEM Education Research Volume 3 represents the practical experience and theoretical insights of education researchers grappling with methodological issues at the cutting edge of the field. The seventeen chapters offer explorations further developed from a long running annual symposium (Contemporary Approaches to Research: CAR) focused on strategically addressing contemporary methodology and methods in STEM education research, distinct from the customary focus on substantive findings that characterise most research gatherings.

Each volume has its own emphases. Volume 3 features a number of chapters that take up a post-qualitative perspective on methodology with two keynote addresses by Emeritus Professor Paul Hart from the University of Regina, and Dr Robin Bellingham from Deakin University. A second theme in the book continues this exploration of self-study, auto-ethnographic methodologies particularly relating to climate change education and sustainability. Further chapters explore diverse topics and perspectives: co-design methodologies; interdisciplinary research method and theory; ethical issues in research; and questions of analysis, data representation and approaches to literature review with contemporary tool application.

The CAR symposium has been an important focal point for researchers in the Deakin University Science, Technology, Engineering, Mathematics, and Environmental (STEME) Education Research Group and a wider set of education researchers who gather annually for the event. The underlying premise driving the symposium is that, as educational researchers, our knowledge and practice of methodologies is critical in framing our research and moving the field forward. Innovation in methodology is part of our brief, and attention to the selection and application of appropriate methodologies and awareness of their underlying epistemologies, their strengths and limitations, is an important aspect of our craft. Changing methodological traditions respond to or drive the theoretical approaches that open up new perspectives on educational practice.

Over the years, the CAR symposium has been a major stimulus for raising the profile of methodological practice and innovation for an increasingly international community, and researchers' awareness of issues and challenges, possibilities and limitations of different approaches to research in education (https://deakinsteme.org/event-category/car-symposiums/). This includes the role of theory and theory-method intersections, and the interrelations between research approaches, research designs, methods and methodology. The symposium represents an annual conversation about a variety of epistemological and ontological topics central to research in education.

The symposium format encourages in-depth engagement with methodologies through extended discussion around grouped presentations. The insights emerging from these discussions in many cases have informed the development and refinement of the chapters in this book. Participants range from experienced researchers to research students grappling with issues of theory and methodology, such that the topics, and the chapters in this book, range in focus from practicalities of methods and their adaptation to different contexts, to sophisticated overviews of theory-methodology interactions, and reviews of trends in theory and method driving rethinking of practice. Unlike formal volumes on methodologies, the chapters attend to context-specific practices, raising questions and speculation about methodological innovation and the challenges of pursuing research in practical contexts. They offer a view of education researchers' practice in framing methods and responding to challenges, which is informative for educational researchers at all levels. It is our intention, through this series, to raise the level of methodological awareness, and methodological debate, within our research community, extending what the symposium itself has done for more than two decades.

Peta. J. White
Russell Tytler
Joseph Paul Ferguson
John Cripps Clark

CHAPTER 1

POST-QUALITATIVE INQUIRY

PAUL HART

Abstract

This chapter is an exploration of the ways that post-qualitative inquiry creates openings for educational research. In moving beyond conceptual limits of qualitative research, it engages and exposes conceptual limitations of qualitative research. It creates conditions for critical engagement with concepts from posthumanist new materialist philosophy. In so doing, post-qualitative inquiry creates a new vocabulary for re-engaging crucial discussions of (new) empirically-based inquiry framed by new ethics of performativity, as naturally occurring and as uninterpreted, within educational and natural settings. Researcher representation is replaced by active conceptual engagements with diffractions, rhizometics, entanglements, assemblages, refrains, affective intensities, as well as a growing range of cartographic mappings of lines of flight, enfoldings and complexities of human-nonhuman intra-actions. At play is the 'politics of engagement' grounded within post-Anthropocene philosophy as a new base for more activist environmental education and political post-humanist critical actions against liberal humanist individualism. When such complexities become entangled with ethics and empirics, they become matters of political agency within post-Anthropocene times. As educational researchers we must decide: Who must we become and what must we do, for the next generation?

Keywords: Post-qualitative, ethics and politics of inquiry, new empiricist inquiry

Speculation

Sometime in the early 21st century, scholars in educational research, and particularly in environmental education research, will encounter issues related to post-qualitative inquiry and its background theory. With the role of science and knowledge production at a crossroads, as societal transformation challenges dominant forms of knowledge production, educational research is challenged (Lotz-Sisitka et al., 2016) to explore transgressive ways of thinking (Temper et al., 2019), and approaches to inquiry prerequisite to societal transformation. Implicated are empirical and ethical theories related to new materialisms and posthumanisms as questions for new empiricisms and new ethics (Clough, 2009; Temper et al., 2019). These impending changes reflect ontological re-orientation which creates conditions for critical bioethical and biopolitical discussions of the nature of being, as well as basic categories of long-term existence. And as we learn to reconceive relations of knowledge and being, relations between human and nonhuman become profoundly ethical-environmental-educational in ways that foreshadow an evolution of thinking about educational inquiry along different planes of thought. St. Pierre et al. (2016) describe this work in terms of an ethical imperative (Braidotti, 2013) that compels educational researchers to explore educational inquiry in terms of alternative futures and to find ways to live a different existence, beyond the Anthropocene. Basic questions concern whether and how it is possible collectively to imagine or speculate on conditions of existence. Perhaps post-qualitative thinking as a beginning, can create points of intersection that compel environmental educational researchers to rethink their responsibilities within the evolving dynamics of social science inquiry?

Introduction: Post-qualitative inquiry within a complex of theoretical starting points

At the turn of this century, Denzin and Lincoln (2002) presumed to introduce the field of qualitative inquiry. They presented a critical framing for interpretive and critical paradigms that displaced quantitative with qualitative inquiry. They described qualitative inquiry within participatory and collaborative processes as grounding for social critique and social action, encompassing a wide range of epistemological viewpoints and research strategies for better understanding people within their natural contexts and in terms of new quality criteria evolving from the 'paradigm wars' and poststructural work of the 1990s. They created openings for scholars such as Elizabeth St. Pierre who described a sixth moment of qualitative inquiry as a rethinking of the ethics of research practices, after Deleuzian ethics principles and figurations provided the grounding for thinking with concepts such as 'rhizome,' 'fold,' 'nomad', and 'haecceity' to question researcher subjectivity. St. Pierre (1997, 2002) and Lather (1997) described this work as mappings for shifts in thinking as concepts/figurations were introduced to challenge inquiry in terms of interpretation and representation. St. Pierre proposed ethical principles from Deleuze and Guattari (1987) to think differently about qualitative methodologies. These philosophical shifts (Braidotti, 1994) enabled mappings of the 'new' empiricism and 'new' ethics as justification for post-qualitative strategies.

In the early years of the 21st century, post-qualitative possibilities have become more common across a wide range of scholarship within educational research, grounded at various levels of thought, within educational inquiry. Jackson and Mazzei's (2012) *Thinking with Theory* challenged simplistic treatment of 'data,' and demonstrated how thinking within conceptual perspectives provides another way of making sense of 'performance' in the material world. These philosophical and conceptual groundings served as underpinnings for educational researchers who began to experiment directly from within 'live' engagements, in educational and field settings.

Certain concepts, within this return to 'realism' (Le Grange, 2018), served as 'starting points' for researchers to think with theory within a philosophy of immanence, framed by new ethics, new empiricisms, posthumanisms, and new materialisms. These changes, particularly in research areas related to environment and environmental education, served as new beginnings for environmental and educational shifts toward what Haraway (2016) describes as 'New Planetary Ethics: Beyond the Anthropocene.'

This chapter focuses on those theoretical and conceptual starting points within what Carol Taylor (2016) calls a cacophonous ecology that responds within an epoch in which humans have become an ecological force with planetary impact. It is manifest, for example, in feminist PhEmaterialisms (Ringrose et al., 2019) which combine realist theories of posthumanisms and new materialisms with questions of ethics and empiricism. It offers new ways of thinking about educational research that cut across issues of education, gender, social justice, and nature/culture/technical interfacts.

St. Pierre (2019) characterises post-qualitative inquiry as setting aside qualitative methodology and instead reading philosophy widely in order to understand the theory-base. What becomes apparent is an immanent philosophy of presence (agential realism) that challenges methodology-based qualitative research practice (Barad, 2007). Following this new theory, educational researchers are working to reconceptualise educational research from the ground up, creating conditions for change within theory in terms of immanent knowledge practices that diverge from 'truth telling' informed by research interpretation and representation.

Post-qualitative inquiry has emerged within background theory as 'transgressive,' as thinking directly from philosophy, and implicates environmental as well as educational theory. For example, Temper et al. (2019) and Elfreich (2019a, 2019b) actively challenge dominant forms of knowledge production in favour of activist approaches to research as transgressive and transformative (see, for example, Hsiung, 2016; Kuby et al., 2016; Lotz-Sisitka et al., 2015, 2016). Theoretical resources include, amongst many others, nomadic theory (Braidotti, 2013), new materialisms (Braidotti, 2002; Coole & Frost, 2010; Dolphijn & van der Tuin, 2012), posthumanisms (Braidotti, 2013), posthuman performativity (Barad, 2003), material feminisms (Alaimo & Hekman, 2008), and bodily nature (Alaimo, 2010, 2016). This literature has become more complex and diverse as background within an emerging post-Anthropocene politics of inquiry and new planetary ethics, beyond the Anthropocene (Haraway, 2016).

This rapidly expanding background for post-qualitative inquiry is beginning to think across transgressive spaces in relation to the politics of nature/planet that follow from post-Anthropocene possibilities. As new grounding for educational/environmental educational research, and within the short span of two decades, theory-savvy researchers, following from Lather, St. Pierre, MacLure, Jackson and Mazzei, Lenz-Taguchi, Taylor, Nordstrom and many others, have engaged a variety of post-qualitative perspectives. This diversity, however, is centered by immanent theory/practice and theory-driven methods, reconceptualised within the conceptual spaces of posthuman and new materialist entanglements and underpinned by new empirical and new ethical thinking.

As an example of how environmental education comes to ground directly from within philosophical perspectives that implicate post-Anthropocene ethics and new empiricisms, the recent *Environmental Education Research* special issue on new materialisms (Clarke & Mcphie, 2020) works toward meetings of politics and ethics. They illustrate how underscoring immanent performativity creates transgressive spaces for thinking differently about research processes as intra-actions captured 'live' from classroom and field experiences. It is in the lived experiences themselves that we can begin to see transgressive flows as human 'becomings' within the entanglements of their educational experiences. Exploring what is really going on within these entanglements of in-school or field experiences as well as speculative engagements and action-oriented, live explorations, the papers in this special issue work to align environmental education research with post-qualitative perspectives.

As a reflection of several discussion sessions at the 2021 Deakin CAR symposium, this chapter focuses on these theoretical and conceptual complexities of thinking with post-qualitative theory to create openings for critical discussion of new conceptual starting points. Concepts such as diffraction, which originate directly within theory, form the 'live' substance for viewing and exploring participants' voices and actions in real time. This process, not unlike dimensions of participatory action research, is intentionally grounded within a complex of theory-based concepts. Barad's (2007) ideas of an immanent philosophy of presence and posthuman performativity challenge educational researchers to work directly with theory outside of pre-existing qualitative research methodologies. There are no pre-existing research methodologies, just certain theoretical and conceptual starting points that illustrate how theory grounds post-qualitative inquiry (see, for example, Kuntz & St. Pierre, 2020).

Post-qualitative possibilities: A complex of new theory bases

Post-qualitative inquiry begins with a return to realism framed by new empiricism within an immanent philosophy of presence (i.e., Barad's [2007] agential realism) at a time when forces in the world (i.e., encounters with difference) require breaking old habits, such as the structures of conventional 'humanist' social science inquiry (i.e., methodology). Following St. Pierre (2002, 2021), after the crisis of representation, researchers' interpretations of the 'real' can only be an ambiguous and unreliable mediation between observers and realities. By engaging an ontology of immanence, researchers are encouraged to work with concepts/figurations (e.g., rhizome, haecceity) to rethink traditional definitions of empirical research practices.

Different ontological presumptions about modes of being implicate a new ethics of engagement by including a nonhuman dimension that questions what counts and who matters beyond the scope of qualitative methodologies and methods. Within these parameters, new inquiry processes are needed to focus on how humans are embedded within the politics and ethics of experiences. This philosophical 'sea change' creates conditions for post-qualitative inquiry to rethink/reworld with nonhuman 'others' and with different assemblages of ecological relations, and decolonial and Indigenous perspectives, as part of the dynamics of shifting forces operating on people (Snaza & Weaver, 2015; Taylor & Hughes, 2006).

New materialisms and posthumanisms (Alaimo & Hekman, 2008; Braidotti, 2002, 2013, 2019; Coole & Frost, 2010; Dolphijn & van der Tuin, 2012; Haraway, 2016; Taylor & Bayley, 2019; Taylor and Hughes, 2016), new empiricisms and posthuman ethics (de Freitas & Truman, 2021) have evolved from philosophical underpinnings for post-Anthropocene inquiry, and in particular for environmentally focused educational inquiry. The response to rapidly emerging concerns about planetary climate change has positioned research in environmental education strategically to adapt into a leading role in this new politics of unsustainability (Blok & Bruun Jensen, 2019).

Gaining understanding of these underpinnings and essences as 'sea changes,' that ground thinking for a post-qualitative turn requires thinking with new frames of inquiry. These new frames have the capacity to articulate conditions for less hierarchical value systems. Such framings refuse the ontological primacy of humans and create openings for thinking differently within contexts that range from micro-classroom episodes to those with planetary implications.

Educational researchers are now actively encouraged to work in nomadic, yet deliberate, ways to explore real conditions and commitments using media such as videos, photos, artworks, and poetry in relation to 'live' enactments and performances, materialised in spaces/places where researchers, along with viewers and readers, can vicariously participate in generating and posing onto-epistemic questions for understanding and perhaps to generate ideas for making a difference. Examples include Malone, Tesar, and Arndt (Book Series); Reinertsen (2016); Ringrose et al. (2019); Ivinson & Renold (2016); Clarke and Mcphie (2020), and the special issue of the *Australian Journal of Environmental Education* (2022).

Post-qualitative inquiry: Creating conditions for new theory/new inquiry

If the overall purpose of critical theory is to create new modes of thought to help us critically engage global education perspectives that serve the dominant (institutional) power structures and control current systems of education and educational research, how can we also engage new theories to illuminate destruction scenarios that continue to emerge in our lives, despite our best efforts to change (Brookfield, 2009)? How can we also learn to disentangle how and what it is that we have learned to accept through educational research and through environmental education research, making it possible to see differently (Hart, 2014, p. 371)?

Some time ago, critical theory was proposed as a means to examine dominant institutional-political power structures that support and maintain the status quo. What seems to have transpired within the movement to post-qualitative inquiry is a more critical evolution of thinking concerning what counts as research in education. An increasing number of forward thinking theory-based researchers, such as Nordstrom and Ulmer (2017), are concerned that researchers might take up post-qualitative inquiry without crucial knowledge about why and how it came to exist as a challenge to interpretive/representational methodology of qualitative inquiry. So how can transgressive thinking, which has created conditions for change in educational research by unmasking the status quo and by raising consciousness concerning material and humanist conditions as manifest in post-qualitative inquiry, work to expose power/knowledge relations between qualitative and post-qualitative which have become political?

- Post-qualitative inquiries are situated as agentic assemblages which implicate agential realism and assemblage thinking as figurations (concepts) that characterise **actual** experiences, each from within the unique politics of their places.

- Reading philosophy is required. For example, following Nordstrom and Ulmer (2017) post-qualitative inquiry engages different theory and theory differently as new ways of thinking and doing research (i.e., 'thinking with theory,' not methodology [Mazzei, 2021, St. Pierre, 2016]). 'Post-qual' is informed directly by philosophy (e.g., Deleuze & Guattari) using concepts as ways of re-orienting thinking, as not interpretive but in concrete encounters with 'the real' (Mazzei, 2021).

- 'Post-qual' is described as ethical inquiry (Lather & St. Pierre, 2013) that embraces the inseparability of ethics, ontology and knowledge (Le Grange, 2018). Concepts, such as haecceity, rhizome, refrain and diffraction, provide different perspectives (i.e., lines of flight/new thoughts and connections in and about 'live' encounters) that enable new thoughts and connections in 'real-time' (Jackson & Mazzei, 2018; St. Pierre, 2013).

- The texts themselves are agentially real (Barad, 2007, p. 712). Each experience, in classroom or field study with human and nonhuman agents forms parts of an overall process as 'affecting or being affected' by human and/or nonhuman relations (Rousell, 2019, p. 897).

- Post-qualitative inquiry resists binaries such as theory/data, researcher/participant/philosopher, writer/reader, so that hierarchies are flattened and the human subject is problematised as the sole source of meaning.

- Relational ways of engaging experiences, from maths lessons to pond studies, can be viewed through multiple conceptual frames, actively engaging the real-lived material complexities and political value of not being able to generalise findings, thus opening to multiple possibilities.

- Concept-as-method turns to theory for creative design experimentation and to focus on engaging 'live' experiences as they lay themselves out naturally in real life.

- In critiquing humanisms, post-qualitative inquiries also raise questions about what counts as knowledge and whose knowledge counts, and about how certain knowledge becomes foundational (i.e., secures truth) within imbrications of knowledge and relations of power; how knowing re-produces being/becoming, produces knowing and so on (St. Pierre, 2013, p. 648).

- Fundamental to this is the Deleuzian concept of immanence (Hein, 2019, 2020) which opens new forms of ethical inquiry as intra-active responsibility. Engaging inquiry immanently also implicates cultural phenomena, including Indigeneity and decolonisation, as essential, albeit political, from various vantage points.

- Working within immanence also has the potential to address affective intensities within immanent material ontologies of field and classroom experiences as transgressive writings and uses of video (Mayes, 2019). These fundamental interrelated concepts are increasingly framed within posthumanist and new materialist perspectives.

Posthumanisms

A growing body of scholarship has taken up the challenge of rethinking the limits of human growth and development in our relationship with earth others. The theoretical core of posthumanism draws attention to life beyond the human self, as immanent forces of life, and to knowledge systems that enable a clearer vision of the three ecologies—environmental, socioeconomic and subjective. Braidotti and Bignall (2019) define a 'posthuman turn' as a convergence of posthumanism with post-Anthropocentrism, as a complex and multidirectional discursive and material event that recognises humans as a geological force capable of affecting all life on the planet. For (environmental) education, it offers a crucial frame for understanding how interlaced assemblages of living things co-exist with nonliving forms and forces within planetary limits. Posthuman perspectives search for ways to reconceive 'humans in relation' with other beings and things, interrogating differences that deterritorialise concepts such as nature/culture, human/nonhuman, body/mind (Hohti, 2018; Holmes & Jones, 2016; Taylor et al., 2013).

Reclaiming human agency with accountability seems crucial for collective resistance to the unprecedented state of social and environmental collapse resulting from contemporary politics and ethics. Engagement in educational research and practice is recognised as crucial in understanding agential responsibilities for engaging critical discussion concerning ethical and political accountability and the tasks associated with framings of 'earth ethics' for geo-ecological and political accountability (e.g., Braidotti, 2013).

A posthuman turn in educational provision implicates ethical and moral issues of social and environmental justice as age-appropriate predicaments acknowledged in relation to common issues of climate change. Whether posthumanist perspectives provide a plausible framework within complex political ecologies of the present education systems seems to rest upon convergences within vital material intersections, as well as educators'/teacher educators' awareness of transformations across new materialist ethical spaces (Kuby, 2017). Questions remain concerning our educational responsibility to translate new ways of knowing/being into (environmental) educational praxis. For example, Haraway's (2016) speculative realist feminism and Barad's (2007) agential realist theory have been generative in creating conditions for human/more-than-human intra-actions with nature/cultures, with techno-bio power questions of local and global dimensions, and with questions that implicate many other dimensions of education that encompass reconceptualised theories of new materialisms within new empiricisms and new ethics.

New materialisms

Fox and Alldred (2015) follow Coole and Frost's (2010) 'New Materialisms': Ontology, Agency and Politics, and van der Tuin and Dolphijn's (2010) 'transversality' to explore underpinnings of new materialism within social inquiry. They found that influential philosophers such as Barad (2007), Hird (2010), and Haraway (2008) were, in different ways, 'always already' framing a 'politics of the future' for educational inquiry. Following Barad's (2007) focus on performativity, and Kirby's (2008) 'there is no outside the text' as 'there is no outside of Nature,' these materialist shifts are grounded within post perspectives on inquiry by Deleuze and Guattari's (1987) refusal of the interpretation/representation of language as the dominating mode of social educational inquiry.

Each of these theorists, in their own way, rethinks educational research as underpinned by engagement with 'live' episodes as research assemblages in what Barad (2007) calls 'intra-actions.' At issue is the new materialist perspective which eschews linguistic translation and searches for new framings within the dynamics of material intra-action. Thus, 'live action,' as already always material/discursive, is what becomes important, rather than the mimicking of a positivist correspondence theory of truth found in the qualitative linguisticism of interpretation/representation. In essence, the philosophy-based grounding by these theorists/researchers is rooted in breaking free from linguistic representationalist theories of matter as movement from epistemology to ontology, or in Kirby's (2008) terms, what if culture was nature all along (Alaimo & Hekman, 2008)? This thinking aligns with van der Tuin and Dolphijn's (2010) notion of the transversality of new materialism, which resonates with Fox and Alldred's (2015) new materialist perspective of thinking about educational and environmental education inquiry as direct viewings of educational experiences.

Twenty years ago St. Pierre and Pillow (2000) in *Working the Ruins* questioned whether humanism's inscription of reality/knowledge and the subject were dangerous fictions and what might a different 'science' look like? After two decades of exploration of this question, Rousell (2019) argued that the field of EE, in view of the evolution of 'thinking with theory,' may now have the political potential to occupy a more central role within the debates about direction in educational research and practice. Many others have worked to integrate the role of feminist, new materialist, and posthuman thinkers to actively engage decolonial and Indigenous reconfigurations of childhoodnatures as groundings for educational inquiry (see Cutter-Mackenzie-Knowles et al., 2020; Malone, Duhn, & Tesar, 2020).

New materialist and posthuman strategies as conceptual starting points

New materialist and posthuman strategies break free from interpretivist research orthodoxies in search of rhizomatic spaces for new conditions of possibility. These movements 'in process,' open inquiries beyond textual representations and focus on 'performance' as enacted and embodied in 'live' experiences. 'Knowing' is the result of direct material engagement in lived experiences, that is, realist intra-actions. Holmes et al.'s (2020) interest in videoing young children's 'games' of play in exploring knotted collections of 'naturecultures' in children's play provides an example.

Fox and Alldred (2015) describe a number of characteristics for a new materialist paradigm as one that explores:

- how desires, feelings, and meanings contribute to social production (Braidotti, 2013);
- breaks in the culture/nature divide and how the culture/nature divide works with 'live' (living evolutions of working with a conception of agency not tied exclusively to human actions);
- how new assemblages can work as relational networks (animate or inanimate);
- research assemblages of bodies, things, territories, abstractions;

- how engaging affective flows can (de) stabilise, (de)territorialise assemblages (territories as the means and places where society and history unfold);
- dynamic materialities filled with affects, forces and desires, flows and intensities, assemblages and disaggregations as 'becomings of' rather than 'beings';
- schizoanalytics as mappings of lines of becomings/affects/flows in schizo- rhizo-nomadilogical methods.

Fox and Alldred (2015) also characterise a posthumanist paradigm as one that explores:

- by engaging critical discourses that interrogate and challenge relations among human/nonhuman Anthropocentrisms and humanist reflexes;
- by reconfiguring how we understand 'politics' and 'knowledge' by troubling humanist concepts such as settler colonialism, racism, sexism, and politics in educational participation in curriculum studies and environmental education;
- by reconsidering humanist politics as incapable of escaping capitalist tendencies and traditions;
- by finding ways to reconceive humans, in relation with other beings and things, as well as interrogating and deterritorialising differences;
- by searching for ways of accounting for things in non-Anthropocentric ways, conscious of the politics of change;
- by engaging in speculations (speculative realisms) with how we must reconceive educational practice and inquiry within new models of the political within communities as relational fields;
- by exploring many curriculum reformations as embedded in post-Anthropocene thinking;
- by engaging in politically sensitive areas of planetary considerations as ecological dimensions of emergent ethics and posthuman politics;

- by engaging politics in research differently in terms of post-Anthropocentric ethics within, for example, feminist, post-colonial, anti-racist projects.

Post-qualitative perspectives have become the focus of special issues in education research periodicals such as the journal *Reconceptualizing Educational Research Methodology*, as well as in such publications as Koro-Ljungberg et al. (2017) and Ringrose et al. (2019). These materials have engaged, in particular, feminist posthumanisms and new materialisms within an entanglement of theory-praxis, as the PhEmaterialism movement that seeks to actualise and account for the layers of complexity in mapping connections and intra-actions across the layers (plateaus) of educational thought.

As a critique of Anthropocentric humanism, posthumanisms have opened inquiry to the more-than-human world with less hierarchical value systems (Braidotti, 2013; Haraway, 2016). The researcher, as 'modest witness,' is not detached or 'objective' and works beyond 'representation' with video, photos, artworks, and poetry, often as performances materialised in spaces where viewers/readers intra-act in generating onto-epistemic questions about these processes to make a difference. Thus, post-qualitative/posthuman methods 'break free' from qualitative research methodological orthodoxies in search of new rhizomatic spaces for conditions of possibility.

The research focus shifts from textual interpretations and representations to postcritical and post-Anthropocentric material engagements with participants' posthuman conceptions of change in empirical educational research (Boden et al., 2021). 'Knowing' is a result of direct engagement (i.e., direct experience) with the material world (at whatever scale). Serious ethical questions about the 'self' within the relational entanglements with human and nonhuman matter (e.g., plants and animals, technologies, nature cultures) are implicated. Inquiries often evolve along 'planes of existence,' for example, in terms of desires, knots, diffractions, political and personal motivations, and more-than-human entanglements.

In fieldwork, researcher interest is immanent. That is, how to capture what is 'going on' in real 'live' situations without recourse to traditional interpretive methodologies. Inquiries into immanent practices begin with perspectives of research 'concepts' such as diffraction, lines of flight, rhizomes, haecceities, folds, affects, and others become useful as ways of viewing, becoming with, and seeing diffractively across complexities of relationships, encounters, emotions and more. These realities are explored as constantly shifting, contingent and uncertain assemblages of relational entanglements that generate questions of how to begin to make sense of these dynamic practices and then to generate arguments for change in educational practice (see, for example, Osgood & Giugni, 2015). These arguments become the 'empirical' starting points for post-qualitative inquiry in rethinking education research, as grounded materially, and as posthuman possibilities.

Rethinking empiricism – The politics of new empiricisms

Drawing from Deleuze's conceptualisation of new empiricisms, Clough (2009) proposed a rethinking of empiricism, as a theory of knowledge, that allowed for a rethinking of bodies, matter and life through encounters with 'transgressive' underpinnings, beyond the positivist, quantitative, statistically oriented method linked to behaviourism, and beyond interpretive methodologies. Clough (2009) traced this evolution of thinking through Deleuze's transcendental empiricism that allows a rethinking of bodies/matter/life through encounters with embodiment, visceral perception, and preconscious affect (Massumi, 2021). Beyond the 1990s, assumptions that social practices could be captured using applied science methods were replaced by qualitative methodologists, questioned by critical theory/research, and refused by poststructuralists within a crisis of representation. More recently, explorations of an expanded (new) empiricism within posthumanist and new materialist theories are argued for agential realist engagements (de Freitas, 2017; de Freitas & Truman, 2021; Jackson & Mazzei, 2018; St. Pierre, 2016).

Where classical empiricism relied on knowledge gained from 'measured' observation, post-qualitative empiricisms are collective, relational, 'knowing together' in working with concepts such as rhizome, haecceity, or diffraction to explore 'live' intra-actions in and out of class situations and field activities. In this post-qualitative shift, the focus is decentering human (individual) agency within relations, across live experiences, where 'becoming' together is engaged as 'live' collectives. Thus infra-empiricism, as new empiricism, is found with realist (authentic) accounts of actual experience, uninterpreted.

Rethinking inquiry beyond interpretation coincides with rethinking empiricisms in terms of immanence. Following St. Pierre (2016) and Mazzei and Jackson (2017), when rethinking matter, we must rethink the empirical. The relation of knowledge and the nature of being are profoundly ethical. St. Pierre (2016) and Mazzei and Jackson (2017) describe this new empirical work as laying out a 'different plane of thought,' assisted by the creation of new concepts and the abandonment of representational thinking (Le Grange, 2018). The structure of humanist research methodologies cannot accommodate this new post-qualitative work where the concept carries the entire structure in which it is imbricated (e.g., St Pierre, 2016).

The profound difference in the shift from transcendental to transgressive empiricism is the difference between metaphysics as transcendental (outside the world), and what exists in the world as transgressive, with immanent, direct connection to lived realities on the ground in 'live actions.' Thus, 'post-qualitative' means abandoning researcher interpretation in favour of agential realist accounts of phenomena from face-to-face intra-actions with research participants (and things) as they make meaning of their experiences, explored directly through diffractive use of concepts (de Freitas & Weaver, 2020).

Rethinking inquiry within new empiricisms and relational ethics

New materialisms/posthumanisms carry new empiricist concepts such as intra-actions and agential realisms into creative and speculative ethical spaces of experimental practice. The material and empirical are so imbricated that they must change together. Where classical empiricism relies on knowledge gained from the senses (observation/measurement), in post-qualitative empiricism, invention is a collective, the relational ability and ethical responsibility to know together via affects and embodiments of experiences that implicate 'relational' ethics. The research focus is on how concepts (e.g., rhizome, diffraction), rather than interpretation, assist researchers in making sense of relations (becoming together) as a joint/collective researcher-participant endeavour.

Rethinking inquiry beyond representation coincides with rethinking empiricism in terms of immanence, that is, in presenting rather than representing/transcending encounters with others as well as in creating openings to more transgressive realist methods of presenting data (Bridges-Rhoads et al., 2016; Bridges-Rhoads & Van Cleave, 2017). The research interest becomes immanent within the realism of 'new empirical' viewers seeing participants in action and/or hearing participants' 'live' speech, with researchers sometimes intra-acting so participants can better articulate what they think they experienced. Listener/viewer/reader hears and/or sees (e.g., video/media) what participants are trying to convey by their own words/actions.

The researcher's role changes from interpreting to inventing ways to give voice to participants, to facilitate viewer access to participants' expressions/actions, voices/passions, and to learn directly from what participants see/hear/feel about their role in participatory experiences. Post inquiry is about creating conditions for direct access to participant voices and performativities within natural and social environments. Thus, new empiricisms, along with new ethics, work to create conditions for a radical break from neoliberal humanism and the dominance of researcher interpretation/reflection. In creating new forms of relational ethics the goal is to create openings so that observers, and participants themselves, can critically examine their systems of thinking (patterns, absences, relations with living and nonliving matter) as habits, views, or lines of force that are often unconscious and/or taken for granted.

New ethics are required to create conditions for forms of radical openness (emergence/immanence) within intra-active human/posthuman entanglements with capacities to read affect and to recognise performances/ voices that play across/among human/nonhuman dimensions, as diffractive of multiple positionings. These inquiries are 'live' at various levels and may be participatory, with or without participation by the researcher, as constitutive enactments. Implicated is the crucial necessity of ontological re-orientations that enable posthuman conceptualisation of matter as agentive.

New ethical thinking is implied as well as biopolitical/bioethical issues in posthumanist assumptions about human agency. Such realist encounters implicate material/discursive, as well as natural and cultural, factors (Alaimo & Hekman, 2008) that act to generate new conditions for new ethics. But it is new empirical foregrounding of 'inner relations of embodied experience' that invites deeper researcher and participant engagement and critical reflection on inner sensations, forces and movements, and impacts relations among individuals (and things) beyond discourse. This process of engagement/entanglement is the manifestation of Deleuzian transgressive empiricism as generated through relational ethics.

Encounters with environmental ethics

Engaging in post-qualitative inquiry, almost by definition, implies working with posthumanist and new materialist philosophies as an ethical endeavour. It implicates new perspectives on ethical responsibilities rooted in relational realist ontology (i.e., knowing through being as a new kind of researcher responsibility and accountability). The responsibilities and politics of ethically grounding 'live' research experiences have generated rethinking of long-held assumptions that have been addressed within posthumanist and new materialist political ecologies, Indigenous relations and animal studies, as dimensions of thought within transgressive, global issues.

As a beginning, the conceptual and material complexities of agential realist theories of post-qualitative inquiry provide grounding for non-representational forms of inquiry (Davies, 2018). Post-qualitative ethics must first acknowledge and then accommodate researcher engagement with concepts as emergent, embodied, material, and posthumanist dimensions of meaning-making within real (lived) experiences.

Agential realist ontology, following Barad (2007) is useful in rethinking ethics in ways that critically engage the complexities of 'live' activities themselves (as uninterpreted) at several levels and in respect of human and nonhuman natures, including environment-related materialities of the planet (Barad, 2007). At one level, research ethics must begin to acknowledge and engage a movement towards a new plane of planetary ethics in terms of realist encounters at all levels of the educational spectrum, including the political-ethical of the administrative apparatus of education. At another level, research ethics must learn how to adapt to new educational and environmental educational inquiry as 'live' encounters with real environments (inside and outside of classrooms). For example, despite frustrations of the PhEmaterialist movement in breaking new ground on post-Anthropocene ethics (e.g., Renold, 2019; Strom et al., 2019), academic ethics committees are beginning to expand criteria to remain credible in new times (see, for example, Davies, 2018; Hultman & Lenz Taguchi, 2010; Zapata et al., 2018).

An ethical challenge for environmental education researchers and teacher educators lies in thinking through how posthumanist and new materialist theory translates ethically (and politically) into new curriculum theory and practice. Ethical questions implicate human responsibility and accountability for new curriculum theory and practice within the geopolitics and political economics of everyday decisions. Such questions have already begun to generate speculative ethics of the future biopolitical landscape of environmental education and environmental and sustainability education in teacher education. Negotiation of the form and content of new ethical futures education, given the forecasts of human-induced planetary change, implicates posthumanist new materialist ethics and biopolitics to the point where scholars such as Braidotti, Haraway, Colebrook, and Alaimo speculate that, because existing knowledge frameworks are likely to create challenges for change within environmental educational research, new beginnings are now required with posthuman ethics and new materialist empirics for environmental education/environmental and sustainability education-based teacher education.

Agential realism and the posthuman 'materialisation' of 'ethics, politics, and care'

Karen Barad (2008) suggests that the 'becoming material' processes that constitute the ontologies of agential realism produce an ethics of mattering. She reasons that 'ethics' is not about the right response to the other but about responsibility and accountability for the lively relationalities of becoming of which we are part. Agential realist ontology reworks ethics, following Barad (2007), as new ethico-onto-epistemology that describes the many ways that our entangled materiality makes each educator responsible for contributing to our collective knowing-in-becoming. Parts of our engaged (educational) intra-actions (as the mutual re-constitution of our entangled agencies and material-discursive praxis) occur among researchers and participants, on the same 'plane of immanence,' in ways that shift the role of the inquirer and the research process from inter- to intra-acting. Anne Reinertsen's (2021) inquiries with pre-school and primary children illustrate how agential realist-based inquiries warrant live video accounts of in-school activities or experiences outdoors, as walking or field episodes.

Given the space devoted to posthumanist and new materialist thinking and practice within educational journals and books, it should not be surprising to find materials for both educational research and teaching such as Malone et al. (2019-2022); Cutter-Mackenzie-Knowles et al. (2020), particularly chapters 3, 6, 29, 58 and 74; Ringrose et al. (2019) and *Environmental Education Research,* 26(9-10). New thinking about research ethics is a component of these exercises and becoming involved in critical discussion with ethics committee members and parent-teacher associations is crucial.

Post-qualitative inquiry assumes that humans have ethical responsibility and accountability in relations which include materialities of the planet (Barad, 2007) at several levels, as practical tasks of research ethics but also in relation to political dimensions (i.e., academic research ethics committees). This is complex, necessary work ranging from school classroom fieldwork to the ecological level of planetary ethics (e.g., Haraway) in terms of preconscious 'affects' as well as realist encounters. University research ethics committees' research approval criteria are in immediate need of updating in view of the evolution of thinking and practice in social and educational inquiry (Davies, 2018) as well as the rapidly expanding literature on post-Anthropocene ethics as posthuman ethical grounding for inquiry (Zapata et al., 2018).

Foreshadowing dramatic changes: Merging theory: Praxis

Bringing embodiment to consciousness beyond the terms of our own subjection (Davies, 2000) is implied by significant changes within research fields where new approaches are underpinned by power and politics. Change is almost always controversial and must be well grounded theoretically. Movement to post-qualitative implies empirical and ethical movement away from interpretive and representational methodologies toward realist performative alternatives. Performances, presented as findings, implicate new technologies that question traditional notions of what counts as data as well as the role of researchers when participants are engaged within natural educational settings, including field experiences. Researchers are working out how many new concepts such as rhizome, diffraction, nomad, haecceity, refrain, and affect work as acts of thought or along planes of thought to make different sense of, for example, the maths lesson or the field trip to the local pond.

Much preliminary theoretical grounding foreshadowed Jackson and Mazzei's (2012) *Thinking with Theory* which created openings for forms of realist inquiry that directly implicated the immanent dynamics of intra-actions. Researchers were encouraged to explore how technologies such as 'live action' videos could create openings for debate about the realities of learning environments. Such practices create openings for application of new empiricisms and new ethics in ways that invite deeper reflection on complexities of teaching/learning situations. For example, inner sensations, forces and movements, participants' thoughts and feelings, as well as overall impact on participants, rather than interpretations by researchers became the focus of these encounters. Theories of transgressive empiricism and intra-relational ethics add depth to debates on curriculum and instruction as well as traditional teacher education.

On another plane, post-qualitative theory is compatible within the deeper philosophical scholarship of post-Anthropocene thinking. For example, Haraway's (2015) new plane of planetary ethics implicates environmental issues within discussions of the materialities of the planet (Barad, 2007). Thus, a new plane of ethical responsibility is embedded within an ethics of change, both internal and external to the academy. These changes directly effect postsecondary education beyond science and environmental education, as well as debate about research criteria within research ethics committees. Examples of these changes are found in several Scandinavian universities where educational research is part of institutional change.

Merging theory with conceptual starting points for post-qualitative inquiry: 'Politics'/ethics 'matter' in the posthuman 'post-Anthropocene'

If impending global issues and philosophy-based theory provide plausible arguments for change within educational inquiry, and if post-Anthropocene politics create openings for posthuman and new materialist thinking, then it could be argued that openings are created for post-qualitative inquiry. Many theorists including Barad (2007), Bennett (2010), Latour (2004), Duhn (2017), and Grosz (2010) urge the legitimation of agential realism and associated new materialist perspectives as grounds to challenge current reduction of these concerns to interpretive methods. Bennett (2010) in particular argues that 'matter' is not simply passive, but vital and that materialities exert affective influences beyond human dimensions.

Matter, in fact, becomes the embodiment of sensations, affects and aesthetics across the more-than-human in the form of new politics/ethics as cultural forces (Malone et al., 2020). Within this frame of thinking post-qualitatively, the concept 'material' means that posthuman systems are underpinned by matter flows such that new materialist-based inquiry questions the privileging of subjectivity and representation and, following Braidotti (2013), replaces textual and other interpretive methods with an ontology of modulated presence (i.e., agential realism).

Margaret Somerville (2017) references Braidotti (Braidotti & Mura, 2014) concerning the concept of the Anthropocene in the face of the advancement of capitalism as a new phase in the history of both humankind and of the

Earth, when natural forces and human forces become intertwined. The fate of one determines the fate of others (Zalasiewicz et al., 2010, p. 2231). As Somerville (2017) indicates, "post-qualitative research methodologies are closely aligned with the posthuman paradigm because" rethinking humanist ontology is key in what comes after humanist qualitative methodology (Lather & St. Pierre, 2013, p. 629). Thus, a new 'starting point' for post-qualitative inquiry is the assumption of human-environment entanglements as already anticipated by several domains of inquiry, including Australian Indigenous studies, animal studies, and new materialism (Diaz-Diaz & Semenac, 2020; Snaza & Weaver, 2015).

The posthuman subject is 'ecological' as embedded within material flows and not only socially discursively but also materially constructed within what Braidotti (2012) calls an ontology of modulated presence. Research praxis, consistent with this theory (as new empiricism) foregrounds/focuses on 'inner relations' of embodied (present) experiences in ways that invite deeper reflection on sensations, forces, and movements that impact intra-relations/intra-actions beyond discourses (i.e., Deleuze's transgressive empiricism).

The shift from qualitative to post-qualitative re-engages the 'real' within non-representational 'live' encounters so that viewers/researchers can experience (through video) with participants and then re-view the sense made by both. Early posthuman/new materialist inquiries broke new ground in struggling with ethics review boards for changes in perspective. They sought, for example, permission for live videos of actual experiences/encounters and permissions to construct, engage and share videos and other live performance media within expanded, rethought empiricisms of inner relations and embodied experiences that invite deeper reflection on 'inner sensation'. At the global level, researcher ethics concerns include political dimensions across relations of human and nonhuman/more-than-human natures which increasingly implicate the materialities of planetary futures (i.e., as in the post-Anthropocene).

Educational researchers, following the (new) ethical imperative to rethink the nature of being on this planet, and the (new) empiricist imperative to engage agential realism within posthuman and new materialist theory-praxis, are beginning to rethink what constitutes inquiry. This ontological re-orientation compels those interested in environmental education to engage posthumanist and new materialist theories as biopolitical and bioethical (St. Pierre et al., 2016). Implicated are socio-materials realities, as geopolitical and as embedded within the socio-economics of everyday life, that challenge our basic humanist assumptions in terms of agential realist intra-actions amongst phenomena that are material, discursive, human, more-than-human, corporeal and technological (Alaimo & Hekman, 2008).

Many applications of post-qualitative inquiry can be found amongst a rapidly expanding trend in environmental education (Cutter-McKenzie-Knowles et al., 2020; Malone et al., 2019-2022) that illustrates environmental education researcher engagements with posthumanist and new materialist theory in relation to 21st century ethics. The Anthropocene, climate and man-made destruction can be studied as entanglements, not as intertwinings of separate entities, but as irreducible relations of responsibility from local to global scales within framings of green political theory, politics of the Anthropocene (Fremeaux, 2019), and the politics of unsustainability. Post-qualitative inquiry is an attempt, on a relatively short timeline, to shift commitments that, beginning with our children and teachers, intend to change mindsets within the local and global political dynamics. Educational inquiries are increasingly found in many educational journals where teacher educators and teachers are experimenting with these new approaches. These inquiries create conditions for re-viewing educational provision and praxis within a different key. The following example is one, among many, in the series edited by Malone et al. (2019-2022).

In this study, Arndt (Arndt et al., 2020) draws on posthumanist and new materialist approaches to subject formation to highlight political and ethical implications of working with others to think with concepts (ethical and moral), such as otherness and interconnectedness (central to Indigenous onto-epistemologies). In working to decentre humanist selves in attending to global relations such as climate change, Arndt, similar to Reinertsen (2016) and Huuki and Lanas (2019), discusses how 'affect' jumps across time and place as affective energy transmitted not only subjectively but intergenerationally through material objects. She characterises childhood as emergent and as constituted by affective assemblages beyond (and critical of) traditional humanism.

Arndt's thinking of children as agentic, with objects such as sand, focuses on children sharing voice and agency and the need for storying the conditions of these relations. She invokes Jane Bennett's speculative ethics, in relation to posthumanist thinking about the impacts of various kinds of vibrant matter (see also Arndt, 2017; Malone et al., 2020; Somerville, 2017; Vladimirova & Rautio, 2020).

In another study, Duhn and Quinones' (2020) speculate with a toolkit of concepts, perceptions and figurations to generate alternative imaginings, that express complexities of subjects they (we) have become. Speculation as method is increasingly used to rekindle imaginations that create spaces for writers to go beyond what they think they know about childhoodnature as a new imagining, beyond representation (pp. 88-89). Every encounter with a figuration, such as childhoodnature, creates possibilities for further 'becoming' (Braidotti, 2011) within contingent assemblages of things/places/texts, as cartographies of power relations. In environmental education, many texts express fleeting moments of emerging ecological sensibilities across embodied matter and materialities—human, more-than-human, and posthuman—as organic/inorganic, vibrant mattered assemblages (Bennett, 2010).

Huuki and Lanas (2019) discuss how 'affects' travel across time-place-age-talk subjectively and intergenerationally, through material objects and posthuman engagements with environments, and challenge readers to take seriously the more-than-human as constituted within complex material/human/posthuman networks of relations (Barad, 2007; Alaimo & Hekman, 2008; Braidotti, 2013). Huuki and Lanas focus on the more-than-present and other-than-conscious in an attempt to understand how affect travels across time-space relations within real relations and environments, as do many post-qualitative studies as complexly composed across material-historical forces. An example of cultural assimilation of the Sami people through intergenerational affective transmission can be viewed as a kind of rhizomatic cartography to explore power, not to explain events, but to understand how childhoods emerge as complexities of historical flows intertwined with multiple affective/force relations.

Post-qualitative advocates continue to engage inquiry within the theory-praxis nexus, and the examples illustrate how this strategy is beginning to make sense. As the critical commentaries within recent volumes of *Qualitative Inquiry* suggest, engaging necessary critical discussion and debate is crucial in creating openings for wider debate and speculation on how these local issues interact with global futures. Environmental education in particular, as already, always political, seems to be repositioning strategically within post-Anthropocene thinking as exemplified in the Springer Series (Malone et al., 2019-2022), specifically Malone et al. (2020), Hargraves (2020), Myers (2019), Crinall (2019), Rautio and Stenvall (2019), as well as Malone et al. (2017).

Postscript

Almost a decade has elapsed since the publication of the landmark special issue of the *International Journal of Qualitative Studies in Education* (2013) on post-qualitative inquiry. Somerville et al. (2020) describe their work in terms of research informed by posthuman and new materialist theorising within the context of human-induced changes to the earth's climate and human entanglement in the fate of the planet (after Zalasiewicz et al., 2010). As I read their paper, amongst many others, as an evolution of my ability to capture the essence of post-qualitative inquiry in this chapter, I have the feeling that one more point needs to be made in relation to the relative space allotted for the opening to the posthuman, and particularly in relation to the space allotted for discussions of ethics.

The turning point becomes obvious for me in the distinction Somerville et al. (2020) identify in relation to new ontologies of entanglement that Lather and St. Pierre (2013) viewed as fundamentally challenging the categories of humanist qualitative research. This challenge is illustrated in the different forms of data between the conceptualisation of entanglement as a methodological project to think data beyond Anthropocentrism and the focus on enacting different forms of data, including data that defies understanding, analysis, and representation. This affirms the importance of placing data in relation to ecosystems and, in particular, within entanglements of biopolitics and global political economies (i.e., planetary well-being).

I believe that the argument above provides justification for giving more research space to ethics which implicates posthuman ethics and therefore arguments for a more intensive engagement with posthuman and new materialist ethics. This is the kind of ethics described by Facca and Kinsella (2021) as 'rhizo-ethics,' cartographic assemblage (Ulmer & Koro-Ljungberg, 2015), and immanent writing (Hein, 2019) as means for reconfiguring research ethics (beyond hierarchies) and to reconsider intra-actions between humans and nonhumans as ethically significant.

References

Alaimo, S. (2010). *Bodily natures.* Indiana University Press.

Alaimo, S. (2016). *Exposed: Environmental politics and pleasures in posthuman times.* University of Minnesota Press.

Alaimo, S. & Hekman, S. (Eds.). (2008). *Material feminisms.* Indiana University Press.

Arndt, S. (2017). (Un)becoming data through philosophical thought processes of past, presents and futures. In M. Koro-Ljungberg, T. Loytonen, & M. Tesar (Eds.), *Disrupting data in qualitative inquiry: Entanglements with the post-critical and post-anthropocentric* (pp. 93-104). Peter Lang.

Arndt, S., Semenec, P., & Diaz-Diaz, C. (2020). Chapter 1: Interview with Sonja Arndt. In C. Diaz-Diaz & P. Semenec (Eds.), *Posthumanist and new materialist methodologies* (pp. 3-11). Springer.

Barad, K. (2003). Posthumanist performativity: Toward an understanding of how matter comes to matter. *Signs, 28*(3). https://doi.org/10.1086/345321

Barad, K. (2007). *Meeting the universe halfway: Quantum physics and the entanglement of matter and meaning.* Duke University Press.

Barad, K. (2008). Posthumanist performativity: Toward an understanding of how matter comes to matter. In S. Alaimo & S. Hekman (Eds.), *Material feminisms* (pp. 120–154). Indiana University Press.

Bennett, J. (2010). *Vibrant matter: A political ecology of things.* Duke University Press.

Blok, A. & Bruun Jensen, C. (2019). The Anthropocene event in social theory: On ways of problematizing nonhuman materiality differently. *The Sociological Review, 67*(6), 1195-1211.

Boden, L., Ceder, S., & Sauzet, S. (2021). Editorial: Posthuman conceptions of change in empirical educational research. *Reconceptualizing Educational Research Methodology, 12*(1). doi.org/10/7577.rerm.4215

Braidotti, R. (1994). *Nomadic subjects: Embodiment and sexual difference in contemporary feminist theory.* Columbia University Press.

Braidotti, R. (2002). *Metamorphoses: Towards a materialist theory of becoming*. Polity.

Braidotti, R. (2011). *Nomadic theory: The portable Rosi Braidotti.* Columbia University Press.

Braidotti, R. (2012). Interview with Rosi Braidotti, In R. Dolphijn & I van der Tuin (Eds.), *New materialism: Interviews & cartographies* (pp. 19–37). Open Humanities Press.

Braidotti, R. (2013). *The posthuman.* Polity.

Braidotti, R. (2019). A theoretical framework for the critical posthumanities. *Theory, Culture & Society, 36*(6), 31-61.
Braidotti, R. & Bignall, S. (Eds.) (2019). *Posthuman ecologies: Complexity and process after Deleuze.* Rowman & Littlefield.
Braidotti, R. & Mura, A. (2014). Indebted citizenship – an interview with Rosi Braidotti. Retrieved from https://www.opendemocracy.net/can-europe-make-it/rosi-braidotti-andrea-mura/indebted-citizenship-interview-with rosi-braidotti
Bridges-Rhoads, S., Van Cleave, J., & Hughes, H. (2016). Complicating methodological transparency. *International Journal of Qualitative Studies in Education, 29*(4), 536-552. http://dx.doi.org/10.1080/09518398.2015.1063733
Bridges-Rhoads, S. & Van Cleave, J. (2017). Writing post-humanism, qualitative inquiry, and early literacy. *Journal of Early Childhood Literacy, 17,* 297-314.
Brookfield, S. (2009). The concept of critical reflection: Promises and contradictions. *European Journal of Social Work, 12*(3), 293-304. http://dx.doi.org/10.1080/13691450902945215
Clarke, D. & Mcphie, J. (2020). New materialisms and environmental education: Editorial. *Environmental Education Research, 26*(9–10), 1255–1265.
Clough, P. (2009). The new empiricism: Affect and sociological method. *European Journal of Social Theory, 12*(1), 42–61. http://dx.doi.org/10.1177/1368431008099643
Coole, D. & Frost, S. (Eds.). (2010). *New materialisms: Ontology, agency, and politics.* Duke University Press.
Crinall, S. (2019). *Sustaining childhood natures: The art of becoming with water.* Springer.
Cutter-Mackenzie-Knowles, A., Malone, K., & Barratt Hacking, E. (Eds.) (2020). *Research handbook on childhoodnature: Assemblages of childhood and nature research.* Two volumes. Springer Nature.
Davies, B. (2018). Ethics and the new materialism: A brief genealogy of the "post" philosophies in the social sciences. *Discourse: Studies in the Cultural Politics of Education, 39*(1), 113-127. http://dx.doi.org/10.1080/01596306.2016.1234682
Davies, B. (2000). *(in)scribing body/landscape relations.* AltaMira.
de Freitas, E. (2017). New empiricisms and the moving image: Rethinking video data in education research. In M. Koro-Ljungberg, T. Löytönen, & M. Tesar (Eds.), *Disrupting data in qualitative inquiry* (pp. 211-223). Peter Lang.

de Freitas, E. & Truman, S. (2021). New empiricisms in the Anthropocene: Thinking with speculative fiction about science and social inquiry. *Qualitative Inquiry, 27*(5), 522–533. http://dx.doi.org/10.1177/1077800420943643

de Freitas, E. & Weaver, J. (2020). Rethinking social inquiry in the wake of science studies: Transdisciplinary pursuits in times of climate change, information flows, and fading democracies. *Cultural Studies <—> Critical Methodologies, 20*(3), 195-202. http://dx.doi.org/10.1177/1532708619883309

Deleuze, G. & Guattari, F. (1987). *A thousand plateaus: Capitalism and schizophrenia*. University of Minnesota Press.

Denzin, N. & Lincoln, Y. (Eds.). (2002). *The qualitative inquiry reader.* Sage.

Diaz-Diaz, C. & Semenec, P. (2020). *Posthumanist and new materialist methodologies: Research after the child.* Springer Nature.

Dolphijn, R. & van der Tuin, I. (2012). *New materialism: Interviews & cartographies.* Open Humanities Press.

Duhn, I. (2017). Performing data. In M. Koro-Ljungberg, T. Loytonen, & M. Tesar (Eds.), *Disrupting data in qualitative inquiry: Entanglements with the post-critical and post-Anthropocentric* (pp. 11-22). Peter Lang.

Duhn, I. & Quinones, G. (2020). Eye-to-eye with otherness: A childhoodnature figuration. In A. Cutter-Mackenzie-Knowles, K. Malone, & E. Barratt Hacking (Eds.), *Research handbook on childhoodnatures* (pp. 87-102). Springer.

Elfreich, A. (2019a). Latina adolescent becomings: Interrupting (un)silenced spaces into entangled modes of resistance. *Cultural Studies <—> Critical Methodologies, 19*(2), 116-125. http://dx.doi.org/10.1177/1532708618784329

Elfreich, A. (2019b). From silence to stillness: Decolonial feminist ontologies and the mattering of history in educational research. *International Journal of Qualitative Studies in Education, 32*(1), 47-66. https://doi.org/10.1080/09518398.2018.1530396

Facca, D. & Kinsella, E. (2021). Emergence, multiplicity and connection: Rethinking ethical discernment in qualitative research through a rhizo-ethics approach. *International Journal of Qualitative Studies in Education.* https://doi.org/10.1080/09518398.2020.1930248

Fox, N. & Alldred, P. (2015). New materialist social inquiry: Designs, methods and the research-assemblage. *International Journal of Social Research Methodology, 18*(4), 399-414. https://doi.org/10.1080/13645579.2014.921458

Fremaux, A. (2019). *After the Anthropocene: Green republicanism in a post-capitalist world.* Palgrave Macmillan.

Grosz, E. (2010). Feminism, materialism, and freedom. In D. Coole & S. Frost (Eds.), *New materialisms: Ontology, agency and politics* (pp. 139-157). Duke University Press.

Haraway, D. (2008). *When species meet.* University of Minnesota Press.

Haraway, D. (2015). Anthropocene, capitalocene, plantationocene, chthulucene: Making kin. *Environmental Humanities, 6*, 159-165.

Haraway, D. (2016). *Staying with the trouble.* Duke University Press.

Hargraves, V. (2020). *Children and the ethics of creativity: Rhythmic affectensities in early childhood education.* Springer.

Hart, P. (2014). Questions of legitimacy and quality in educational research. In A. Reid, P. Hart, & M. Peters (Eds.), *A companion to research in education* (pp. 363-374). Springer.

Hein, S. (2019). Deleuze, immanence, and immanent writing in qualitative inquiry: Nonlinear texts and being a traitor to writing. *Qualitative Inquiry, 25*(1), 83-90. https://doi.org/10.1177/1077800418784328

Hein, S. (2020). Deleuze and absolute immanence: Achieving fully immanent inquiry. *Qualitative Inquiry,* 1-10. https://doi.org/10.1177/1077800420935929

Hird, M. (2010). Indifferent globality: Gaia, symbiosis and 'other worldliness.' *Theory, Culture & Society, 27*(2-3), 54-72. https://doi.org/10.1177/0263276409355998

Hohti, R. (2018). Siiri and the "bag lady": Analysing the material entanglements of special needs. *Reconceptualizing Educational Research Methodology, 9*(1), 6-16. https://doi.org/10.7577/rerm.2698

Holmes, R. & Jones, L. (2016). Flickering, spilling and diffusing body/knowledge in the posthuman early years. In C. Taylor & C. Hughes (Eds.), *Posthuman research practices in education* (pp. 108-127). Palgrave Macmillan.

Holmes, R., Jones, L., & Osgood, J. (2020). Mundane habits, ordinary affects, and methodological creations. In A. Cutter-Mackenzie-Knowles, K. Malone, & E. Barratt Hacking (Eds.), *Research handbook on childhoodnature* (pp. 259-275). Springer Nature.

Hsiung, P. (2016). Teaching qualitative research as transgressive practices: Introduction to the special issue. *Qualitative Inquiry, 22*(2), 59-71. https://doi.org/10.1177/107780041567204

Hultman, K. & Lenz Taguchi, H. (2010). Challenging anthropocentric analysis of visual data: A relational materialist methodological approach to educational research. *International Journal of Qualitative Studies in Education, 23*(5), 525-542.

Huuki, T. & Lanas, M. (2019). Sámi child-adult/past-present entanglements in a painful lecture in university. In P. Rautio & E. Stenvall (Eds.), *Social, material and political constructs of Arctic childhoods: An everyday life perspective* (pp. 135-148). Springer.

Ivinson, G. & Renold, E. (2016). Girls, camera, (intra)action: Mapping posthuman possibilities in a diffractive analysis of camera-girl assemblages in research on gender, corporeality and place. In C. Taylor & C. Hughes (Eds.) (2016). *Posthuman research practices in education* (pp. 168-185). Palgrave.

Jackson, A. & Mazzei, L. (2012). *Thinking with theory in qualitative research.* Routledge.

Jackson, A. & Mazzei, L. (2018). Thinking with theory: A new analytic for qualitative inquiry. In N. Denzin & Y. Lincoln (Eds.), *The SAGE handbook of qualitative research* (5th edition, pp. 717-737). Sage.

Kirby, V. (2008). Natural convers(at)ions: Or, what if culture was really nature all along? In S. Alaimo & S. Hekman (Eds.), *Material feminisms* (pp. 214-236). Indiana University Press.

Koro-Ljungberg, M, Löytönen, T., & Tesar, M. (Eds.). (2017). *Disrupting data in qualitative inquiry.* Peter Lang.

Kuby, C. (2017). Why a paradigm shift of "more than human ontologies" is needed: Pulling to work poststructural and posthuman theories in writers' studio. *International Journal of Qualitative Studies in Education, 30*(9), 877-896. hptts://doi.org/10.1080/09518398.2017.1336803

Kuby, C., Aguayo, R., Holloway, N. Mulligan, J., Shear, S., & Ward, A. (2016). Teaching, troubling, transgressing: Thinking with theory in a post-qualitative inquiry course. *Qualitative Inquiry, 22*(2), 140-148. https://doi.org/10.1177/1077800415617206

Kuntz, A. & St. Pierre, E. (2020). Proclaiming the new: An introduction. *Qualitative Inquiry,* 1-5. https://doi.org/10.1177/1077800420939125

Lather, P. (1997). *Troubling the angels: Women living with HIV/AIDS.* Westview.

Lather, P. & St. Pierre, E. (2013). Introduction: Post-qualitative research. *International Journal of Qualitative Studies in Education, 26*(6), 629-633. https://dx.doi.org/10.1080/09518398.2013.788752

Latour, B. (2004). *Politics of nature: How to bring the sciences into democracy.* Harvard University Press.

Le Grange, L. (2018). A (re)turn to realism(s). *Educational Philosophy and Theory, 50*(14), 1594-1595. https://doi.org/10.1080/00131857.2018.1461424

Lotz-Sisitka, H., Ali, M., Mphepo, G., Chaves, M., Macintyre, T., Pesanayi, T. (2016). Co-designing research on transgressive learning in times of climate change. *Current Opinion in Environmental Sustainability, 20,* 8—87. https://doi.org/10.1016/j.cosust.2016.04.004

Lotz-Sisitka, H., Wals, A., Kronlid, D., & McGarry, D. (2015). Transformative, transgressive social learning: Rethinking higher education pedagogy in times of system global dysfunction. *Current Opinion in Environmental Sustainability, 16,* 73-80. http://dx.doi.org/10.1016/j.cosust.2015.07.018

Malone, K., Duhn, I., & Tesar, M. (2020). Greedy bags of childhoodnature theories. In A. Cutter-Mackenzie-Knowles, K. Malone, & E. Barratt Hacking (Eds.), *Research handbook on childhoodnatures* (pp. 19-30). Springer.

Malone, K., Tesar, M., & Arndt, S. (Eds.) (2019-2022). *Children: Global posthumanist perspectives and materialist theories* (Series). Springer Nature.

Malone, K., Tesar, M., & Arndt, S. (2020). *Theorising posthuman childhood studies.* Springer Nature.

Malone, K., Truong, S., & Gray, T. (Eds.). (2017). *Reimagining sustainability in precarious times.* Springer.

Massumi, B. (2021). *Parables for the virtual.* Duke University Press.

Mayes, E. (2019). The mis/uses of 'voice' in (post)qualitative research with children and young people: Histories, politics and ethics. *International Journal of Qualitative Studies in Education, 32*(10), 1191-1209. https://doi.org/10.1080/09518398.2019.1659438

Mazzei, L. (2021). Postqualitative inquiry: Or the necessity of theory. *Qualitative Inquiry, 27*(2), 198-200. https://doi.org/10.1177/1077800420932607

Mazzei, L. & Jackson, A. (2017). Voice in the agentic assemblage. *Educational Philosophy and Theory.* http://dx.doi.org/10.1080/0013857.2016.1159176

Myers, C. (2019). *Children and materialities: The force of the more-than-human in children's classroom lives.* Springer Nature.

Nordstrom, S. & Ulmer, J. (2017). Curations, creations, and other post-qualitative possibilities. *Reconceptualizing Educational Research Methodology, 8*(3), 1-15.

Osgood, J. & Giugni, M. (2015). Putting posthumanist theory to work to reconfigure gender in early childhood: When theory becomes method becomes art. *Global Studies of Childhood, 5*(3), 346-360.

Rautio, P. & Stenvall, E. (Eds.) (2019). *Social, material and political constructs of Arctic childhoods.* Springer.

Reinertsen, A. (Ed.) (2016). *Becoming earth.* Sense.
Reinertsen, A. (2021). Activist ESD pedagogies and the end of critique: An edu-poetic attempt to bring in the missing child—becoming child. *Qualitative Inquiry, 27*(8-9), 1114-1122. https://doi.org/10.1177/1077800420978744
Renold, E. (2019). Becoming AGENDA: The making and mattering of a youth activist resource on gender and sexual violence. *Reconceptualizing Educational Research Methodology, 2,3*(2), 208-241.
Ringrose, J., Warfield, K., & Zarabadi, S. (2019). Introduction feminist posthumanismsm/new materialisms & educational research: Response-able theory-practice-methodology. In J. Ringrose, K. Warfield, & S. Zarabadi (Eds.), *Feminist posthumanisms, new materialisms and education* (pp. 1-15). Routledge.
Rousell, D. (2019). Inhuman forms of life: On art as a problem for post-qualitative research. *International Journal of Qualitative Studies in Education, 32*(7), 887-908.
Snaza, N. & Weaver, J. (Eds.) (2015). *Posthumanism and educational research.* Routledge.
Somerville, M. (2017). Thinking critically with children of the Anthropocene. *International Review of Qualitative Research, 10*(4), 395-410. https://doi.org/10.1525/irqr.2017.10.4.395
Somerville, M., Woods, A., Duhn, I., Rautio, P., Powell, S., McConnell-Imbriotis, A., & Galvez, S. (2020). Big data: And the micropolitics of entanglement in the Earth's becoming. *International Journal of Qualitative Studies in Education.* https://doi.org/10.1080/09518398.2020.1753848
St. Pierre, E. (1997). Methodology in the fold and the irruption of transgressive data. *International Journal of Qualitative Studies in Education, 10*(2), 175-189.
St. Pierre, E. (2002). Comment: "Science" rejects postmodernism. *Educational Researcher, 31*(8), 25-27.
St. Pierre, E. (2013). The posts continue: Becoming. *International Journal of Qualitative Studies in Education, 26*(6), 646-657.
St. Pierre, E. (2016). Rethinking the empirical in the posthuman. In C. Taylor & C. Hughes (Eds.), *Posthuman research practices in education* (pp. 25-36). Palgrave Macmillan.
St. Pierre, E. (2019). Post qualitative inquiry in an ontology of immanence. *Qualitative Inquiry, 25*(1), 3-16.
St. Pierre, E. (2021). Post qualitative inquiry, the refusal of method and the risk of the new. *Qualitative Inquiry, 27*(1), 3-9.

https://doi.org/10.1177/1077800419863005

St. Pierre, E., Jackson, A., & Mazzei, L. (2016). New empiricisms and new materialisms: Conditions for new inquiry. *Cultural Studies <—> Critical Methodologies, 16*(2), 99-110.

St. Pierre, E. & Pillow, W. (2000). *Working the ruins: Feminist poststructural theory and methods in education.* Routledge.

Strom, K., Ringrose, J., Osgood, J., & Renold, E. (2019). Editorial: PhEmaterialism: Response-able research and pedagogy. *Reconceptualizing Educational Research Methodology, 2,3*(2), 1-39.

Taylor, C. (2016). Educ-crafting a cacophonous ecology: Posthumanist research practices for education. In C. Taylor & C. Hughes (Eds.), *Posthuman research practices in education* (pp. 5-24). Palgrave Macmillan.

Taylor, C. & Bayley, A. (Eds.). (2019). *Posthumanism and higher education: Reimagining pedagogy, practice, and research.* Palgrave Macmillan.

Taylor, C. & Hughes, C. (Eds.). (2016). *Posthuman research practices in education.* London, UK: Palgrave Macmillan.

Taylor, C., Blaise, M, & Giugni, M. (2013). Haraway's "bag lady story-telling": Relocating childhood and learning within a "post-human landscape." *Discourse: Studies in the Cultural Politics of Education, 34*(1), 48-62. https://doi.org/10.1080/01596306.2012.698863

Temper, L., McGarry, D., & Weber, L. (2019). From academic to political rigour: Insights from the 'Tarot' of transgressive research. *Ecological Economics, 164.* https://doi.org/10.1016/j.ecolecon.2019.106379

Ulmer, J. & Koro, M. (2015). Writing visually through (methodological) events and cartography. *Qualitative Inquiry, 21*(2), 138-152. https://doi.org/10.1177/1070800415442706kk

van der Tuin, I. & Dolphijn, R. (2010). The transversality of new materialism. *Women: A Cultural Review, 21*(2), 153-171. https://doi.org/10.1080/09574042.2010.488377

Vladimirova, A. & Rautio, P. (2020). Unplanning research with a curious practice of methodology. In A. Cutter-Mackenzie-Knowles, K. Malone, & E. Barratt Hacking (Eds.), *Research handbook on childhoodnature: Assemblages of childhood and nature research* (pp. 335-360). Springer

Zalasiewicz, J., Williams, M., Steffen, W., & Crutzen, P. (2010). The new world of the Anthropocene. *Environmental Science and Technology, 44*(7), 2228-2231. https://doi.org/10.1021/es903118j

Zapata, A., Kuby, C., & Thiel, J. (2018). Encounters with writing: Becoming-with posthumanist ethics. *Journal of Literacy Research, 50*(4), 478-501. https://doi.org/10.1177.1086296X18803707

Paul Hart (BSc (Hons), BEd, MEd, PhD) - University of Regina, Regina, Saskatchewan, Canada. Paul.hart@uregina.ca
Paul Hart is Adjunct Professor and Professor Emeritus of Science and Environmental Education in the Faculty of Education at the University of Regina where he has taught graduate and undergraduate courses. He has authored, edited and co-edited many publications, including *Teachers' Thinking in Environmental Education* and the *Companion to Research in Education* as well as serving as an editor for major environmental education journals. His research interests focus on post-qualitative research with special applications for environmental education. His current research explores critical and post-informed theories for research practice in writing beyond human-environment boundaries with particular emphasis on ethics and politics of inquiry. https://orcid.org/0000-0002-5747-1722

CHAPTER 2

DIFFRACTIVELY READING 'SUSTAINABILITY' AND THE VIOLENCE OF MODERN THOUGHT IN CURRICULUM AND DISCOURSE

ROBIN BELLINGHAM

Abstract

The term 'sustainability' is an integral part of the contemporary lexicon used in the Australian Curriculum and in wider discourse to discuss responses to ecological crises. In this chapter I consider the term 'sustainability' from a decolonial perspective, and to do so I enact a diffractive reading to generate insights about its ethico-onto-epistemological entanglements and their implications. I provide an outline of a diffractive and decolonising reading methodology. I then apply this in a reading of sustainability discourse and curriculum through Ferreira Da Silva's conceptualisation of the 'violence of modern thought'. Ferreira Da Silva contends that the separability, determinacy, and sequentiality that are the ontological foundations of modernity enable and cocreate the conditions for forms of violence. In light of Ferreira da Silva's critique I consider the entanglement of the term 'sustainability' with the violence of modern colonial-capitalist thought, how these entanglements also emerge in Australian sustainability curriculum, and subsequent implications for reimagining eco-cultural relations. Last, because a diffractive reading involves critical participation in experimenting with what texts do and can do, I take the opportunity to suggest a small number of alternative curriculum concepts that might be used to support the decolonisation of curriculum and understanding of more-than-human relationships.

Keywords: Environmental education, diffractive reading, decolonising education, sustainability, curriculum

Introduction: The onto-epistemological problem

It is not only what we think, learn and teach *about*, but the way many of us think, learn, teach, live, and relate; our habits-of-being (Stein, 2019) and the systems that support these, that are implicated in the transformations necessary to shift the patterns of damage to the Earth that we are entrenched in. This paper poses the problem of particular words we use to discuss ecological crises, issues and actions, and how these might continue to support, undermine, or reimagine, these complicit habits-of-being and systems. The dominant words and concepts we have for thinking about the challenges the earth faces and the shifts we need to create stem from the classical onto-epistemology of the "representationalist triadic structure of words, knowers, and things'' (Barad, 2003, p. 813), that co-produces the nature-culture divide and that is fundamental to the ontology of colonial-modernity. This onto-epistemology is entangled with and strengthened by the supremacy of science. Science is understood and presented as emblematic of a superior onto-epistemology, one that is objective and value-neutral, the pinnacle of thought and the driver of progress through which nature becomes known, quantifiable, generalisable, and predictable (Higgins, 2016).

The establishment of science as representing truth and certain knowledge is evident in mainstream conversations about global crises such as the COVID-19 pandemic and climate change. The dominant rhetoric about these issues is to 'listen to the scientists' and to 'trust the science'. 'The science' is understood within the context of a debate in which trust in science and science denial are discussed unproblematically and reductively, as absolutely polarised, and as the only possible positions. Nuanced ideas such as the fact that science has made possible diverse inventions that have added to the magnitude of greenhouse gas emissions are typically left out of mainstream debates (Sjögren, 2020). The polarised nature of these debates forecloses a range of other possible ecocultural responses and futures (Lövbrand et al., 2015). That science is "a powerful and self-interested political and philosophical force" (Houston, 2021, p. 51) in itself, which often works to stifle and dismiss critique of the contribution it makes to the problems it seeks to solve, is a view that is suppressed.

Mainstream education has for some time been a key agent in cementing, universalising, and acclaiming the thought forms of science, its methodologies and habits-of-being, so as to assist in placing them largely beyond critique. Science, as a force in education, has involved the presentation of the scientism of colonial-modernity as the truest and most valuable knowledge system, and the ongoing erasure of other knowledge systems (Higgins, 2016). Science education has evolved various movements in response to recognition of some of the limitations of the traditional rationale, goals, and methods of the subject, for example the socio-scientific issues agenda (SSI) (Prain, 2012) and the scientific inquiry approach (Jadrich & Bruxvoort, 2011). These advocate the exploration of complex real-world problems over the re-confirmation of established theories, concepts and processes, but tend to uphold the scientific onto-epistemology of the world as knowable and controllable through the extraction and use of data to identify problems, methods and solutions.

Because the legitimacy and power of the modern, mainstream education system are historically invested in and reinforced by ongoing forms of colonisation (Stein, 2019), in order to sustain itself education cannot acknowledge the deeper implications of the confusion of our times; that our systems devoted to mastery, quantification, competition, progress, and standardisation are the conditions for the socio-ecological violence, neglect and trauma that are all around us. Education must leave its more profound responsibilities in world-making largely unexamined. For students whose experience, identity, family, culture, language, resources, or homelands are in ongoing ways negatively impacted by the above habits-of-being of colonial-modernity this educational experience is in many senses one of violent epistemic assimilation (Higgins, 2016).

A diffractive and decolonising reading methodology

In a diffractive methodology, texts are entanglements of more-than-human relationships (Adema, 2021; Barad, 2007; Van der Tuin, 2016). Rather than an interpretation or a critique of the supposedly determinate and stable knowledge texts contain and represent, a diffractive reading is a reading of how entanglements of texts do something, where the reading produced is part of the entanglement and the ongoing doing of something (Povinelli, 2016; Van der Tuin, 2016). More-than-human textual relations might involve texts and other texts; selves and other selves (including authors and readers); lexicons and textual genres as material-discursive phenomena; disciplinary fields; the political economy; technology including processes of text production; history; future; time; space; place; animals; plants; rocks; biology; geology; chemistry; and any other more-than-human phenomena. The emergent reading is a contingent entity; an enactment and reading of patterns particular to the entangled relationships that are attended to and/or prominently manifest themselves.

A diffractive reading and a decolonising reading can coalesce around the key intention to direct attention to the ethico-onto-epistemological entanglements that co-constitute phenomena including knowledge and texts, and to the implications of these for being, agency, possibilities, and vitality for the spectrum of more-than-human entities. This shared intention implicates, among other possibilities, a further common intention to recognise the damage of the competitive hierarchy and eco-cultural boundaries that are features of ontology in modernity, and to re-imagine or re-constitute these features. In Barad's words, a diffractive methodology is "a commitment to understanding which differences matter, how they matter, and for whom" (Barad, 2007, p. 90). Both Barad (2007) and (Haraway, 1994, p. 62) emphasise that a diffractive methodology necessitates the reader's critical participation in the reading; that "[t]he point is to get at how worlds are made and unmade, in order to participate in the processes, in order to foster some forms of life and not others."

A lexicon is a material-discursive agent that is actualised in habits of thinking and doing, including pedagogy and curriculum, exercising significant structures and constraints. In this chapter I explore the relations between the term sustainability, central to our lexicon for talking and thinking about ecological responsiveness, and the colonising onto-epistemology of modernity. That is, for example, how the term might be used in contexts that work to destabilise, or to support, the notion of "a single, universally-relevant rationality and set of values that offer certainty, predictability, and consensus"; the equation of aspiration and success with wealth and resource accumulation; worth determined by economic utility and value; or the belief in the independence and autonomy of (certain) humans (Stein, 2019, p. 3). How it might disrupt or reinforce the treatment of knowledge as a means to "index, control, and order the world" (Stein, 2019, p. 3); the invisibility of the self and subjectivity; the privileging of argument, authoritative voice, reductivism, and auditing; disconnection from Country; and blindness to the trauma that all of these co-constitute.

Selection of texts in this reading is not an exercise of unidirectional agency from myself as researcher for representative coverage of viewpoints. Texts bring themselves to my attention, are selected and select themselves, by provoking thinking about sustainability in modernity. I read historical and contemporary sustainability policy, mainstream sustainability publications, interdisciplinary academic scholarship on education and sustainability in the global and Australian contexts, and existing and proposed Australian Sustainability Curriculum, through Ferreira Da Silva's (2016) philosophical essay "On difference without separability" (2016), in which the author conceptualises grammars of violence in modern thought. Through this reading I posit that the term sustainability is closely entangled with colonial-modern thought, and that it does particular things in support of its endurance. I provide an interpretation of this activity, in the spirit of resistance, ethical concern, and experimentation, with the intention that this interpretation will also do something.

Grammars of violence in modernity and eco-cultural relations

In the essay "On difference without separability" (2016), academic and artist Ferreira Da Silva conceptualises the "violence of modern thought" (2016, p. 59), arguing that the separability, determinacy, and sequentiality that Barad (2007) has also critiqued as limited ontological constructions, cocreate conditions for forms of violence that emerge in modernity. Separability and determinacy are the assumptions that allow modernity to conceive of the social as pertaining to humans, distinct from nature and a whole in itself, but also made up of further properly separable and determinable components, for example gender, race, sexuality, nationality, and ethnicity (Ferreira da Silva, 2016). Sequentiality rests on the notion that events occur in a linear, one-directional series. Sequentiality also rests on separability and determinacy. Without the identification of entities and phenomena separate from one another, it would not be possible to determine sequences. Sequentiality, determinacy, and separability underlie Newtonian causal relations; that is, the notion that the effect of an action of one entity on another can be determined, and this occurs in a linear sequence.

These constructs naturalise opposition and alienation. Where separability, determinacy, and sequentiality are the basis of ontology, difference emerges as a manifestation of an "unresolvable estrangement" (Ferreira da Silva, 2016, p. 66). This notion underlies Darwinian evolution as it is commonly understood and has been put to use in strengthening and reproducing the "violence of modern thought" (Ferreira da Silva, 2016, p. 59). The wisdom that Natural Selection operates as a forceful struggle for existence and competition for resources between innately differentiated groups, resulting over time in further differentiation and hierarchies of those who are more or less successful, underlies colonialism and capitalism and has gone relatively unquestioned as a principle of nature in modernity. In humans a "racial grammar" (Ferreira da Silva, 2016, p. 57) is produced, naturalising and justifying struggle, conflict, hierarchy, and violence between groups. In the more-than-human world, this violence includes the accepted exploitation of more-than-human entities as resources, and capture, enslavement, experimentation and manipulation, torture, burning, fracking, ripping, drilling, mining and polluting, as forms and by-products of this exploitation.

Carr and Milstein (2021) contend that it is a self-preservation mechanism for humanity in modernity to live in purposeful inattentiveness to our inter-relationships with and spectacular harms to the ecosystems we are dependent on. The resulting sanctioned ignorances and disappearances are also forms of colonial-capitalist violence. Maintaining these ignorances removes the obligation for responsibility or action and is essential in staving off "ontological crises resulting from the implosion of dominant socioculturally constructed boundaries between human and 'natural' realms" (Carr & Milstein, 2021, p. 191).

Education institutions, pedagogy, and curriculum play a significant role in such obscuring practices. By upholding the privilege of scientism and colonial-capitalist thinking including the nature-culture binary, education disavows connections with land, nature and Country, and disables interest in, critique of, and responsibility for the connections, relationships, implications, and ethics associated with living in ecologies. The result is normalised silence and inactivity on issues and instances of violence that are all around us. The invisibility of trauma in education has recently been highlighted by school students striking for climate change action, who explicitly name this normalised silence and inactivity as a source of anger. However, many other traumas remain largely invisible in education, including the depth of the impact of ongoing damage to culture and Country experienced in particular ways by Indigenous people in Australia. These spaces of invisibility signal phenomena that cannot be acknowledged lest this topple the integrity of the core beliefs of mainstream society, and of the system. This invisibility also serves to keep us from the understanding that our inseparability from and entanglement with all else means that the violence we do to others and to nature we also do to ourselves (Stein, 2019).

Ferreira Da Silva (2016) argues that modern critical tools cannot support an undermining of constructions of modern thought because these tools themselves rehearse the imaging of the World as an "ordered whole composed of separate parts relating through the mediation of constant units of measurement and/or a limiting violent force" (Ferreira da Silva, 2016, p. 58). The author gives examples of lexical terms based in separability, determinacy, and sequentiality and emerging from wars for control of natural resources, including "refugee crisis", "Muslim Terrorist", and "starving African" (Ferreira da Silva, 2016, p. 57). These terms are agents in reinforcing assemblages of fear and uncertainty and undermining efforts to address inequities in human rights protections. The lexical term of interest in this chapter is 'sustainability' and next I attend to this term and its onto-epistemological premises, through the above analyses of violence and sanctioned ignorance. Specifically, I consider the entanglement of sustainability with notions of time, hope, progress, and violence in the Australian context, its entanglement with the imperative to find 'solutions', and the scientific and militaristic implications of this. Following this I draw the Australian Curriculum Sustainability Cross-Curriculum Priority (ACARA, 2015b) into the diffractive reading and consider its use of the term and its disturbance and reinforcement of these onto-epistemological patterns.

Sustainability, time, hope, progress, and violence

While the definition of the term 'sustainability' continues to be ambiguous and controversial (Eckert & Kovalevska, 2021), the term is entrenched in the lexicon of responses to climate change and ecological crisis. In general terms it suggests the capacity to endure in a relatively ongoing way. Kuhlman and Farrington (2010) note the evolution of a common conceptualisation of sustainability through two key reports. The first, a report of the Club of Rome (Meadows et al., 1972), anticipated that many natural resources crucial to our survival would be exhausted within one or two generations. The second report, from the UN World Commission on Environment and Development (WCED) in 1987 presented an explicit response to this dire message and "was welcomed for showing a way out of impending doom. It was this report which adopted the concept of sustainability and gave it the widespread recognition it enjoys today" (Kuhlman & Farrington, 2010, p. 3437). The WCED (1987) report specified the dimensions of economic viability, environmental protection, and social equity as the components requiring balance for sustainability.

Kuhlman and Farrington's (2010) critique suggests that a function of the term sustainability in public policy has historically been to divert public attention from the realistic consequences of modern ways of life and to help set an agenda, generally regarded to be appropriate and necessary, for an optimistic discourse about the state of the planet and the future of humanity. The WCED (1987) report explicitly shackled sustainability and development together in a coupling that has taken a firm hold in modernity. Kuhlman and Farrington (2010, p. 3438, italics in original) note that in 1987:

> The question which Brundtland and her colleagues posed themselves was: how can the aspirations of the world's nations for a better life be reconciled with limited natural resources and the dangers of environmental degradation? Their answer is sustainable development, in the Commission's words: *development that meets the needs of the present without compromising the ability of future generations to meet their own needs.*

The above definition of sustainable development implicates several assumptions premised on separability, determinacy, and sequentiality, including separate nature as resource and asset for human use, and the necessity of ongoing sequential accumulation and progress for humanity. These are global norms today. For example this definition has been adopted by the European Union in their Green Deal policy (Eckert & Kovalevska, 2021), and is also used in the Australian Curriculum (Australian Curriculum Assessment and Reporting Authority (ACARA), 2015b), to define sustainability itself. This view of sustainability positions technological solutions and economic growth as the substantive constitutive ingredients and typically involves measurement and quantification of environmental impacts, development of suitable regulatory apparatuses, and use of market mechanisms to encourage modernisations of technologies with the goal to improve efficiency, i.e., to use fewer resources but increase production (Eckert & Kovalevska, 2021). This interlinking of sustainability with economy and development is a firm principle of modernity. In this framing, sustainability is tied to a particular form of hope, that is, a hope premised on the assumption of a future in which humans as an individuated species and as individuated selves can continue to accumulate wealth and resources (Lobo et al., 2021, p. 1497).

Counter views and uses of the term have emerged, for example, the concept of 'strong sustainability' (Kuhlman & Farrington, 2010; O'Neil, 2018). In this conceptualisation nature is "intrinsic and basic to human survival" and "a deep ecological view of the world that supports bio-equity of species and where humans cannot stand apart from nature" is espoused (O'Neil, 2018, p. 370). There is value in this kind of ongoing debate and reconceptualisation, however it is simultaneously essential that we question the extent to which the term sustainability can really be helpful in challenging the status quo from which it has emerged. As Stein argues: "from a decolonial perspective, the very idea of sustainability is suspect, and it therefore becomes necessary to ask: What are we trying to sustain, and why?" (Stein, 2019, p. 201).

Coloniality provides the forms of ordering, relating, calibrating, and governing of modern existence to such a deep extent (Stein, 2019) that these endure in spite of contestations about policy directions. For example, the message of Greta Thunberg, an iconic voice in climate change activism, strongly critical of political inaction for sustainability, is presented as a counter-discourse and yet draws on traditional colonial-capitalist ontology in a discourse of sustainability that retains human privilege and separation from nature (Sjögren, 2020). The unfairness of leaving the beautiful planet spoiled for children and grandchildren is a key part of the Thunberg message. It does not in essence deviate from colonial values for the Earth as a place primarily for humans to continue to thrive unencumbered and to be recipients of the same kinds and the same, or greater, degree of possibilities as those that have characterised the past. Sustainability is not now a concept that lends itself to disturbance of the thinking of modernity. It has "become a call to conceive a mode of (multi)existence that is pliant to our desires even as political alliances become very confusing" (Povinelli, 2016, p. 12).

In Australia, the conditions with which a particular understanding of sustainability co-emerge, include the heritage of a particular strain of ruthless colonialism (Bird Rose, 2004). Australian settler ontology, co-emergent with, among other things, a marked dehumanising disdain for Indigenous peoples and cultures, involved a particular future-facing orientation and individualised ambition that took brutality for granted. We might ask whether this ethic may have contributed to a deeply engrained cultural respect for those who can exploit contexts in the service of finding hyper-competitive strategies for further accumulation.

While the elevation of this future-facing, individualistic ambition to a cultural aspiration may seem to have worked to the advantage of many in the short-term, it is not grounded in logic. The linear, sequential forward movement of time this is premised on is problematic in many Indigenous knowledges. For example, for the Indigenous Action group (2020, p. 3) this thinking is embedded in an "(a)pocalyptic logic", a compulsion with a self-fulfilling prophecy of extermination. Where everything is subject to the rhythm of this forward march of progress (Tsing, 2017), the end is not a future event, but a constant inherent part of the march because 'progress', that is, the accumulation of knowledge, wealth and resources, dictates that everything around us is drawn with us into daily participation in its own demise. So ingrained are our understanding of time as linear and our aspiration of continued accumulation and longevity, we cannot imagine a concept of sustainability from which these are disentangled.

The powerful influence of the Fossil Fuel Order is a related condition from which a particular notion of sustainability emerges in Australia, intensifying the hold of coloniality over the concept. The Fossil Fuel Order is a central global system in the maintenance of coloniality. In Australia, the growing power of the coal, oil, and gas industry is supported in its representation from both big businesses and the labour unions that represent the workers in the industry (Ritter, 2021) helping it to continue both to shore up its future even as many other nations move away from fossil fuels, and to continue to wield strategic power over and maladjust other sectors, including education. Sustainability discourse in Australia is subject to this influence, assisting to sustain the notion of fossil fuel use as the inevitable normal (Ritter, 2021).

The Fossil Fuel Order as a hegemonic system, with its vested interests in our democratic system, bolsters particular discourses of sustainability and climate change. For example, the Fossil Fuel Order has assisted in promulgating the general view that responsibility for action to address these issues is an individual matter (Ritter, 2021). Agency is a matter of self-determination, of individuals acting to reduce their carbon footprint, with a supposedly quantifiable effect. This is evident in the Australian Curriculum (ACARA, 2015a), as is discussed later in this article. The cognitive dissonance created by this belief in individual responsibility alongside colossal institutional and business interests and the complexity, scale, and time-frames of climate change, leaves people unmoored and inert, and many young people struggling for hope. It enables sanctioned ignorance of the role and responsibility of institutions of power in ethical relations with the planet and denies the insight that our best opportunity for action is to participate in collective challenges and transformations to overturn the status quo.

'Sustainability', solutions, and the hold of science

Separability, determinacy, and sequentiality also underlie the modern claim that knowledge, including knowledge of environmental crisis, can be produced with certainty by the establishment of efficient causes and mathematical demonstration (Ferreira da Silva, 2016). This, along with the colonial habit-of-being that idealises unrestricted autonomy and conceives responsibility and relationships as matters of utility-maximisation, helps to reduce responses to environmental crises such as climate change to the finding and applying of solutions (Stein, 2019). Our over-reliance on 'solutions thinking' is essentially flawed, for at least two reasons. First, the finding of solutions upholds the human-nature divide and the notion of humans, and especially of the human capacity for science, as possessing the capacity to control and master nature. Second, in the context of ecological crisis, solutions thinking tends to be dependent on extreme kinds of reductionism, resulting in misleading forms of knowledge and strengthening the sanctioned ignorance of ethical responsibility in the relations between places and times and emergent issues.

The finding of solutions is dependent on the identification of specific and separate problems, and so 'climate change' has come to stand for the main problem of relevance to sustainability and the overarching problem we face in the world today. The term is both reductive and nebulous. On the one hand it identifies the problem as pertaining specifically to the climate and positions it as something external to ourselves. On the other hand, there is little public understanding as to the range of issues that the term implicates. Because of the primacy of science, education, research and communication about climate change are constructed primarily as matters of establishing the sources of authoritative knowledge, disseminating information from these sources about what is certain and predictable about it, and using this information to process and ascertain solutions to master it.

Mastery, solutions, and science remain firm in the public mind as appropriate sustainability responses. For example, this is evident in the message of sustainability activist Greta Thunberg (Sjögren, 2020). Thunberg's message emphasises the choices of human civilization in managing future events and further "evokes a grand modern worldview in which change can be made from a point of unity" (Sjögren, 2020, p. 11); the basis of that unity being science. We need, above all, to listen to "united climate science" (Thunberg, 2019, p. 55). Thunberg is specific about the parts of science that we should attend to: "…the only thing that we need to look at is the emission curve. And I'm sorry, but it's still rising. That curve is the only thing we should look at. Every time we make a decision we should ask ourselves: how will this decision affect that curve? We should no longer measure our wealth and success in the graph that shows economic growth, but in the curve that shows the emissions of greenhouse gases" (Thunberg, 2019, p. 64). The message is suggestive of what Hulme (2011, p. 245) has called "climate reductionism", wherein a specific and measurable phenomenon is attributed a significance that makes it an inarguable focal point and a measure against which all decisions should be made, significantly narrowing discussion and foreclosing other perspectives that may not arise from science, or from that particular scientific standpoint.

The extraction of information for the identification of solution strategies, the application of these strategies and the quantification of their effectiveness is the normalised rational approach. This imperative justifies and obscures the forms of violence inherent to the systemic way the method is operationalised. Situated human decision making is replaced "with an impersonal, unquestionable, un-controllable" (Gordon, 2008, p. 167) and decontextualised logic. Because the identification, deployment, and impact measurement of solutions does not attend to encounters and relationships, bodies, experiences, narratives and relations, disappear.

Market-based solutions such as carbon trading, policy solutions such as emission reduction targets, and science and technology solutions such as carbon sequestration (Stein, 2019) and other forms of geoengineering all appear to represent the main kinds of rational and feasible collective actions that are possible. In the context of sustainability and climate change the truth that a cycle dependent on strategy application followed by measurement of effect demands stability and controlled conditions to be meaningful is unaddressed. While the processes and knowledge that uphold colonial order remain largely disavowed and mainstream education and media do little to assist in understanding the entanglement of the colonial order with ecologies and the current instabilities, it continues to appear to make sense for people to look to existing systems and institutions of science, technology, and politics for collective solutions (Stein, 2019).

'Sustainability' in the Australian cross curriculum priorities

Next I read the previously discussed entanglements of sustainability with the colonial-modern ontology of separability, determinacy, and sequentiality, through the Australian Sustainability Curriculum (ACARA, 2015b). Curricula are potentially potent texts, as educators and learners interact closely and explicitly with their vocabularies and concepts, strengthening their power. Curriculum specifically aims to delineate what we should know and how we should come to know it, and in so doing plays a role in co-constituting what it is possible to know. I consider the extent to which evolving ideas of sustainability in the Australian curriculum preserve or depart from the constructs of colonial-modernity I have discussed, including the human-nature binary, the primacy of development and progress via technology, science and solutions thinking, the dysfunctional individualistic future-focussed orientation, and the sanctioned ignorance of institutional power and past and ongoing violence.

In the Australian Curriculum the notion of sustainability is most prominently represented in the Sustainability Cross Curriculum Priority, where it “places emphasis on Sustainability as a priority for study that connects and relates relevant aspects of content across learning areas and subjects” (ACARA, 2015b). The dominant scientific worldview of a separate nature requiring human management and control manifested in the Sustainability Cross-curriculum Priority at its inception. The Melbourne Declaration on Educational Goals for Young Australians (Ministerial Council on Education Employment Training and Youth Affairs (MCEETYA), 2008), made the recommendation that sustainability be an addition to the curriculum, emphasising scientific problem solving to meet challenges such as climate change:

> Complex environmental, social and economic pressures such as climate change that extend beyond national borders pose unprecedented challenges, requiring countries to work together in new ways. To meet these challenges, Australians must be able to engage with scientific concepts and principles, and approach problem-solving in new and creative ways. (MCEETYA, 2008, p. 5)

In the proposed revisions to the Sustainability Cross Curriculum Priority (Australian Curriculum Assessment and Reporting Authority (ACARA), 2021), due to be confirmed in 2022, some shifting views around separability and determinacy thinking are evident. These include the explicit connection between humans in earth systems, and identification of interdependence and connections as central organising ideas, for example, in the statement that: “All life forms, including human life, are connected through Earth’s systems on which they depend for their wellbeing and survival” (ACARA, 2021, p. 5). These inclusions can be interpreted as recognition of, and explicit attempts to make space for educators to address, the problematic human-nature division.

At the same time, the curriculum continues to define sustainability itself in the colonial-capitalist terms of ‘sustainable development’ that were outlined in the 1987 Brundtland report: That is, that: “(s)ustainability addresses the ongoing capacity of Earth to maintain all life. Sustainable patterns of living seek to meet the needs of the present generation without compromising the ability of future generations to meet their needs” (ACARA, 2021, p. 3). Other statements continue to echo the existing dominant managerial approach to nature-as-a-resource, for example that: “Sustainable patterns of living require the responsible use of resources, maintenance of clean air, water and soils, and preservation or restoration of healthy environments” (ACARA, 2021, p. 4).

A revitalised effort to reinforce industry and technology supported by processing of information about impacts as the major drivers and enablers of change, rather than changes to habits-of-being or the political-economic order, is evident. In the proposed curriculum revisions, a new organising idea entitled 'Design' aims to "expand the focus on sustainable design of products, environments and services" (ACARA, 2021, p. 2). The revisions state that "(c)reative and innovative design is integral to the identification of new ways of sustainable living", and "(s)ustainable design requires an awareness of place, past practices, research and technological developments, and balanced judgements based on projected environmental, social and economic impacts" (ACARA, 2021, p. 4). Escobar (2018, p. 25) problematises the design field as also closely entangled with colonial-capitalism, in that it is integral to the structures of unsustainability that hold the modern world in place, and much of what occurs in the name of design "involves intensive resource use and vast material destruction."

The view presented in the Sustainability curriculum revisions also tends to uphold rather than enable shifts in the idea that responsibility and action towards sustainability must be matters for individuals and communities to negotiate for themselves. Several mentions of individuals and communities in terms of actions and responsibility, and the invisibility of politics and institutions in this regard sustain this pattern. For example, the revisions aim to: "provide clearer support to explore how individuals and communities can take action and effect positive change" (ACARA, 2021, p. 2). As was also noted by Smith and Gough (2016) of the 2015 version of the curriculum, no space is created for individuals to explore or understand political constraints on the agency of individuals in achieving these ends.

In its explicitly stated future-orientedness, the curriculum "is designed to encourage students to reflect on how they interpret and engage with the world and support them to take informed action to create a more environmentally and socially just world" (ACARA, 2021, p. 3). Institutional invisibility is preserved in the failure of the curriculum to address history, past planetary violence and its ongoing consequences, or the responsibility for these. Responsibility for the past is made immaterial, as what is important is that individuals in school should now come to understand their own responsibility to the Earth, to support their own future (Smith & Gough, 2016). Where reflection on the past is mentioned it is in the context of "(c)ommunity action for sustainable futures (which) requires individuals with shared beliefs and values to collaborate, seek information, identify solutions, influence others and reflect on and evaluate past actions as they work towards a desired change" (ACARA, 2021, p. 6). Human activity is acknowledged as an agent in relation to the imbalance and depletion of the Earth's interconnected physical and biological systems. However the nature of these activities remains unspecified and mention of historical, ethical, economic, and political dimensions of these is absent. As in the 2015 version of the curriculum, the links between an absence of understandings of interdependence and reciprocity in dominant systems and thinking, and current conditions, are denied (Smith & Gough, 2016). Nowhere in the Sustainability Cross Curriculum Priority is it suggested that issues of unsustainability might have a direct relationship to the ongoing intensification of colonialism, or any other aspects of systems of modernity.

In summary, the 2015 version and proposed revisions of the Sustainability Cross Curriculum Priority can be interpreted as aiming to address an absence of attention to ecological challenges and to address, to some extent, the nature-culture divide. Simultaneously the curriculum assists in sustaining the violence of modern thought in an anthropocentric and instrumental worldview, an onto-epistemology of scientism, and the invisibility of institutional power. In these ways the Curriculum demonstrates the inextricability of sustainability from the thinking of colonial-modernity.

Given these entanglements of sustainability with the foundations of colonial-modern thought, the use of the term sustainability may be of limited value in facilitating more connectedness and responsiveness to ecological systems and crises. While we continue to use the term sustainability to help frame our responses, a raft of deep and complex denials and sanctioned ignorances are allowed to stand. These include denial of the illogicality and the ongoing damage of a global capitalist economic system of continuous growth without consequence; of institutional and state-led violence toward, and expropriation, instrumentalisation and exploitation of, the more-than-human world on the basis of othering; of the gifts and understandings of other human and more-than-human knowledge systems; and of interdependence and our responsibilities to each other and the more-than-human world and its ecological metabolism. Sustainability in the Cross Curriculum Priorities (ACARA, 2015b) has been captured by coloniality; perhaps it is too steeped in colonial ordering to escape.

Possibilities for a lexicon for decolonising curriculum

If the project of decolonisation is, in part, a questioning of the onto-epistemological dimensions of modern thought, to recognise, understand, and address the ethico-political-material challenges this raises in the global present, I ask the question: what role does the notion of sustainability play in this project? "Thought does something; it assembles and correlates" (Povinelli, 2016, p. 137). This chapter posits that the assemblages and correlations entangled with the concept and term sustainability are closely entangled with and support the preservation of the power, thought and systems of colonial modernity and its ethic of accumulation, making the term at best highly problematic for use in discourse, curriculum, learning, and action toward transformative change. In this reading I have considered some of the ways that sustainability reinforces the desirability and necessity of ongoing progress for human groups, and its role in constructing ecological crises as phenomena external to ourselves and to institutional power. In education, this denies us opportunities to engage with the onto-ethico-political dimensions of these issues in their entanglement with ourselves, our histories, political-economies, and our habits-of-being.

We need a range of different terms to support us to recognise and understand our failures and to help to create conceptual, ontological, and actionable change (Ross, 2021). Ferreira Da Silva suggests that the situation demands "that we release thinking from the grip of certainty" (Ferreira da Silva, 2016, p. 58). To do so is a significant ontological and affective challenge, necessitating that we relinquish our investment in the image of mastery of the future. Aspects of this challenge are articulated by both Barcan (2020) and Whyte (2020), who note the ontological paradox of our possible responses to climate change. It may be possible to take large scale, immediate action drawing on our current tools, in which case we may achieve something in terms of lessening the force of pending ecological events, but we will leave the habits-of-being that are at the heart of these conditions untouched. Alternatively, we might patiently address our relationships with the Earth and those most damaged by coloniality, and work to transform our habits-of-being. In this case we may learn something about living with care and co-responsibility but it will be too late to avoid many ecological catastrophes. The lesson is that we need to confront and deeply consider the fact that the future is radically uncertain and uncontrollable, but that we have a responsibility to transform our responses and relationship with the Earth regardless.

Coping with the knowledge that we will not master the Earth and the future requires intellectual, affective, and relational preparation. In education, such an approach would understand learning as an ongoing, open-ended relational process rather than a finite phase of acquisition of knowledge. It would also understand learning as inseparable from being; that is, learning and being are intertwined experiences integral to everyday life (Higgins, 2016). Barad (2000, p. 237) proposes that in such a shift, "scientific literacy becomes a matter of agential literacy – of learning how to intra-act responsibly within the world". The literacy that is needed should account for the ways in which phenomena are always already natural-cultural. It should account for the ways in which phenomena are not produced in cause and effect sequences, but co-produce and are co-produced by, other phenomena, and it should account for the always entangled ethics of the relations of phenomena, that implicate forms of co-responsibility in the world's becomings (Higgins, 2016).

Because a diffractive reading involves critical participation in experimenting with what texts do and can do toward the fostering of some forms of life, and not others (Haraway, 1994), below I identify a small collection of terms which could assist us in shifts towards curriculum and pedagogy that support the above aims, the valuing of multiple knowledges, and the opening up for examination the existing, violent forces of modern thought that have cocreated the unequally distributed exploitations and damage to ecosystems we see today. There is a wide range of terms with such potentials, prominently including many from Indigenous knowledge systems. I identify the few I outline here as a brief experiment in supporting other assemblages of thought forms for curricula, and other possibilities.

Ecology of knowledges

The concept of an ecology of knowledges (Santos, 2007) is based on recognition of the plurality and interconnectedness of knowledges. Knowledges emerge with certain conditions, and all knowledges are recognised as both insufficient and indispensable. In curriculum and pedagogy, this conceptualisation enables destabilisation of the hierarchy of knowledges in which science sits at the top. It enables both the valuing of, and critique of the limits of, all knowledge systems, including knowledges of the more-than-human. It sets the expectation of opportunities for engaging with different knowledge systems for learning and responding. It implicates learning from difference, rather than learning to master difference (Ahenakew, 2016) as the relational process at the heart of ongoing understanding.

Pedagogy of porosity

A pedagogy of porosity (Bellingham, 2021) is a conceptualisation of pedagogy not for assisting to know the parts of the world, and to utilise them, but as space for facilitation of "the flow of living knowledge" with "as many other agents as possible" (Yunkaporta, 2019, p. 99). Porosity in coral provides a demonstration of the commitment every entity already has to this flow of living knowledge. Coral porosity is the crucial space through which nutrients flow, connections are made, and communication throughout the ecosystem occurs (McCalman, 2013). Porosity, like the void in quantum field theory, tells us that the universe is "teeming with the full set of possibilities of what may come to be" (Barad, 2007, p. 354). In pedagogy and curriculum, porosity suggests the sharing and co-creation of learning and the strengthening of relationships, and sets the expectation of supporting dynamic spaces for many voices, and for plurality, difficult conversations, resistance, and contestation to emerge (Bellingham, 2021). This pedagogy implicates ongoing recognition of and accountability to the regeneration and sustainment of ecological relationships (Higgins, 2016). This requires enriching our capacity to attend and listen, to refocus on our reciprocal and mutually generative relations with others (Taylor, 2017).

Ecocultural identity

Ecocultural identity recognises the entanglement of ecologies with selves (Milstein, 2020). The concept assists in understanding how and to what extent people and cultures separate or connect themselves with their ecologies in their concept of themselves, the way boundaries between these are regulated and controlled, and how othering occurs. In pedagogy and curriculum, the concept can assist in helping us recognise the normative separation of identity from ecology in modernity, and the ways this has a constraining impact on our own experience, and on the possibilities of the culture for responsive connection-making. It can enable exploration of a range of forms and experiences of eco-cultural identity, for example, knowing-in-being with nature, finding a passion for what is at stake in our relations with nature, and facing the fact that complexity and complicity are the constitutive conditions of these relations. This means accepting that those of us who are invested in and gain from modernity are implicated in the harms it produces, while also accepting that we have no choice but to engage as ecocultural selves in the rectification of these harms and ongoing care of the planet.

Negative capability

Negative Capability is the poet Keats' (1925) term for a capacity to rest in uncertainty, mystery, and doubt without striving toward factual or philosophical certainty or definitive answers. A state of negative capability allows for pondering, imagination and reception, without the constraint of a need to resolve or produce, enabling openness to a wider sphere of perception, recognition, and attention. It is in essence a rejection of preconceived ideas about the world. The valuing of such a capacity in curriculum and pedagogy would help to cultivate our openness and receptivity to a broader range of connections with the world we exist in and to tolerate and generatively engage with the uncertain.

References

Adema, J. (2021). The commodification of the book and its discursive formation *Living Books*. MIT Press. Retrieved from https://livingbooks.mitpress.mit.edu/pub/qq1wouku/release/1

Ahenakew, C. (2016). Grafting Indigenous ways of knowing onto non-Indigenous ways of being: The (underestimated) challenges of a decolonial imagination. *International Review of Qualitative Research, 9*(3), 323-340.

Australian Curriculum Assessment and Reporting Authority (ACARA). (2015a). Australian Curriculum. Retrieved from https://www.australiancurriculum.edu.au/

Australian Curriculum Assessment and Reporting Authority (ACARA). (2015b). Sustainability. *Australian Curriculum Version 8.4.* Retrieved from https://www.acara.edu.au/curriculum/foundation-year-10/cross-curriculum-priorities/sustainability-ccp

Australian Curriculum Assessment and Reporting Authority (ACARA). (2021). *Australian Curriculum Review. Cross Curriculum Priorities. Sustainability*. Retrieved from https://www.australiancurriculum.edu.au/media/7018/ccp_sustainability_consultation.pdf

Barad, K. (2000). Reconceiving scientific literacy as agential literacy. In R. Reed & S. Traweek (Eds.), *Doing Science+Culture*. New York, NY: Routledge.

Barad, K. (2007). *Meeting the universe halfway: Quantum physics and the entanglement of matter and meaning*. Durham: NC: Duke University Press.

Barcan, R. (2020). The campaign for legal personhood for the Great Barrier Reef: Finding political and pedagogical value in a spectacular failure of care. *Environment and Planning E: Nature and Space, 3*(3), 810-832. https://doi.org/10.1177/2514848619886975

Bellingham, R. (2021). Reef pedagogy: A narrative of vitality, intra-dependence, and haunting. *Educational Philosophy and Theory*, 1-23. https://doi.org/10.1080/00131857.2021.1917365

Bird Rose, D. (2004). *Reports from a wild country: Ethics of decolonization.* Kensington, Australia: University of New South Wales Press.

Carr, J. & Milstein, T. (2021). "See nothing but beauty": The shared work of making anthropogenic destruction invisible to the human eye. *Geoforum*, *122*, 183-192.
https://doi.org/10.1016/j.geoforum.2021.04.013

Eckert, E. & Kovalevska, O. (2021). Sustainability in the European Union: Analyzing the discourse of the European Green Deal. *Journal of Risk and Financial Management, 14*(80).

Escobar, A. (2018). *Designs for the Pluriverse (New Ecologies for the Twenty-First Century)* [Kindle version]. Duke University Press.

Ferreira da Silva, D. (2016). On difference without separability. In J. Volz, J. Rebouças, I. Rjeille, P. Fundação Bienal de São, & M. Pavilhão Ciccillo (Eds.), *Incerteza viva: 32nd Bienal de São Paulo: 7 Sept-11 Dec 2016 Catalogue* (pp. 57-65). Retrieved from
https://biennial.com/files/pdfs/7879/d-ferreira-da-silva-on-difference-without-separability.pdf

Gordon, A. (2008). *Ghostly matters: Haunting and the sociological imagination*. Minneapolis, MN: University of Minnesota Press.

Haraway, D. (1994). A game of cat's cradle: Science studies, feminist theory, cultural studies. *Configurations, 1*, 59-71.

Higgins, M. (2016). Decolonizing school science: Pedagogically enacting agential literacy and ecologies of relationships. In C. Taylor & C. Hughes (Eds.), *Posthuman research practices* (pp. 267-289). Basingtoke, UK: Palgrave Macmillan.

Houston, C. (2021). Should we 'follow the science'? *Arena,* (7), 51.

Hulme, M. (2011). Reducing the future to climate: A story of climate determinism and reductionism. *Osiris, 26*(1), 245-266.

Indigenous Action. (2020). Rethinking the apocalypse: An Indigenist anti-futurist manifesto. Retrieved from

http://www.indigenousaction.org/rethinking-the-apocalypse-an-indigenous-anti-futurist-manifesto/
Jadrich, J. & Bruxvoort, C. (2011). *Learning & teaching scientific inquiry: Research and applications*. Arlington, VA: NSTA Press.
Keats, J. (1925). *Letters of John Keats to his family and friends*. London, UK: Macmillan and Co.
Kuhlman, T. & Farrington, J. (2010). What is sustainability? *Sustainability, 2*(11), 3436-3448. Retrieved from https://www.mdpi.com/2071-1050/2/11/3436
Lobo, M., Bedford, L., Bellingham, R. A., Davies, K., Halafoff, A., Mayes, E., . . . Lucas, C. (2021). Earth unbound: Climate change, activism and justice. *Educational Philosophy and Theory*. Retrieved from https://www.tandfonline.com/doi/full/10.1080/00131857.2020.1866541
Lövbrand, E., Hedrén, J., Beck, S., Chilvers, J., Forsyth, T., Hulme, M., . . . Vasileiadou, E. (2015). Who speaks for the future of Earth? How critical social science can extend the conversation on the Anthropocene. *Global Environmental Change, 32*, 211-218. https://doi.org/10.1016/j.gloenvcha.2015.03.012
McCalman, I. (2013). *The reef: A passionate history*. [Kindle version]. Retrieved from Amazon.com.
Meadows, D., Meadows, D., Randers, J., & Behrens, W. (1972). *The limits to growth*. Washington, DC: Potomac Associates, New American Library.
Milstein, T. (2020). Ecocultural identity boundary patrol and transgression. In T. Milstein & J. Castro-Sotomayor (Eds.), *Routledge handbook of ecocultural identity*. Milton, UK: Taylor and Francis Group.
Ministerial Council on Education Employment Training and Youth Affairs (MCEETYA). (2008). *Melbourne declaration on educational goals for young Australians*. Melbourne, Australia: Ministerial Council on Education Training and Youth Affairs.
O'Neil, J. K. (2018). Transformative sustainability learning within a material-discursive ontology. *Journal of Transformative Education, 14*(4), 365-387. https://doi.org/10.1177/1541344618792823
Povinelli, E. (2016). *Geontologies: A requiem to late liberalism*. Durham: NC: Duke University Press.
Prain, V. (2012). Acting on sustainability. *Research in Science Education, 42*(1), 149-154.
Ritter, D. (2021). Empire of the dead. *Arena* (6 Winter).
Ross, D. (2021). Dare to think past the Anthropocene. *Arena, 6* (Winter).

Santos, B. (2007). Beyond abyssal thinking: From global lines to ecologies of knowledges. *Review (Fernand Braudel Center), 30*, 45-89.

Sjögren, H. (2020). Longing for the past: An analysis of discursive formations in the Greta Thunberg message. *Culture Unbound.*

Smith, W. & Gough, A. (2016). Critical discourse analysis of sustainability curriculum. *Curriculum Perspectives, 36*(2), 28-40.

Stein, S. (2019). The ethical and ecological limits of sustainability: A decolonial approach to climate change in higher education. *Australian Journal of Environmental Education, 35*(3), 198-212. https://doi.org/10.1017/aee.2019.17

Taylor, A. (2017). Beyond stewardship: Common world pedagogies for the Anthropocene. *Environmental Education Research, 23*(10), 1448-1461. https://doi.org/10.1080/13504622.2017.1325452

Thunberg, G. (2019). *No one is too small to make a difference*. London, UK: Penguin.

Tsing, A. (2017). *The mushroom at the end of the world: On the possibility of life in capitalist ruins*. Princeton, NJ: Princeton University Press.

Van der Tuin, I. (2016). Reading diffractive reading: Where and when does diffraction happen? *Disrupting the Humanities: Towards Posthumanities, 19*(2). http://dx.doi.org/10.3998/3336451.0019.205

Whyte, K. (2020). Too late for indigenous climate justice: Ecological and relational tipping points. *WIREs Climate Change, 11*(1), e603. https://doi.org/10.1002/wcc.603

World Commission on Environment and Development (WCED). (1987). *Our common future*. New York, NY: Oxford University Press.

Yunkaporta, T. (2019). *Sand talk*. London, UK: The Text Publishing Company.

Robin Bellingham (B Appl Sc (Health), Dip Teach, M HSc (Hons), MCW, PhD) - Deakin University, School of Education, 221 Burwood Highway, Burwood, VIC 3125, Australia. r.bellingham@deakin.edu.au

Robin Bellingham is a Senior Lecturer in Pedagogy and Curriculum at Deakin University, Melbourne. Her research explores how education and methodologies can respond ethically to pressing problems of modernity such as educational and political disempowerment and disengagement, the ongoing effects of colonisation, and ecological crisis. She draws on different writing genres, posthumanism, new materialism, decolonial theory and critical theory.

https://orcid.org/0000-0002-0425-6770

CHAPTER 3

A MOVEMENT, A SYSTEM, A MOUNTAIN, A CHORUS: ART PRACTICE WITHIN ENVIRONMENTAL EDUCATION AS A TRANSDISCIPLINARY MODE OF POST-QUALITATIVE INQUIRY

CASSANDRA TYTLER

Abstract

The way truth and knowledge are produced is steeped in power relations and privilege of access. As researchers working within the academy, this poses an interesting task for research outcomes where objectivity becomes problematised due to the social positions (St. Pierre, 2021) and hierarchical processes born from being within the institution (Whelan, 2015). Post-qualitative inquiry is an approach to research that works to disrupt these hierarchies. This paper argues that embracing art practice within environmental education research is a mode of post-qualitative inquiry that decentres the human subject researcher. I turn to Barad's analysis of diffraction (2007), to argue for research that is transdisciplinary, post-qualitative, and ethical. I use an analysis of two public artworks that were exhibited as part of *Treatment: Flightlines* (2017). Turning to a reading of affect created by artworks exhibited *in situ* as an entangled *ethico-onto-epistem-ology* (Barad, 2007) is a way of activating inquiry so that participants are responding to multitudinous conditions that are separate from representationalism, set methods, and the purely linguistic. I conclude that there is the opportunity in environmental education to use art practice for ethical and political relations, and material-discursive metamorphoses of understanding.

Keywords: Post-qualitative, transdisciplinary, public art, post-humanism, environmental education

A chapter in a book

The outcome of this paper, a chapter in a book, creates its form in advance. The language used, its structure and even its audience is pre-set. Communication of knowledge is expected within the mode of the academic voice; one that is neutral and learned. The framework of the text is defined by its mode of production, in this case the academic paper, above that of the author, whose independence is skewered in service to the form the work is presented in and the expectations of the reader/audience (Benjamin, 1970). The critical canon–and indeed, my citing of Benjamin was calculated–that is inferred has social, cultural, and intellectual value and legitimacy in this academic realm (Whelan, 2015). Truth, knowledge, and the way it is produced in this sense is steeped in power relations and the privilege of access. To quote Richardson: "Wherever truth is claimed, so is power; the claim to truth is also a claim to power" (2002, p. 415). Here, a singular 'truth' is problematised. As researchers working within the academy, this poses an interesting task for research outcomes where objectivity becomes difficult due to the social positions (St. Pierre, 2021) and hierarchical processes born from being within the institution (Whelan, 2015).

Post-qualitative inquiry is an approach to research that works to disrupt the hierarchies mentioned above. It aims to de-centre the researcher, to work alongside theory (most specifically post-structuralist thinkers Foucault, Deleuze, and Derrida), and to reject a humanist qualitative methodology, where the subjectivity of the researcher is never objective (see Tytler, 2022, p. 1). Post-qualitative inquiry is always becoming and has no end. In St. Pierre's words, "[t]his is truly experimental inquiry—attending to the surprises that point to difference and refusing the impoverished answers we've given to the questions the world has posed" (St. Pierre, 2018, p. 605). Entanglements form between theory, the researcher, the text produced and the witness/participant, where relationality, affect, more-than-human intra-actions (Barad, 2003), and diffractive encounters (Haraway, 2018; Barad, 2007) take place in an ever-evolving rhizomic multiplicity (Deleuze & Guattari, 1987).

From the outset, as researcher and as writer it is necessary to place myself up front to try to re-position myself to the side. Post-qualitative inquiry tells us that we cannot escape our social positions. They inform our analysis, and even before that, our methods within research. We researchers are social beings, hailed as subjects through the call of power (Althusser, 2001).[1] I, a white Australian settler on Aboriginal land (specifically in my case, the Wurundjeri lands of the east Kulin nation), write this text with the advantages that this social position has given me in a colonised country. My academic training gives me the language to express myself within this forum, however, I am aware that this position follows specific dualistic lines of legitimacy promoted within Western thought. Indeed, as noted by Nkosinkulu (2021), following the work of Fanon (2008), the curriculum has often been used as a racist and ideological tool in dismissing knowledge coming from outside the Western education system and dismissing Indigenous knowledge systems. In her keynote address for the CAR Symposium, 2021, Bellingham (see Chapter 2) also pointed to these hierarchies of knowledge accumulation and their situatedness through relations in politics. In this sense, the doing of post-qualitative inquiry is an act informed in the politics of postcolonialism, feminism, queerness, posthumanism, disability studies and intersectionality, all of which reject the idea of the 'universal' (male, white, heterosexual, and able-bodied) human citizen.

Within environmental education (EE) where does this approach leave us, especially as educators, where clear *answers* are so often sought? Kerasovitis (2020) describes the process of post-qualitative inquiry, with its embrace of the "not knowing," the "chaotic in itself" to a piece of "art that momentarily escapes comprehension, cocooning itself in its chaotic nature in order to render a concept sensory" (p. 66). I take this as my cue going forward in proposing a transdisciplinary approach to EE and art practice and its affective potential within research.

From a posthumanist perspective, Barad (2007) has taught us that agency does not rest in the individual, but takes place through intra-actions, that affect humans and matter in ever-evolving encounters within material conditions and with each other: "neither the discursive practices nor material phenomena are ontologically or epistemologically prior" (p. 152). As one example, in research, creating encounters rather than interviews allows the

---

[1] Louis Althusser argued that subjects are formed through answering (or being recognised within) the call of figures and institutions of power. Our subjecthood is reinforced and regulated through these institutions and the ideologies that they promote.

investigator to de-centre both themselves and the participant *in situ*.[2] The *answers* sought in research are entangled in these intra-actions. What may be more pertinent in taking a poststructuralist and posthuman approach to post-qualitative inquiry, is that there is no singular *question*, but instead multiple modes of becoming in knowing, or as Barad would call it and which I will explain further below, *knowing in being* (ibid).

Going forward I argue that embracing art practice within EE research is one mode of post-qualitative inquiry that can work towards a shifting from the centre and unbinding of the human subject researcher. I turn to Barad's analysis of diffraction, influenced by Haraway (2018), to argue for inquiry that is transdisciplinary (as opposed to interdisciplinary), post-qualitative, and ethical.

Diffractively transdisciplinary

Rather than acting as discrete discourses influencing one another in reductionist terms, as is the case with approaches to interdisciplinary inquiry, this paper is an embrace of transdisciplinary research across EE, Philosophy, and Art. It works in a diffractive manner, where the beginnings and ends are unmarked, embedded, and attuned to variance. Haraway (2018) reads diffraction against reflexivity as a space where interferences trace the effects in difference. She explains that "[d]iffraction is the production of difference patterns in the world, not just of the same reflected— displaced— elsewhere" (p. 268). To Haraway, reflexivity is caught up in finding the *real* or the *original* and reflecting it back, whereas diffraction finds the points of difference and recognises the disparate effects created in coming together. Haraway's proposition is a strategy for working in the world that is anti-reductionist. She sees diffraction as "an optical metaphor for the effort to make a difference in the world" (p. 16). Importantly, an embrace of diffraction from Haraway's perspective develops an approach to the relations between humans and non-humans, considering it an open possibility of the non-binary, the fluid and the relational enmeshment of all things.

Barad (2007) extends Haraway's metaphor or ethics of diffraction to a physical occurrence that happens in practice. Her theory of diffraction

---

[2] See Mazzei's extension of the Deleuzian Body Without Organs to a Voice without Organs in an analysis of the interview in posthumanist research: "A voice that does not emanate from a singular subject but is produced in an enactment among research-data-participants-theory-analysis" (Mazzei, 2013, p. 732).

comes from Bohr's understanding of wave behaviour in quantum physics from sound and light waves to the waves in water. She gives the example of diffraction occurring in water waves, "the way [they] combine when they overlap and the apparent bending and spreading out of waves when they encounter an obstruction" (p. 28). This happens when waves lash over a barrier or go through a hole, or intersect with one another, creating diffraction patterns with the change of movement, twisting, and broadening out. This application of the physical phenomenon of diffraction extends it from the metaphorical proposition conceived by Haraway, to a methodology that builds on Haraway's positive engagement with difference. Barad's reading of difference combines quantum physics with an attention to feminist and queer theory as an opening up of discourses around difference, and as a way of broadening perceptions and finding innovative understandings.

Reading discourse diffractively where the difference in difference is celebrated, and where processes, subjects, and apparatuses change and adapt in reference to one another invites a transdisciplinary tactic. As Barad notes, "diffraction involves reading insights through one another in ways that help illuminate differences as they emerge: how different differences get made, what gets excluded, and how those exclusions matter" (p. 30). This is pertinent in discourse that is transdisciplinary, and works from a posthuman, queer, and post-qualitative perspective because rather than looking at discrete divisions of knowledge from, as an example, the visual arts versus the sciences, a diffractive methodology opens new possibilities across disciplines through the differences they share.

This tactic also works to de-emphasise a singular, humanist, Western knowledge system. Not only is the subject/researcher de-centred by looking for difference through diffraction across disciplines, but this also encourages us to look beyond the human subject alone. In other words, the distinction between subject versus object, human versus non-human or nature versus culture is dissipated, and the 'I' of the researcher is dissolved. In an interview with van der Tuin and Dolphijn, (2012) Barad explains that she sees this notion of difference that comes from reading diffractively, not as a subtraction or as a critique but "in the sense of it being suggestive, creative and visionary" (p. 50).

Diffraction is a part of our worlding through the entanglements created by the constant communication between matter intra-acting. Elements or entities are not independent from one another but materialise in their intra-action (see, Barad, 2003, pp. 814-815). The sense or meaning we make from matter, as well as its measurement and the apparatuses used to measure, all contribute to what Barad (2007) calls "*mattering*" or "*material discursive*" (pp. 146-153). In an interview in *Mousse*, she explains that matter is not bodies nor objects alone, but a:

> …substance in its iterative intra-active becoming—not a thing, but a doing, a congealing of agency. It is morphologically active, responsive, generative, and articulate. Mattering is the ongoing intra-active differentiating of the world. (Kleinman, 2012, p. 80)

Here we see that the material and discursive are not removed from one another but change and adapt according to different configurations within the world. "*Onto-epistem-ology*—the study of practices of knowing in being" is how Barad (2007) describes the in-separability of ontology and epistemology where "knowing is a matter of part of the world making itself intelligible to another part" (p. 185).

Barad's theory is a deeply ethical one, where exclusions and inclusions within world-making must be considered conscientiously. Here, matter is seen as ethical engagement, ontological and epistemological, or what Barad (2007) names *ethico-onto-epistem-ological*: "an appreciation of the intertwining of ethics, knowing, and being" (p. 185). We are not separate from the world we exist in, rather we are affected by or affecting matter in its constant intra-activity and therefore can never be exterior in analysis because we are always entangled *within* it. This understanding frames post-qualitative inquiry discussed above, where the researcher-subject accepts they can never be objectively outside or separate from their investigation. Going forward, my approach to post-qualitative inquiry harnesses Barad's *ethico-onto-epistem-ological* methodology in arguing for a transdisciplinary, non-hierarchical, and ethical approach to EE that uses artworks in place.

Affected by or affecting matter

As I have argued elsewhere (Tytler, 2022), a turn to affect guides an approach to the interrelations between the human body, the social, the cultural, as well as the pre-linguistic embodied connection to things. Using artwork as a research tool for participants to encounter in place, decentres the researcher as well as the human subject participating, and creates new modes of activation and understanding. Turning to a reading of affect as an entangled ethico-onto-epistem-ology, or *ethically knowing in being*, is a way of activating inquiry so that participants are responding to multitudinous conditions that are separate from representationalism, set methods and the purely linguistic. Artwork used within EE has the potential to turn this attention in different ways where outcomes are not set through inflexible, researcher-centred methodologies. This is where force and intensity can instigate new understandings of human, the more-than-human and environment, and the constant and changing dynamic flows taking place between them. Representationalism is displaced through the sensation of affective encounters by an artwork and an embodied knowing is taking place beyond what can be cognitively communicated, or a *knowing in being*. Using artworks that participants engage with in an environment as part of a research inquiry to engage with the more-than-human/more-than-researcher-participant is a way forward in harnessing post-qualitative inquiry as ethico-onto-epistem-ological. The examples from the public art project *Treatment: Flightlines*, discussed further below lend themself to greater benefits in the transdisciplinary field of EE and art research.

Following Spinoza, Massumi (2002) describes affect as the body's capacity to act and to be acted upon or affect and be affected. Affect is not fixed, but moves in ebbs and flows at different intensities, where the body acts as a passage. As Massumi says, "[t]here is no body. There is a continuous body*ing*" (Massumi's emphasis, p. xxxii). From moment to moment, a body is characterised by the capacities it holds.[3] Building artworks *bodying* the space of the researcher and participant creates an encounter where affect and therefore embodiment takes place according to the varying intensities of the moment. Following a politics of affect, situations can be remodulated, and participants respond according to speculative experiments in research. Massumi gives the example of an alarm going off. Whether one decides that there is a danger or not, they are nevertheless coming to that conclusion in a space that was for an instant, one of danger. All other bodies too, are reacting to that alarm in that space at the instance it goes off, "as inhabitants of the same affective environment" (xxxviii). Therefore, there is the space for a guiding principle when placing participants within the same environment. For example, an artist's sonic intervention made within a natural environment that pans sound from left to right can direct participants' focus from one point to another. If the sound is electronic amongst trees, participants recognise the interference as mediated by the artist. Here, focus is not upon the researcher and a strict methodology. To quote Barad (2007), "the point is not merely that knowledge practices have material consequences but that *practices of knowing are specific material*

[3] See (Tytler, 2022) for a discussion of the intra-body created through intermedial practice as EE research, where human, technology and environment are connected in affective encounters.

engagements that participate in (re)configuring the world" (Barad's emphasis, p. 91). The bodies that become body*ied* are not human bodies alone but are material-discursive entanglements of phenomena in the world.

Treatment: Flightlines, 2017 – Like a murmur of birds

As an illustration of the varying ways that art practice can be used from a post-qualitative position of inquiry within EE, I would like to write about my experience as a participant/audience member of the public artwork, *Treatment: Flightlines* 2017, curated by Cameron Bishop. This was a project featuring eight artworks (by ten artists), that were responding to the 10,000 hectare Western Treatment Plant south of Werribee, Victoria that treats more than 182,500 million litres of sewage a year and produces 40 billion litres of recycled water (Melbourne Water, 2022). The art was experienced by visitors on the site across the vast landscape of the treatment plant. To get from place to place the participants were transported on a mini-bus. Certain works would be experienced through the windows of the bus, some within the natural environment, and others inside the plant or in the abandoned town of Cocoroc on the site.

In the exhibition essay, Bishop (2017a) explains that artists were directed to choose an exhibition site intuitively. One that "might be thought of as a space in-between, or a site overlooked … hard to fix and territorialise … following an indeterminate pattern impacted by a variety of unseen forces – like a murmur of birds" (p. 1). My own responses to the artworks, written here, rely mostly on memory, alongside the affective resonance that certain creations gave me, and which still sit with me today.[4] In this sense, the 'truth' of these artworks that I write about is the truth of my affective engagement with them as I remember the experience. In the spirit of post-qualitative inquiry, I am rejecting the representational and working in experimentation alongside the artists' work that I experienced four and a half years ago. As St. Pierre (2021) explains, "[r]epresentation is not the goal of post qualitative inquiry. … The 'what is' question denies immanence, the not yet, the what is 'to come' of poststructuralism" (p. 6). I am also following Riley and White's (2020) interpretation of Massumi's pedagogies of 'attuning with', where affective encounters do not discretely exist and then depart but change in context and over time. As they say,

[4] I made contact with the two artists, Lyndal Jones and Zoe Scoglio, whose work I write about here. They clarified small recollections that I wanted confirmed, and also provided me with further documentation of their work to refer to as I wrote about them.

"bodies are 'marked' through nonconscious experiences of pre-personal affective intensities, bodies are continually undergoing a dynamic and reiterative actualising of virtual possibilities within their assemblage of relations" (p. 264).

My own body dynamically remembers the artworks that I write about here, while also recognising that a description of my (and the artists') exact actions in 2017 may be difficult to put into words (representation) due to a faded memory (my own and the artists') as well as the very nature of 'truth', knowledge, and representation discussed above. I consider this propositional research towards what will serve as post-qualitative inquiry. I take Bishop's (2017a) call in the exhibition essay as part of my *attuning with* and critical reflection:

> *Treatment: Flightlines* is a public art project that does not seek to compete with the wildlife, landscapes, histories or technologies of the WTP [Water Treatment Plant]. What it does do is offer a variety of public artworks to channel those aspects of the plant that at first encounter, go unnoticed, in the hope that they might be experienced in affective, sometimes spectacular, and critical ways. (p. 1)

Bishop makes clear that the artworks do not function above the site they take place on. They are immanent, and in turn have embedded me, an audience participant, into them through my affective and now critical *becoming with* these artworks.

This is my memory

The mini-bus bumps and grinds along the dirt road coming from the old town of Cocoroc, where we have seen archival photographs in old-abandoned buildings of the townspeople who once lived and worked there. I sit at the back of the bus, trying to keep my motion sickness at bay. The heat is not helping, and I regret my sensitive ears/eyes/stomach. A pang of relief as we stop and are allowed outside to experience some more art (even though being transported across the site is a part of the art experience itself). Lyndal Jones' (2017b) work, *Walking with pencil pines* begins at the opening of the suggestion of a path marked through the bush. We learn that the previous townsfolk of Cocoroc would spend relaxation time in this area, tracing paths through the bush forest. Jones addresses us as a group. She seems kind and speaks in a way that feels personal and present. I feel comfortable, which is not always the case when asked to interact live when in a group setting. I appreciate the fresh air, and luckily my motion sickness has not rested with me. (Having felt it highlights the bloom of the air I am now breathing). We are instructed that to experience this walk we are required to wear fluorescent vests and to walk in single file, leaving a large space between the person in front, and by extension behind us. We are also given the choice of three stories on a small piece of paper to choose from: '*A Casual Geography*', '*An Archaeology of the Recent Past*', and '*The Literary Path.*' I select '*The Literary Path.*' The story we choose should inform the walk we make through the bush forest and Jones has made small interventions where objects have been placed within the paths, on branches, or on the ground. In recent communication Jones explained that she tied certain objects to specific stories such as a deer hide draped over a tree referring to the literary path, but participants were not directly told this, so they were presumed to find their own meanings from what they encountered. The act of walking through the bushland path stays with me and my memory is that my response to the site changed when I picked a second story and walked a slightly different path. I also encountered the placed objects differently.

I stroll at a slow pace, with enough distance between myself and the other participants so that I feel uninterrupted by them yet can still see their movements thanks to the fluorescent vest they wear. Walking through a tight path, I am suddenly aware of the smell of the flowers, mixed with eucalyptus. Did Jones plan the walk so that I would have this response suddenly on this spot? Or is it the guided story that opens my senses? I enter a shrouded section of bushland where there are wooden box-like chairs to sit on. I believe they are placed there by the artist, so I purposely sit down. The wind moves in the trees. I look up and notice the movement. Having the affordance to be alone in this section, as instructed through the distance we keep with other participants, the sound of the world has opened itself to me in a way it hadn't during that day. Again, I wonder whether this attunement would have occurred if the artist hadn't placed me here, actively sitting in this circular section where the sound of the environment becomes more apparent. Again, my breath is attuned to its relationship with the trees, leaves and flowers rather than the less-pleasant bus interior. I think to my body's senses and consider smell also. I can see other participants tracing their way on different paths. People are everywhere, yet my interaction is directed towards myself in nature, listening, smelling, feeling, and breathing alongside all that is around me. I am attuned to the aliveness of the place as well as to the stories embedded within it. I think to the townspeople of Cororoc, who I have just seen pictures of as well as the architectural remnants such as the empty swimming pool of their disbanded environment. The Pencil pine or Cypress pine, which the artwork is named after was introduced during post-colonisation. The deer hide draped across a branch I walk past, makes me think of peoples and stories from Europe. Moving through the bush with my senses opened in this directed way, where people walk with me, following stories that become sense-stories, takes me further back to the land before white settlers.

Indigenous and other knowledges

Indigenous Australians also have stories that connect them to the land. These stories share knowledge. In *Sand Talk*, Yunkaporta (2019) tells us that all things are connected, and all knowledge has value with Indigenous thinking. He instructs us not to put ourselves above others, otherwise we cut ourselves off from knowledge. In this way Indigenous knowledge is a "living knowledge" where "the answers are everything, all of us, and anything" (p. 42). In conversation in *Artlink*, Sarah Lynn Rees (Briggs, Delany & Rees, 2021) explains the embedded knowing that comes from Indigenous knowledge:

> [I]f we're talking about a river red gum, we understand it as a tree that is physically tangible, but it's also a marker in the landscape, a marker of what kind of flora and fauna are likely to be nearby, a marker of seasons, it forms part of stories that inform navigation, it has its own story, family, experience of time, it's a home and even maybe a supermarket to fauna, it is so much more than its potential commodity as a building material. (p. 15)

This approach to knowledge flattens the hierarchies of matter and meaning where nothing stands above or takes a privileged position in reference to anything else. This is an immanent philosophy where the borders between matter, organisms and things are in flow with one another, or following Barad (2007), are intra-acting and continually transforming in diffraction with one another. Moving forward with post-qualitative inquiry is a step towards decolonising research and listening with Indigenous and other knowledges. This is a form of ethico-onto-epistem-ology. By 'attuning with' Jones' artwork, my sense of these interconnections between environment, stories, and politics is awakened through this art encounter of felt observation and knowing in being.

Walking with pencil pines – Storying and bodying me

'The Literary Path', the story that I chose as part of *Walking with pencil pines*,[5] refers to the pencil pine, the international symbol of grief, as being native to the eastern Mediterranean region. It recounts a story by the Roman poet Ovid. Here, a beautiful young prince Keos accidentally kills Apollo's favourite stag. Keos prays to Apollo to grieve for eternity to which Apollo turns him into a Cypress tree. However, Jones notes that while this is the well-known version, she learnt a variation to the tale, where the stag is in fact Keos' lover, and he kills him because he is unable to restrict his love for him. It is then that he asks Apollo to allow him to grieve forever. Having learnt the variation on the original, Jones states that she cannot look at a pencil pine without thinking of grief and passion, pulling this past story into her present. She asks, "[w]hat then of the grief necessarily attached to the history of Eucalypts and saltbush you may well ask" (Jones, 2017b, para. 6). As a white settler Australian on unceded Aboriginal land, feeling the histories and connection of others to land, as well as affectively feeling my own connection, heightened thanks to this artwork; I am affectively attuned to a knowing of grief and passion of this moment through story*ing* my position and in that act body*ing* it also. This is a practice of a *knowing in being* that is reconstituting my knowledge of the world and creating an assemblage of relations between matter, story, history, and politics that act diffractively in reference to one another thanks to Jones' treatment of varying elements.

This is a form of ethico-onto-epistem-ology, activating and educating me further (in body and mind) than I had previously been before this intervention. As Riley and White (2020) attest, "assemblages of relations in environmental science education provide opportunities to illuminate the lived existence of 'Other(s)' through ourselves, in ways that discursive practices within the discipline could not" (p. 268). Jones' artwork expanded my knowledge and position in this environment in a lived and embodied mode of being, thinking and doing. This also incorporated stories, belongings, the more-than-human and histories of different peoples. Jones' artwork was placed within a larger intervention of public art made on this site, all of which intra-acted and affectively created an ethico-onto-epistem-

[5] The Literary Path story can be found here:
https://www.theavocaproject.com/_files/ugd/fd6324_f67f8622ca994285bcaf127cd4828d6d.pdf

ology of sensation, connecting the human body, the social, the cultural, and the pre-linguistic.

In her review of *Treatment: Flightlines, Our Collective Waste*, Boland (2017) claims the event made her re-consider her place within the Anthropocene and as a human on Earth. "I thought about the human impact on the environment, of my own impact, and of how each of us successfully ignores the refuse left behind by our own body" (para. 3). By placing artwork in a waste treatment plant, diffraction illuminates differences, allowing Boland to be engaged in place, the environment, and in an entangled relation to others bodily and metaphorically.

Where do you end and begin? Zoe Scoglio's The Sedimentary Collective

Scoglio's (2017) work *The Sedimentary Collective*, also a part of the exhibition, extends this connection between bodies and the geological formations that are produced from human waste from the Werribee Treatment Plant. Scoglio deploys the technology of audio headsets to guide participants on a collective walk at the bio solids precinct. The presentation of this piece is mediated in a different way from Jones' work, where participants are not listening to sounds produced in an environment, but instead are directed by Scoglio's voice as they sit in the bus, and then exit, walking along a directed path. Scoglio speaks in the slow, relaxed tones of a meditation instructor: "I am speaking to you from my gut. From one gut to another. Welcome to the wetlands. The bio solid mounds where our bodies rest" (Scoglio, 2017, audio, provided by the artist). She instructs the participants to walk close to one another in single file along a gravel path. We can see that the path leads to a number of chairs placed in a circle with coloured bunting flags shaped like a teepee over them. The audio of Scoglio's voice, and the accompanying music is synched so that we are all listening to the same instructions at the same moment, while we jointly walk, sit, or stand. We move in a line, instructed by Scoglio to stay close to one another. She pushes this collectivity further in her voice over, where she says:

> Notice the others in the group. We are each a colony of cells. A community of microbes. Together we are a movement, a system, a mountain, a chorus. We are the sedimentary collective. We will move as one, rest as one. A digestion stroll in single file. (Scoglio, 2017)

The act of being together, in unison, walking on mounds made of the waste of others, and ourselves, creates a being of interconnection and exchange. Scoglio asks, "So where is the edge of your system? Where do you end and begin?" Indeed, this rhetorical question is answered in our collective presence, where in this moment, the boundaries between bodies are blurred intra-actively activating us into an assemblage of relations. As Melbourne residents, have we already been present here, with our waste contributing to the ground we are walking upon? As Scoglio has told us we are "the sedimentary collective," therefore we are entangled with these interconnected sedimentary forms, in place and in time. Experiencing this work tells us that there are no fixed boundaries between humans and matter. Instead, here, we walk in tandem, a voice in our shared ear, the rocks crunch beneath our feet, and flesh bodies move alongside us. We are here in a celebration of immanence, shared intra-action and diffraction with the world. This is a moment where we recognise our mutual and constant communication with the ecosphere and the biosphere. The participant subject 'I' does not exist, but they instead connect and diffract through the production of intra-actions. This diffractive approach challenges the binary and oppositional separate researcher. Through the diffractive elements of this artwork, Scoglio has positioned the participants as fluid, entangled, interdependent in co-constituted existence, where through the affective encounter of being, touching, moving and listening, we are affectively transformed beyond a singular human vessel. As Scoglio tells us, "Here we become geological" (2017).

In her article, Boland (2017) quotes Bishop, who says:

> Waste is a by-product of living, and at this site, … works like Zoe Scoglio's, … connect us to what we are – not who we are. The operations, histories and practices of the sewage plant are amplified by the artists to expose us as collective, habitual, cyclic, and ultimately, as beings just passing through. Contemporary art, in this sense, becomes the vehicle through which we make these realisations. (para.16)

The Sedimentary Collective connects participants in an embodied, affective, and relational entanglement, which in turn links us to the site, and further, to the Earth we inhabit in *ethico-onto-epistem-ology*.

Treatment: Flightlines artworks – An analogue to denial of waste

Unfortunately, there is not enough space to write about all eight artworks in depth here, however, they all intersect and diffract with one another, affectively influencing the pieces that come before or after in the tour. As discussed above, the photos and abandoned town of Cocoroc informed my interaction with Jones' work, just as the tour in the mini-bus did. Other artworks in the exhibition included the aerial photography of the ponds precinct by Louis Helbig. The site, viewed from above is visually reassessed as an abstraction of colour and line, flattening anthropomorphic hierarchies. Amanda Shone's work *View des Ponds*, used performing chefs, and a *performance* of a restaurant. Business cards and a menu of said establishment were made, waiters pushed trolleys of food that were grown using recycled water from the Western Treatment Plant, yet when participants went to order, they learnt there was no food to eat. In communication with me, Bishop stated that, "the denial was an analogue to the kind of denial we give to our waste" (email communication). The affective rumbling of tummies, with the fancy looking menu and chefs promoting further enticement, over-rode the disgust often placed upon recycled water. Here was a message of sustainability promoted through denial. Zoë Bastin's work used an improvisational dance performance amongst left-over industrial objects. The bodies dressed in beige and brown, and contorting into the shapes within the junkyard resembled both flesh and concrete, where the beginnings and ends of each form were confused. In *About Now*, the collaborative group comprising of Peter Fraser, Shaun McLeod and Olivia Millard, also used dance, but instead of meshing with the place they danced in (three water treatment ponds) they wore the orange uniforms of treatment plant workers. Their work was experienced by participants from the bus window, where in the distance and then closer, the orange outfits could be seen moving in nature. Olivia Millard danced with a metre long bird head attached to her skull, making one think of the rich diversity of bird species present on the plant, while noting the human intervention of fake trees that Millard danced amongst that the birds were not roosting upon. Fiona Hillary's immersive sonic and light inflected work was situated in the architecture of the historic Water Tower, which acted as a guiding invitation, where the sound echoed out into the atmosphere, thanks to the shape of the performance space/tower. Maree Clarke's work featured six images from her series, *Ritual and Ceremony* (2012): photographs of 84 Aboriginal men and women who represent the Aboriginal language groups of Victoria. They are dressed in black with white ochre masks painted on their faces that represent mourning and grief for lives of ancestors lost. As part of *Treatment: Flightlines* the images were displayed upon a shining gold background that reflected in the sunlight so that the eye was drawn to the black and white photographs of mourning, reminding us of the harm done to Aboriginal peoples in the name of colonisation. The intra-action of light on gold, pulls our vision, reminding us of shameful histories, but also telling us that Aboriginal people are the original inhabitants of the country we live on, and they must be counted.

An assemblage of relations forms an intra-body – guided by artworks in place

In *Treatment: Flightlines* artists found the difference in difference, activating the treatment plant in new ways for a participant so that *knowing in being* took place with the ethical implications of sustainability, Indigenous knowledge and belonging, stories across place and time and the echoes still felt as part of this, the more-than-human and a recognition of our own connection to each other and all things, were communicated.

Each artist remodulated a site so that participants could respond through the affective encounter created within a set space. Artist-led reshaping of an environment such as those discussed, are material-discursive entanglements of phenomena. Affective encounters are created by the guiding principle of the artworks. Here, planes of difference are flattened, but intra-act through the diffractive processes that the artwork introduces *in situ*. The insertion of elements, which would not otherwise be there but for the artworks, produces diffraction. It is in the differences within difference as participants affectively encounter in unison that a new form or intra-body is created. The blurring of boundaries between bodies creates an assemblage of relations in place, where we walk in unison listening to the same voice, the sediments beneath our feet previously from the human form; or where stories link environments of belonging, where the bodies of others perform future or past paths for us through the bush.

Art practice and interventions *in situ* such as *Treatment: Flightlines* offer a transdisciplinary path to EE research when approaching it through a diffractive and affective ethico-onto-epistem-ology. As we have seen these artworks respond to a site in a reciprocal recognition of *indeterminate patterns* or *unseen forces* (Bishop, 2017b). The works are not necessarily research unto themselves but can count as or contribute to inquiry in its multiple forms, philosophising the formation of new models and ways of investigating where the researcher is not central. As Barad (2007) aptly notes, "[d]iscourse is not a synonym for language. … Discourse is not what is said; it is that which constrains and enables what can be said" (p. 146). This is a propositional inquiry in thinking and feeling through how *knowing in being* is enacted, in relation, in the environment, and how material-discursive intra-actions take place and change through the intervention of art practice. Art in its very existence is evocative, poetic, relational, and revelatory. It works beyond a reductive cognition and perception. It is often non-representational, and its use within EE is experimental inquiry in the guise of the post-qualitative. There is the opportunity in EE to use art practice such as that discussed here for ethical and political relations, and material-discursive metamorphoses of understanding and learning. We must look here for the differences, similarities, and echoes across the material, the written, and actions in place.

In a STEM teaching and learning context, where so often the guiding principles of education are policy-driven and celebrate determinant practices and fixed categories, post-qualitative and transdisciplinary research such as this acts as an interruption. Investigating *through*, *with* and *alongside* artworks, people, and place, is one practical example that EE research can use as embodied learning. Celebrating the decentring of the researcher in EE in this way promotes ethical, non-hierarchical entanglements of becoming in knowing through the experiential, intellectual, and sensory. There is no singular answer and not one question but immanent possibilities of ethical worlding for the researcher.

References

Althusser, L. (2001). *Lenin and philosophy, and other essays*. Monthly Review Press.

Barad, K. (2003). Posthumanist performativity: Toward an understanding of how matter comes to matter. *Signs: Journal of Women in Culture and Society*, *28*(3), 801-831. https://doi.org/10.1086/345321

Barad, K. (2007). *Meeting the universe halfway: Quantum physics and the entanglement of matter and meaning*. Duke University Press. https://doi.org/10.1515/9780822388128

Bellingham, R. (2021, November 22-23). *Reading diffractively, decolonising methodology* [online presentation]. Contemporary approaches to research in mathematics, science, health and environmental education symposium, online. Retrieved from https://deakinsteme.org/event/2021-contemporary-approaches-to-research-in-mathematics-science-health-and-environmental-education-symposium/

Benjamin, W. (1970). The author as producer. *New Left Review*, *62*, 83-96.

Bishop, C. (2017a). On golden ponds. *Exhibition Essay*. Retrieved from https://www.publicartcommission.com/project/treatment-flightlines-2017

Bishop, C. (2017b). *Treatment 2017: Flightlines*. A temporary public art project. Western Treatment Plant Werribee, Melbourne.

Boland, B. (2017 June 2). *Our collective waste*. Overland. Retrieved from https://overland.org.au/2017/06/our-collective-waste/

Briggs, C., Delany, M., & Rees, S. L. (2021). Ngargee Djeembana: On the materiality of public space and country. *Artlink*, *41*(2), 12-17.

Deleuze, G. & Guattari, F. (1987). *A thousand plateaus capitalism and schizophrenia*. University of Minnesota Press. Retrieved from https://ebookcentral.proquest.com/lib/rmit/detail.action?docID=310185.

Dolphijn, R. & Tuin, I. van der. (2012). *New materialism: Interviews and cartographies*. Open Humanities Press Imprint. https://doi.org/10.3998/ohp.11515701.0001.001

Fanon, F. (2008). *Black skin, white masks*. Pluto. Retrieved from https://ebookcentral.proquest.com/lib/rmit/detail.action?docID=3386656

Haraway, D. (2018). *Modest_witness@second_millennium. FemaleMan_ Meets_OncoMouse: Feminism and technoscience* (2nd ed.). Routledge. https://doi.org/10.4324/9780203731093

Jones, L. (2017a). On cupressus semperiverens and grief The literary path. *Story Artwork*. Retrieved from https://www.theavocaproject.com/_files/ugd/fd6324_f67f8622ca99428 5bcaf127cd4828d6d.pdf https://www.theavocaproject.com/walking-pencil-pines-avoca

Jones, L. (2017b). *Walking with pencil pines. Treatment 2017: Flightlines*, Western Treatment Plant Werribee, April 27th and 29th, 2017.

Kerasovitis, K. (2020). Post qualitative research - Reality through the antihierarchical assemblage of non-calculation. *Qualitative Report*, *25*(13), 56-70. https://doi.org/10.46743/2160-3715/2020.4756

Kleinman, A. (2012). Published in Special dOCUMENTA (13) Issue of Mousse Magazine. 34. (Milan, Italy), Summer.

Massumi, B. (2002). Parables for the virtual: Movement, affect, sensation. Duke University Press. https://doi.org/10.1215/9780822383574

Mazzei, L. A. (2013) *A voice without organs*: Interviewing in posthumanist research. *International Journal of Qualitative Studies in Education*, *26*(6), 732-740. https://doi.org/10.1080/09518398.2013.788761

Melbourne Water. (2022) *Western treatment plant*. Retrieved from https://www.melbournewater.com.au/water-data-and-education/learning-resources-and-tours/guided-tours-and-virtual-excursions/western

Nkosinkulu, Z. (2021). A decolonial curriculum is everything: An Afrocentric approach. In N. Mthembu (Ed.), *Ethical research approaches to Indigenous knowledge education*. IGI Global. https://doi-org.ezproxy.lib.rmit.edu.au/10.4018/978-1-7998-1249-4.ch003

Richardson, L. (2002). Writing sociology. *Cultural Studies Critical Methodologies*, *2*(3), 414-422. https://doi.org/10.1177/153270860200200311

Riley, K. & White, P. (2020). 'Attuning-with', affect, and assemblages of relations in a transdisciplinary environmental education, *Australian Journal of Environmental Education*, *35*, 262-272. https://doi.org/10.1017/aee.2019.30

Scoglio, Z. (2017). *The sedimentary collective* a binaural guided audio walk. Treatment 2017: Flightlines, Western Treatment Plant Werribee, April 27th and 29th.

St. Pierre, E. A. (2018). Writing post qualitative inquiry. *Qualitative Inquiry*, *24*(9), 603-608. https://doi.org/10.1177/1077800417734567

St. Pierre, E. A. (2021). Post qualitative inquiry, the refusal of method, and the risk of the New. *Qualitative Inquiry*, *27*(1), 3-9. https://doi.org/10.1177/1077800419863005

Tytler, C. (2022). We found a body: The intrabody of human, technology, narrative and environment as postqualitative inquiry. *Australian Journal of Environmental Education,* 1-9. https://doi.org/10.1017/aee.2021.32

Whelan, A. (2015). Academic critique of neoliberal academia. *Sites: A Journal of Social Anthropology and Cultural Studies*, *12*(1), 130-152. http://dx.doi.org/10.11157/sites-vol12iss1id258

Yunkaporta, T. (2019). Sand talk: How Indigenous thinking can save the world / [eReserve] (Third edition.). Text Publishing.

Acknowledgements

Many thanks go to Lyndal Jones and Zoe Scoglio who corresponded with me about their artworks, and Cameron Bishop, who also communicated with me on the exhibition as a whole. Their generosity is appreciated.

Cassandra Tytler (BA, BFA (Hons), MFA, PhD) - Independent Scholar.
cass@cassandratytler.com
Cassandra Tytler is an artist and academic. She works across single-channel video, performance, and video installation. She completed her practice-led PhD within the Faculty of Art (Theatre Performance) at Monash University in 2021. Her research examines the potential of video performance to create a relational and aware politics of resistance. She has exhibited, screened, and performed work nationally and internationally. She has also spoken at conferences and published papers in academic journals. She has gained funding from organisations such as the Australia Council, The Ian Potter Cultural Trust and The City of Brimbank Council. https://orcid.org/0000-0003-0357-7123

CHAPTER 4

(RE)FIGURING SELF-STUDY OF PRACTICE: STORYLINES OF TRANSFORMATION IN THE SPACES BETWEEN OF CLIMATE CHANGE EDUCATION

BRONWYN A. SUTTON

Abstract

In this chapter I propose a (re)figuring self-study of practice, involving multispecies collaborative inquiry, as a productive approach for attending to the posthumanist challenge of decentring the human in climate change education research and practice. I draw on situated knowledge of practice and a year-long experimentation with rethinking my education practice in the area of environmental public pedagogy, within the liminal, messy, affective, in between spaces of practice transformation. I discuss the productive nature of the challenges involved in developing a methodology that seeks to disrupt humans from their centred position and explain how I have created and enacted a (re)figuring self-study of practice that takes up curious practices of multispecies collaborative inquiry. I diffractively read *Encountering Air,* a storyline created for a climate change education workshop, through the methodology to discuss the development of curious practices for multispecies collaborative inquiry. In keeping with the methodological focus of self-study, this chapter is an instance of sharing research findings. I also offer the chapter as an invitation for researcher/educator bodies like mine to think with what may surface as they read this as a storyline of transformation in the spaces between climate change education.

Keywords: Posthumanism, affect, practice transformation, climate change education, self-study

An opening

In a performative enactment of methodology, this chapter is a (re)figuring self-study in and of itself. I set out to write about decentring the self in self-study, which I have found to be a productive assemblage because I thought that what I had to say about methodology might be helpful for other researcher-educator bodies like mine as they similarly rethink practice. Instead, I found the writing process disruptive, productive, generative, and transformative as chapter writing became a textual and embodied storyline of feminist, posthumanist, new materialist pedagogy, practice and research with the affects of climate change education.

Enacting this methodology reveals the possibilities in shifting towards multispecies forms of inquiry that allow us to open to the "shimmer all around us" (Gan et al., 2017) and the ever-present affects in more-than-human, multispecies entanglements. It reminds me of the productive nature of responding to the posthumanist demand to disrupt the human-self from its privileged position as the only body who knows, feels, and is in control of what is taught and learnt. Challenging assumptions is productive in self-study, because it creates possibilities for rethinking, refiguring, and reshaping practice in ways that can remain hidden, even in reflective forms of inquiry. In introducing a (re)figuring self-study of practice I highlight the ways multispecies forms of collaborative inquiry might be significant in transforming how we rethink climate change education.

I open by briefly framing my situated knowledge of education practice, and trace the development of a (re)figuring self-study of practice. I explain how I have been attending to the posthumanist challenge of decentring the human with this self-study, and introduce the language-terms *bodies like mine* and *this body* as devices for troubling the binaries of self/other, us/them, human/nonhuman that often surface in my writing, speaking, education practice, and research. I then share *Encountering Air*, a storyline created in 2020, and diffractively read it with curious practices (Haraway, 2015, 2016) for multispecies collaborative inquiry. These are 'knots' in a somewhat string figured (Haraway, 2016) research assemblage (Barad, 2007) that threads feminist, posthumanist scholarship with new materialist research and knowledge production practices.

Throughout the chapter I highlight the dissonances and dilemmas that have surfaced in taking up risky practices (Haraway, 2016) to point out the productive nature of research that inhabits the messy, liminal, affective spaces between of practice transformation and climate crisis. I highlight how a (re)figuring self-study has become a research approach that is ethical, response-able and systematic, and aligned to a feminist ethico-onto-epistemology.

Coming to self-study: Situated knowledge of climate change education practice

We live in a timespace described by Crutzen (2021) as 'Anthropocene', "the proposed name for a geologic epoch in which humans have become the major force determining the continued liveability of the earth" (Gan et al., 2017, p.1). The World Wildlife Fund, in the 2018 Living Planet Report, explain burgeoning human consumption as a force that is transforming the planet through increased demand for natural resources, land, water and energy, speaking a storyline of the Anthropocene through a scienced-based assessment of the health of the earth. The report points out,

> There cannot be a healthy, happy and prosperous future for people on a planet with a destabilized climate, depleted oceans and rivers, degraded land and empty forests, all stripped of biodiversity, the web of life that sustains us all. (Grooten & Almond, 2018, p. 4)

Globally, food security is declining, sea levels are rising, there is increasingly significant biodiversity loss, natural resources are dwindling, waste is burgeoning, there is continued widespread environmental devastation. Climate change drives increasingly variable weather patterns that impact water resources, ecosystems and the health of all living species. While this situation continues, countries like Australia that are prone to greater variability in weather patterns are likely to face a higher likelihood of national natural disasters like drought, bushfire, and flood (Commonwealth of Australia, 2020). In the last three years, Eastern Australia has experienced the Black Summer bushfires of 2019-20 and the 2022 flood disaster.

Education and lifelong learning have long been recognised as critically important in leading the global changes necessary to address such ongoing multiple ecological crises (Wals & Benavot, 2017). The sustained, productive efforts of generations of environmental educators across a variety of fields and disciplines (Sauve, 2005; Walter, 2009) are challenged by multiple calls to rethink environmental education practice (Lotz-Sisitka et al., 2015; Selby & Kagawa, 2018). I have participated in these efforts since 2004 through my work in public sphere environmental communications, waste education and community engagement, through volunteer climate change activism and more recently in higher education. I frame my climate change education practice as public pedagogy, because it focuses on forms, process and locations of learning that occur in informal, non-formal, every day and beyond school spaces (Sandlin et al., 2011). The choice to situate my practice in these spaces is in response to a deep intra-connection with Earth and its related planetary systems, and knowledge that our choices matter.

I became interested in affects and the ways more than human encounters shape pedagogy and practice in response to dissonances and resonances I felt when encountering feminist scholarship, posthumanist philosophy and new materialist thinking in the early stages of PhD research. Thinking with ontologies of relationality (Lange, 2018) and affect theory (Seigworth & Greg, 2010) was unsettling and provocative, in that it surfaced assumptions I had about what was shaping my practice. I became curious about how affect and encounters with the environment were shaping my pedagogical choices but did not have language to explain what I was noticing. I wondered how I could understand more deeply what was happening in the messy, liminal spaces between transformation, and why the affective dimensions of learning, which felt somehow significant, seemed to be missing from research into transformative learning. I felt that a practice led research methodology would allow me to consider multiple ways of knowing, while arts-based inquiry would enable me to explore embodied and sensory aspects of learning in deeper and different ways.

I understood that rethinking education practice would mean acknowledging response-ability (Haraway, 2016) for the choices I make about research, teaching, and learning. As Haraway (2016) reminds us, "It matters what thoughts think thoughts. It matters what knowledges know knowledges. It matters what relations relate relations. It matters what worlds world worlds. It matters what stories tell stories" (Haraway, 2016, p. 35). Each decision I make, each knot in the research methodology shaped and was shaped by what became possible within a feminist posthumanist ethico-onto-epistemology and a (re)figuring self-study of practice assemblage.

Self-studies enrich and affirm the experiences of the educator because "at the very heart of self-study is the radical idea that those who practice have something important to say about that practice" (Taylor & Coia, 2019, p. 2). What we say about our practice is relevant for our professional development and for others who encounter our research, practice and writing. Self-studies recognise that teachers and researchers are continually shaping, and are shaped by, the worlds of which they are part and support educators to take an analytical stance (Samaras & Freese, 2006) towards practice. Taking up self-study meant sitting with discomfort as assumptions I held about education surfaced through post-graduate study and stretching the boundaries of what and how it is possible to research, teach, and learn.

Self-study was an entry point for this research because it allowed me to interrogate my practice in relation to "learning theory, beliefs about the nature of teaching, and moral, ethical and political values regarding the means and ends of education" (LaBoskey, 2004, p. 818). In keeping with the methodological focus of self-study, this chapter is an instance of sharing research findings. I also offer it as an invitation for researcher/educator bodies like mine to think with what might surface as they read this storyline.

A (re)figuring self-study of practice assemblage

Self-studies are qualitative research methodologies often used to research teaching and teacher education practice within education settings (LaBoskey, 2004). They help us interrogate our own practices in relation to "learning theory, beliefs about the nature of teaching, and moral, ethical and political values regarding the means and ends of education" (LaBoskey, 2004, p. 818). As ethical and rigorous approaches for "investigating and reframing practice over time" (Samaras & Freese, 2006, p. 42), self-studies demand the use of open-ended, systematic and transparent research processes (Samaras, 2011). Framing and reframing of approaches and findings during the investigation is a core practice of self-study, as is applying improved learning about teaching practice as they occur (Samaras, 2011).

Iris van der Tuin's (2019) insistence that "assembling, disassembly, and re-assembling may happen at the same time and in multiple ways…(a) machine in, and with which, a great number of antagonistic, different or differing movements happen simultaneously" (p. 22) resonated with the generative and recursive nature of self-study. The use of bracketed (re) preceding the word figuring in a (re)figuring self-study of practice (a (re)figuring self-study) therefore denotes a simultaneous making, unmaking and remaking, and assembling, disassembling and reassembling that occurs as a body like mine, this body, performatively enacts a (re)figuring self-study. It becomes a generative, emergent, collaborative, creative assemblage that honours the five foci of self-study (Samaras, 2011). It is personally situated and involves critical collaborative inquiry that has an intention of improving education practice; uses transparent and systematic research processes and aims to generate and present knowledge. In a (re)figuring self-study I think with situated knowledge as I change my approach to climate change public pedagogy. I enact multispecies forms of collaborative inquiry, taking up curious practice (Haraway, 2015, 2016) as systematic approach for researching the affects and processes of practice transformation.

Decentring the human may seem counter-intuitive in self-study, which ostensibly is personally situated and focused on practice improvement. However, a (re)figuring self-study shifts this body towards relational, response-able (Haraway, 2016) research that acknowledges the intra-active (Barad, 2003) and affective nature of entanglements between bodies. I looked to posthuman performativity and attempted to respond to Barad's (2003, 2007) argument that language and representation have "more power than they deserve" in shaping ontologies. Barad's insistence that a posthuman performative account involves "understanding thinking, observing, and theorizing as practices of engagement with, and as part of, the world in which we have our being" (Barad, 2007, p. 133) resonated with my sense that affects that surface within entanglements and encounters with multispecies and more-than-human kin are important in shaping education practice and research choices.

Thinking with posthuman performativity pushed this body to consider forms of collaborative inquiry that acknowledged the agency multispecies and more-than-human kin. Experimenting with new materialist concepts and approaches that acknowledge the agency of matter (Barad, 2007) revealed on one hand the potential and possibilities of posthumanist perspectives for understanding affects of transformation, and on the other the difficulty in responding to demands to decentre the human if I only draw on qualitative research methods.

A (re)figuring self-study therefore recognises multiple ways of knowing; the importance of embodied, curious and creative practices; more-than human and multispecies agency; the significance of affect; and the deep, rich, and diverse insights produced in heterogeneous spaces of living, learning, practicing, and researching. It insists I recognise the agency of more-than-human bodies and demands insights be read through the body (Lenz Taguchi, 2012).

Bodies like mine

Bodies like mine are multispecies, more-than-human, material-discursive figurations (Schadler, 2019, p. 216), entanglements entangled with a multiplicity of bodies like mine. They are agentic, and can affect and are already, always affected, shaped by and shaping immanently through endless multispecies encounters and entanglements between bodies, always becoming with. 'Human' bodies are bodies like mine; sensory, thinking, creationary assemblages (Barad, 2007) in ongoing entanglement with multispecies, more-than-human bodies and different worlds of living, research, and practice. 'Educator' bodies like mine live, research, and practice education in ways that shape and are shaped by multispecies encounters that shape how they perform research and respond to knowledge that unfolds, surfaces, emerges and is produced with/in a situated inquiry. We are in constant relation, entangling with bodies, encountering and inseparable, in ongoing being and becoming, regardless of the human-made stories we tell. Given matter and meaning are mutually constituted (Barad, 2003) bodies like mine are produced intra-actively through encounters with bodies like mine. They have capacities to learn, be, become, practice, and transform through living entanglements with similarly made, more-than-human and multispecies bodies. Knowledge can be produced through encounters between bodies like mine.

The use of terms bodies like mine, a body like mine, and this body, signify my attempts to decentre the human 'self' as the only creature, being or species with intelligence, the only ones who know, or produce knowledge, the only beings who affect and can be affected, and whose experiences are important or significant, the species who transform, change, shapeshift, or respond. It signifies a recognition that humans are not the only bodies with agency. Rather, multispecies, more-than-human bodies are bodies like mine who shape and are shaped by entanglements with bodies and worlds. In a (re)figuring self-study, this body decentres itself in an effort to think differently about what shapes the performance of education practices, acknowledging that teaching and learning is already, always relational. The framing of this body as a body like mine acknowledges that multispecies and more-than-human encounters between bodies are relational, affective and agentic. The 'I' writing this chapter, is a body like mine, named *this body*.

This body is a body like mine

This body thinks, creates, feels, is agentic, affected, and affects bodies like mine. It is neither and both object and subject, rather an intra-active part of a worldly whole. It is named *this body* to differentiate it as the body like mine performing research and writing a chapter. It reminds me *I am* a body like mine that feels, senses, affects, and is affected. This body becomes with and through entanglements and encounters with bodies like mine. This body does not have a separate voice that speaks for or about other bodies. Instead, it is a co-constituted assemblage produced through entanglements with bodies like mine, inseparable from the enactments in which it was produced (Mazzei, 2013). This body was made, and is unmade and remade simultaneously (van der Tuin, 2019), through lifelong relational encounters with bodies like mine. This body knows, and produces knowledge, and is produced by multispecies entanglements with bodies like mine, is different in different entanglements, and doesn't only thinking-know.

As an apparatus for making meaning and knowledge production, this body is inseparable from the assemblage, is inextricably, materially, discursively 'fused together' (Barad, 2007) with all aspects of the assemblage. In a (re)figuring self-study, this body is part of the production of data, the findings and the research storylines created, which are similarly bodies like mine. Because this body cannot disentangle from the apparatus, its embodied thoughts and senses and practices are part of how it produces produce knowledge. Multiple 'selves' are fused into becoming with as this body enacts a (re)figuring self-study. Embodied, spiritual, creative, sensory, immanent, and relational ways of knowing become possible through this body, though they often remain elusive and difficult to pin down or articulate textually and artistically. Embodied ways of knowing, which are only possible if matter is acknowledged as having agency, allow this body to learn and become with bodies like mine.

Performing research curiously and responsively with and through this body involves opening to different possibilities and different ways of knowing. It involves allowing this body to affect and be affected by (Seigworth & Gregg, 2010) bodies like mine in multispecies collaborative inquiry. The framing of this body as a body like mine acknowledges that multispecies and more-than-human encounters between bodies are relational, affective, and agentic. A (re)figuring self-study demanded this body open to allowing encounters and entanglements to 'do their work' (Barad, 2003). Insights produced are read diffractively through a body like mine. In (re)figuring self-study there is a shift from reflective and reflexive practice towards diffractive practices that allow this body to trouble the boundaries of how and what is possible to know through collaborative inquiry.

Displacing reflection with diffractive practice

Diffraction is practiced in a (re)figuring self-study as a counterpoint to reflection (Haraway, 2016), in that it does not repeat the past or represent faithfully and truthfully the meaning of an encounter as being the same or similar for the bodies involved. Diffractive practices shift self-study away from being a reflective or reflexive practice undertaken by an individual on practice, towards a collaborative, intra-active, and productive practice in which bodies like mine entangle and encounter each other, create interruptions and disruptions, and look for what might be different from what was previously understood. Reading data diffractively 'through' rather than 'reflecting on' experiences reveals difference patterns rather than simply displacing the same elsewhere (Haraway, 2016).

Diffraction can be a practice, a process and a result, or a being and a becoming which is productive (Bozalek & Zembylas, 2017) when enacted through embodied entanglements, or encounters, between bodies like mine, and read through this body. Diffractive practices involve reading insights through each other and mapping where differences appear (Bozalek & Zembylas, 2016), and then exploring the patterns of differences and understanding the ways these differences matter. Threading data, storylines, theory, and methodology through each other produces patterns that sometimes glowed (MacLure, 2013) to reveal insights that might otherwise remain hidden from the self performing the research.

Diffractive practices are affirmative and productive approaches to producing and reading patterns in data that highlight differences that might otherwise remain hidden. They emphasise difference and support the decentring and destabilising of the self, pushing a (re)figuring self-study beyond the structures of qualitative research (Mazzei, 2014). They have allowed this body to follow threads that run in multiple directions, read embodied insights through theory, and think with what matters. In response, this body produces storylines.

Storylines

The storylines created with multispecies forms of collaborative enquiry are expressions of the palpable, visceral, and powerful affects that surface within this body in entanglements with bodies like mine. They speak to what this body feels, rather than what it thinks. In a (re)figuring self-study, storylines are embodied and performative material-discursive expressions of knowledge produced with encounters between bodies like mine. This body produces storylines in response to these encounters, giving form to them through writing, poetry, drawing, photography, music, video, digital art, and spoken word. Storylines give voice to what feels important and what affects this body, and to what matters. They are experimental presentations of research, pedagogy, and methodology. In enacting a (re)figuring self-study I have come to know that this body's creative practice is inseparable from response-ability (Haraway, 2016; Lobo et al., 2021) to different ways of knowing, and entanglements with the climate crisis and affects of climate change that are ever present, intra-active forces always already shaping bodies like mine.

While the storylines are produced through curious, creative practice rather than thinking-knowing as is common in scholarly research and writing, they are revealing in that they allow for recording and presenting different forms of knowledge that shape and are shaped by a research assemblage. Multiple storylines were created during the (re)figuring self-study, not all of which are in response to the research questions, and not all of which can be represented artistically or textually. Many simply remain woven into this body, part of becoming with multispecies collaborative inquiry. In presenting research, I choose the storylines that matter to me (this body) and those that highlight or reveal patterns and traces that reveal aspects of practice that would otherwise remain hidden.

Encountering Air

Storylines like *Encountering Air* become bodies like mine, more-than-human entanglements with capacities to affect and be affected. This body produced the text, created the performative storyline, and reads the storyline diffractively with multispecies collaborative inquiry as a way of explaining methodological development. I invite you to read *Encountering Air,* through your body like mine with what your body knows, as it reads the remainder of this chapter. https://video.deakin.edu.au/media/t/1_cx0a9ken

#

Figure 4.1: Making a storyline: Encountering Air

I am air. Enduring, alive and complex.
A delicate balance of multiple elements
Nitrogen and oxygen mainly
and a handful of others
I breathe life into all beings on earth
My sisters and my brothers

I support plants like moss to grow
Help birds like the Owlet-nightjar take flight
I'm in the soil and waters as they ebb and flow

Without me, no living being could take a breath
Even my relationship with fire is life or death

You see me as movement through the leaves
Feel me on your lips as you breathe
Hear me in winds that howl and scream
A mushroom, a fish, pollution, a rose
Are simply suspended particles traveling on me
Until your senses awaken, as smell, when I touch your nose

I carry the effects of climatic patterns
An invisible presence
Dancing on the skins of all beings
Protecting earth from the sun, and the solar system
I bring our planet alive
I am everywhere, inviting you to listen

But now I am becoming dirtier and heavier with carbon
Tainted by smoke from industrial works
Human ways are destroying my health
You are literally taking your own breath away
And it's getting harder to sustain your needs
I can't regenerate, rebalance or support my health
While deforestation continues to destroy ecologies and trees

I invite you to take a breath

And live in harmony with me

Offer your gratitude for the life I bring

Take up the call to care

Living lightly on our planet

Is giving thanks to air.

Making Encountering Air

Encountering Air is an artistic storyline of an encounter with air, a body like mine, that was produced with curious practices of affective attunement (Willink & Shukri, 2018), sitting with discomfort and noticing (Harrison et al., 2017; Klein, 2018). It reveals this body's response to discomforts and resonances that surfaced when I was asked to 'give voice' to a more-than-human aspect of this world for a workshop based on the *Council of All Beings* (Work That Reconnects Network, 2017) created with my colleagues from the Climate Change Education Network (Sutton et al., 2020). What began as an opportunity to model a process became an expression of an encounter with air, and was significant in shaping the curious practices I needed to experiment with multispecies forms of collaborative inquiry. The storyline that remains woven into this body is an entanglement of embodied knowledge of air, climate change, the *Council of All Beings,* the affects of climate change felt by this body, a body that cares deeply for its multispecies kin, and multispecies collaborative inquiry.

Curious practices and Multispecies Collaborative Inquiry

Multispecies collaborative inquiry is a curious practice of "attunement to multispecies entanglement, complexity and the shimmer all around us" (Gan et al., 2017, p. 11) that opened this body to response-ability (Haraway, 2016) through encounters with bodies like mine. Curious practices demand this body think-practice-know differently, troubling the limits of what is known, what can be known and whether we need to know. Curious practices are performed by this body as it notices, explores, affects, and is affected by encounters with bodies like mine. It is a high stakes approach to encountering difference, or something new, while not always knowing how this body will respond or know, though it must, as Haraway (2016, p. 28) explains,

> Despret trains her whole being, not just her imagination, in Hannah Arendt's words, to go visiting…venturing off the beaten path to meet the unexpected, non-natal kin, and to strike up conversations, to pose and respond to interesting questions, to propose together something unanticipated, to take up the unasked-for obligations of having met.

This body has 'gone visiting' many times, in multiple ways in enacting a (re)figuring self-study. Intentionally immersing this body in 'in between' spaces of transformation, encountering familiar and unfamiliar bodies like mine, and responding in embodied, performative, artistic, creative, and scholarly ways. The right curious practices for multispecies collaborative inquiry emerged as this body encountered bodies like mine, practiced diffraction and made storylines like *Encountering Air*. I found the curious practices that are productive for this body are affective attunement, sitting with discomfort and diffractive reading because they enact noticing, wondering, creating, questioning, disrupting, diffracting, patterning possibilities, and response-ability (Haraway, 2016) through encounters with bodies like mine.

Encountering involves intentionally entangling with bodies like mine, and with ideas, theories, concepts, and thinking that are different and challenging to this body; and sitting with discomfort that surfaces or created through these encounters with the familiar and unfamiliar. It becomes an uncomfortable yet generative and creative process for rethinking and refiguring practice, where I think with, feel with, 'try out' and diffract different approaches, practices by entangling with multispecies and more-than-human bodies that are different to mine. Encounters that helped me think differently as I enacted a (re)figuring self-study were many. Encounters with creatures and plants in my garden and the park that surrounds it. Peers and colleagues in conversation, my PhD supervisors and co-creators of workshops and public pedagogies. White sage, butterflies, nettle, K'gari, plants, earth, air, ocean ripples and waves at Sandringham beach, fire, the light and shade of shadows. Memories, timespace, stories, spirituality, camera, video, phone, poetry, computer, social media. Many pens and a lot of notebooks. Textual representations of research, the words of scholars and philosophers too many to name here. Posthumanism, new materialisms, post qualitative research. Indigenous and first nations knowledges. A multitude multispecies kin including plants, herbs, trees, and insects.

More-than-human encounters were more traditionally academic, as I read scholarly and research literature across different fields and disciplines;

thought with situated knowledge of practice; shared what I was learning; collaborated with human colleagues with similar and different interests to me; created and enacted climate change public pedagogies (including blog and Instagram posts, poetry, songs, video, conference sessions, workshops, events, websites, and publications); and wrote about climate change education. Through these encounters I learnt about affective attunement (Willink & Shukri, 2018) and practiced applying it in interviews about practice transformation with my peers. Attempting to collect data in this way, I raised more questions than I was answering about researching affect and transformation.

The affects that arise in these encounters are often difficult to represent in language, powerful in shaping education practice, and challenging to research using traditional methodologies. I noticed that affects seemed to be present, though not explicitly stated, in as I read the storylines (like *Encountering Air)* I made in response to encounters with more-than-human bodies. Affects are visceral forces which are generally other than conscious knowing, vital forces beyond emotion that drive us towards movement, thought, and action. Affects are revealing and powerful, but are often ephemeral, hard to describe and articulate using human language, and cannot be fully represented with spoken or textual language. Affects are "always beyond our control, beyond our grasp, elusive, and nonrepresentational, but very real and very impactful" (Willink & Shukri, 2018, p. 187).

Affective attunement (Willink & Shukri, 2018) is a performative practice for noticing affect through presence to the sensations that surface "in the midst of *in-between-ness*: in the capacities for bodies to act and be acted upon" (Seigworth & Gregg, 2010, p. 1). I have found it to be a generative curious practice of being present to what is happening in an encounter, for noticing movements, intra-actions, and vital forces through sensory engagement. It provides access to sensory ways of knowing and brings forward the feelings, sensations and affective responses of bodies like mine within the encounter. Affective attunement enables this body to research, respond and learn about affects that arise in the messy, liminal spaces between of professional and practice transformation. It reminds me that bodies like mine often respond in ways that cannot be controlled, and the knowledge that is created through encounters cannot always be represented.

Affective attunement allowed this body to feel the affects of air and climate change as I produce, and re-encounter *Encountering Air* as a body like mine in a different timespace as I write this chapter, even though the affects are not, and cannot be, explicitly spoken. In this way the encounter between bodies like mine produces embodied, intra-active, situated knowledge of affect and the entanglements of air and climate change. I understand that this body is affected by multispecies collaborative inquiry and by writing about producing *Encountering Air*. Sitting with the discomfort, attuned to the affects surfacing of the multispecies and more-than-human encounters across multiple timespaces, I allowed affect to do its work on me, noticing the surfacing sensations and responding by writing what was there, within this body, to be expressed about air. There were many questions, some of which are answered in *Encountering Air,* and many which are not. Why does air matter, why should we care? How can I give voice to an invisible force that we know exists, that we know is life force, yet that we don't question? How does air, also body like mine, respond to climate change?

Encountering Air in a (re)figuring self-study

Encountering Air was produced with curious practices of multispecies collaborative inquiry that 'work' for this body. At the time it was written I was still experimenting with which curious practices worked within a (re)figuring self-study about the processes and affects of practice transformation. Thinking through how I could differently know a more-than-human aspect of the world involved a shift from my knowing, thinking self to a practice of stillness and affective attunement to what was happening within this body in encounters with air. The deep connection I have with earth and its related planetary systems, the sensations that surface as affects through encounters with multispecies kin (in particular, plants and trees) are evident to me as I encounter *Encountering Air* as a body like mine.

I chose to present Air in this chapter because it is an immanent life force that is already, always present in every encounter, much like my *self* in teaching and learning. As the storyline reads, "without air, no living being could take a breath." Air is a body like mine. The discomfort I felt as I encountered bodies like mine–air, pollution, climate change, human action, environmental devastation, clouds–in artistic production and again reading *Encountering Air* highlight that bodies like mine, and this body, are complicit, a part of it all and response-able. It reveals a wider and deeper knowledge that I feel, but cannot always find language to express, about why it matters how we practice climate change education.

Curious practices of encountering, affective attunement, sitting with discomfort, and diffractive reading provide access to different ways of knowing for this body. The production of an artistic storyline in response to encounters with these bodies like mine reveals the affects, for example, in the changes in my voice as I perform the poem in the recorded video. The practice, the performance of the methodology, is important in helping me understand affective learning in a deeper and different way. The practice of diffractive reading reveals hidden or unseen aspects of our 'selves' by offering a method of reading our own experiences, our production, our pedagogy, and our creative practices through and against each other. I notice through doing this that air is intra-active, and that I can write affective storylines that become bodies like mine–intra-active, co-constituted, always being and becoming, and open to interpretation and reinterpretation over time. The implication for my own practice is that it offers me a way to surface what matters in my practice by giving me something to read with and through. It's only productive when you put the methodology to work. Reading *Encountering Air* with methodology to write the chapter became another opportunity to learn with climate change, the affective connection I have with earth, and my practice.

Thinking with *Encountering air* on multiple levels as I read it through this body, and with the assemblage that produced it, nearly two years later, revealed the possibilities and potential of posthumanist and new materialist approaches to self-study. Diffracting humanist assumptions through posthumanist theories and new materialist practices brought my attention to the ways humanist assumptions are woven into and shape my practice, and how these assumptions often remain hidden, even in reflective forms of inquiry into self and practice. For example, I named the entanglement of curious practices as 'multispecies collaborative inquiry', to give a textual form to a collection of practices within a self-study methodology. The intention was to acknowledge the entanglements and encounters with multispecies, more-than-human kin that have been integral, and again an attempt to decentre the human. However, the discomfort in using the term collaborative surfaced as I thought with the notion of intra-action. I realised that the word collaboration implies a separateness that posthumanist thinking seeks to trouble. For now, I use the word collaborative between multispecies and inquiry to acknowledge the critical collaborative foci of self-study (Samaras, 2011), and note that this may shift and change as I continue to enact this form of inquiry. This reminds me that (re)figuring self-study, and transformation, are immanent practices and process of becoming with.

Reading *Encountering Air* through this body for the purpose of describing methodological development for this chapter reminded me of the ways curious practices of multispecies collaborative inquiry enabled its production. As a body like mine (reader, viewer, listener) encounters *Encountering Air* in a different timespace, it takes on a different meaning. And a different meaning again when read diffractively with methodology and theory. *Encountering Air* read with posthumanist theory highlights that air, like humans, are intra-actively, immanently, entangled with climate change, and our past, present, and future. It is all inseparable; we are always already becoming bodies like mine. There are multiple implications for rethinking climate change education, and perhaps this chapter writing is another beginning. Another opening. Another suggestion of future directions for research and practice, or production of storylines, beyond the boundaries of this chapter.

(Re)figuring possibilities

Doing self-study in a way that acknowledges the agency of multispecies and more-than-human bodies pushed me to trouble both the boundaries of self-study and my own assumptions about what it means to research, practice, know, transform, and write a book chapter. Writing about my attempts to disrupt the human from its centred position in this chapter, at the same time as writing about methodology, and reading an artistic storyline created by this body with the methodology, became a performative enactment of self-study. Threads were pulled together, woven, dropped, picked up again. Knots were made and unmade simultaneously. Points were underscored, patterns revealed, and questions asked. I reiterated and returned to what I spoke of earlier, diffractively reading as a practice of surfacing what this body felt was significant and learning more deeply about practice. I made, unmade and remade, picked apart and picked up where I left off, looked to the past and the future, thought with, remained present, noticed and made decisions about what was written and what left out. I recognised the affective potential of rethinking climate change education practice in ways that were previously hidden and noticed the possibilities that were ever present when we dare to trouble the boundaries of what research and education practice looks like.

Decentring the human in educational research involves taking up practices that go beyond thinking knowing, to open to the possibilities of something different, and continuing to trouble the boundaries of what we think we know. Chapter writing has been a response-able, ethical, productive process for disrupting assumptions (again) about what it means to do academic research as I grappled with expressing the embodied knowledge that continues to be produced with a (re)figuring self-study. Through this, I developed a deeper understanding of how this body works, and its responses, within a feminist posthumanist ethico-onto-epistemology. I wonder what will shift if I move to more holistic, embodied teaching and learning practices that acknowledges intra-action, response-ability, and that bodies are affected by encounters with bodies like mine. I wonder what will happen as I continue in my attempts to decentre my self in self-study and take up curious practices and multispecies forms of collaborative inquiry. Will it become less about being a reflection of the world in which we learn, of recreating the same elsewhere (Haraway, 2016) and more about collapsing the boundaries around what is possible to know, to express, to teach, and learn?

There is never a conclusion to practice transformation, only openings to 'what's next' for this body as it refigures its approach to climate change education in an immanent climate crisis. Producing this chapter has revealed the productive possibilities of sitting with the discomforts and dissonances that surface in the liminal, messy spaces between of becoming with. However, discomfort, disruption, and dissonance are just some of the threads woven into this storyline of practice transformation. Experimenting with curious practices and multispecies forms of collaborative inquiry has revealed the potential and possibilities that are always, already present in encounters between bodies like mine. The implication is for bodies like mine to continue to question, challenge and disrupt, to take up curious practices and multispecies forms of collaborative inquiry, to continue to embody response-ability within research and practice, as we rethink and refigure climate change education in these troubled and troubling times.

References

Barad, K. (2003). Posthumanist Performativity: Toward an understanding of how matter comes to matter. *Signs: Journal of Women in Culture and Society*, *28*(3), 801-831. https://doi.org/10.1086/345321

Barad, K. (2007). Meeting the universe halfway: Quantum physics and the entanglement of matter and meaning. Duke University Press.

Bozalek, V. & Zembylas, M. (2016). Critical posthumanism, new materialisms and the affective turn for social just pedagogies in higher education. *South African Journal of Higher Education*, *30*(3), 193.

Bozalek, V. & Zembylas, M. (2017). Diffraction or reflection? Sketching the contours of two methodologies in educational research.

International Journal of Qualitative Studies in Education, *30*(2), 111-127.
https://doi.org/10.1080/09518398.2016.1201166

Commonwealth of Australia. (2020). Royal Commission into National Natural Disaster Arrangements – Report. Canberra, ACT: Commonwealth of Australia.

Crutzen, P. J. (2021). We live in the anthropocene, so will our grandchildren. *Herald of the Russian Academy of Sciences*, *91*(1), 13. https://doi.org/10.1134/s1019331621010020

Gan, E., Tsing, A., Swanson, H., & Bubandt, N. (2017). Introduction: Haunted landscapes of the anthropocene. In A. Tsing, H. Swanson, E. Gan, & N. Bubandt (Eds.), *Arts of living on a damaged planet*. University of Minnesota Press.

Grooten, M. & Almond, R. E. A. (Eds.). (2018). Living Planet Report - 2018: Aiming Higher.

Haraway, D. (2015). A curious practice. *Angelaki: Journal of the Theoretical Humanities*, *20*(25), 5-14.
https://doi.org/https://doi.org/10.1080/0969725X.2015.1039817

Haraway, D. (2016). Staying with the Trouble: Making kin in the Chthulucene (Kindle ed.). Duke University Press.

Harrison, N., Bodkin, F., Bodkin-Andrews, G., & Mackinlay, E. (2017). Sensational pedagogies: Learning to be affected by country. *Curriculum Inquiry*, *47*(5), 504-519.
https://doi.org/10.1080/03626784.2017.1399257

Klein, S. R. (2018). Coming to our senses: Everyday landscapes, aesthetics, and transformative learning. *Journal of Transformative Education*, *16*(1), 3-16. https://doi.org/10.1177/1541344617696969

LaBoskey, V. K. (2004). The methodology of self-study and its theoretical underpinnings. In J. J. Loughran, M. L. Hamilton, V. K. LaBoskey, & T. L. Russell (Eds.), *International handbook of self-study of teaching and teacher education practices* (pp. 817-878). Springer.

Lange, E. A. (2018). Transforming transformative education through ontologies of relationality. *Journal of Transformative Education*, 16(4), 280-301. https://doi.org/10.1177/1541344618786452.

Lenz Taguchi, H. (2012). A diffractive and Deleuzian approach to analysing interview data. *Feminist Theory*, *13*(3), 265-281.
https://doi.org/10.1177/1464700112456001

Lobo, M., Bedford, L., Bellingham, R. A., Davies, K., Halafoff, A., Mayes, E., ... & Lucas, C. (2021). Earth unbound: Climate change, activism and justice. *Educational Philosophy and Theory*, *53*(14), 1491-1508.

Lotz-Sisitka, H., Wals, A. E., Kronlid, D., & McGarry, D. (2015). Transformative, transgressive social learning: Rethinking higher education pedagogy in times of systemic global dysfunction. *Current Opinion in Environmental Sustainability*, *16*, 73-80.

MacLure, M. (2013). The wonder of data. *Cultural Studies – Critical Methodologies*, *13*(4), 228-232. https://doi.org/10.1177/1532708613487863

Mazzei, L. A. (2013). A voice without organs: interviewing in posthumanist research. *International Journal of Qualitative Studies in Education*, *26*(6), 732-740. https://doi.org/10.1080/09518398.2013.788761

Mazzei, L. A. (2014). Beyond an easy sense: A diffractive analysis. *Qualitative Inquiry*, *20*, 742-746. https://doi.org/10.1177/1077800414530257

Samaras, A. P. (2011). *Self-study teacher research*. Sage Publications Inc.

Samaras, A. P. & Freese, A. R. (2006). *Self-study of teaching practices primer*. Peter Lang Publishing, Inc.

Sandlin, J. A., O'Malley, M. P., & Burdick, J. (2011). Mapping the complexity of public pedagogy scholarship: 1894–2010. *Review of Educational Research*, *81*(3), 338-375. https://doi.org/10.3102/0034654311413395

Sauve, L. (2005). Currents in environmental education: Mapping a complex and evolving pedagogical field. *Canadian Journal of Environmental Education*, *10*(Spring 2005), 11-37.

Schadler, C. (2019). Enactments of a new materialist ethnography: Methodological framework and research processes. *Qualitative Research*, *19*(2), 215-230. https://doi.org/10.1177/1468794117748877

Seigworth, G. J. & Gregg, M. (2010). An inventory of shimmers. In M. Gregg, G. J. Seigworth, S. Ahmed, B. Massumi, E. Probyn, & L. Berlant (Eds.), *The affect theory reader*. Duke University Press.

Selby, D., & Kagawa, F. (2018). Teetering on the brink: Subversive and restorative learning in times of climate turmoil and disaster. *Journal of Transformative Education*, *16*, 302-22. https://doi.org/10.1177/1541344618782441

Sutton, B. A., Blades, G., Upton, M., & White, P. J. (2020). A storyline: Deepening our commitment to our world through collaborative practice during Covid-19. *Eingana, The Journal of Environment Education Victoria*, *43*(2), 17-20.

Taylor, M. & Coia, L. (2019). Co/Autoethnography as a feminist methodology. In J. Kitchen, A. Berry, S. M. Bullock, A. R. Crowe, M. Taylor, H. Guojonsdottir, & L. Thomas (Eds.), *2nd International*

handbook of self-study of teaching and teacher education. Springer. https://doi.org/https://doi.org/10.1007/978-981-13-1710-1_18-1
van der Tuin, I. (2019). Deleuze and diffraction In B. Rosi & B. Simone (Eds.), *Posthuman ecologies: Complexity and process after Deleuze*. Rowman & Littlefield Publishers.
Wals, A. E. J., & Benavot, A. (2017). Can we Meet the Sustainability Challenge? The Role of Education and Lifelong Learning. *European Journal of Education Research, Development and Policy*, 52, 404-413. https://doi.org/DOI: 10.1111/ejed.12250
Walter, P. (2009). Philosophies of adult environmental education. *Adult Education Quarterly*, *60*(1), 3-25. https://doi.org/10.1177/0741713609336109
Willink, K. G. & Shukri, S. T. (2018). Performative interviewing: Affective attunement and reflective affective analysis in interviewing. *Text & Performance Quarterly*, *38*(4), 187-207. https://doi.org/10.1080/10462937.2018.1526409
Work That Reconnects Network. (2017). *Council of all beings*. Retrieved from https://workthatreconnects.org/resource/council-of-all-beings/

Acknowledgements

Bronwyn would like to thank her supervisory team of Peta White and Robin Bellingham for their support and guidance throughout the PhD journey, and during the writing of this chapter. She would also like to thank the reviewer and the numerous colleagues, peers and networks for their part in shaping and guiding her thinking.

Bronwyn Sutton (BA, MEd) - Deakin University, School of Education, 221 Burwood Highway, Burwood, VIC 3125, Australia.
bsutton@deakin.edu.au
Bronwyn Sutton is an educator, activist and early career researcher living, working and researching on the unceded Country of the Kulin nations, otherwise known as Australia. She respectfully acknowledges the traditional custodians of the lands on which she continues to research, learn and teach, including Butchulla, Boon Wurrung, Wurrundjeri, Taungurung, Yorta Yorta, Dja Dja Wurrung and Wauthurong people, and their elders past, present and future. Bronwyn's education practice and research focuses on environmental public pedagogies, climate justice activism, transformative pedagogies, materiality, and affect. She has a particular interest in teaching and learning that occurs in informal, nonformal,

community, and beyond school spaces. As a PhD student at Deakin University Bronwyn is working on a research project that explores professional and practice transformation using self-study, through which she is experimenting with feminist, posthumanist, and new materialist approaches.
https://orcid.org/0000-0001-7190-9092

CHAPTER 5

UNDERSTANDING PEDAGOGIES FOR CLIMATE CHANGE EDUCATION THROUGH COLLABORATIVE ARTS-BASED AUTOETHNOGRAPHY

JO RAPHAEL, SHELLEY HANNIGAN AND PETA J. WHITE

Abstract

Climate change education focuses on understanding the complex interactions of social, economic, and environmental systems and our role in negotiating uncertain futures. As educators we embody the responsibility to support our students to learn about the climate crisis in ways that enable actions even though the future remains uncertain. Collaborative Arts-based Autoethnography (CABAE) is a methodology employed to enable reflective practice through arts-based strategies in a context of collaboration. Three education academics from different disciplines brought perspectives and practices together to interrogate and understand climate change education. We investigated how we adapt our practices, the challenges we experience and why. We confront some issues for climate change education and come up with ways forward to address our climate crisis through our teaching practices. We experience a sense of solidarity in working on this together, reminding each other of the human impacts that devastate our environments and the importance of rallying youth to act in ways that encourage governments to listen and act. We re-affirm our roles to play as educators and inevitably turn to deepen our climate change education practice.

Keywords: CABAE, arts-based inquiry, autoethnography, transdisciplinary pedagogy, climate change education

Uncertain times call for innovation in practice

We are living in a period of rapid human induced climatic change (IPCC, 2021) and complex uncertainty which we claim inevitably impacts education systems (White et al., 2021). Teachers who are already working with complex curriculum demands and increasingly diverse classrooms, are also challenged by the pedagogical pivot brought on by the global pandemic. Teachers are striving to meet the educational needs as well as the wellbeing needs of students who are experiencing precarity in relation to their social and economic futures, employment, and the future of the planet in the face of human induced climate change (Diaz et al., 2019; IPCC, 2021). As teacher educators we are concurrently impacted and strive to pre-emptively prepare preservice teachers to cope with the challenge of teaching in these times (Mulvihill & Martin, 2021). As teacher educators, we acknowledge the importance of making time to listen, share and dialogue with each other about these challenges to bring new insights and ideas into our teaching practice.

We are three academics in a School of Education from different discipline areas across the arts and sciences. We began our collaboration in 2014 adopting a methodology of self-study of teacher education in a larger community of practice originally comprised of eight teacher educators (White et al., 2020). We adopted self-study research as a personally situated critical and collaborative inquiry (Samaras, 2011), aiming to improve learning, deepen scholarship of teaching practice and professional lives. We employed elements of critical friends (Lomax, 1991), friendship as method (Tillmann-Healy, 2003), and intimate scholarship (Hamilton & Pinnegar, 2015). A self-study invites the application of diverse methods (Tidwell et al., 2009), and we soon moved to include arts-based inquiry (Hannigan et al., 2016; Raphael et al., 2016) and collaborative autoethnography (Chang et al., 2016, Palmer et al., 2018).

We set ourselves provocations of teacher education practice to explore collectively, placing high value on the opportunity to collaborate and to benefit our personal wellbeing. Our approach has since evolved into what we now call collaborative arts-based autoethnography (CABAE) (Hannigan et al., in press). We suggest this innovative methodological approach as a supportive and congenial process of inquiring into teacher education practice that provides the conditions for deep reflection, careful listening, and productive dialogue set within the challenging context of our current global crises. It is also a practice that is embodied, creative, and personally encouraging, and as such, a practice we always manage to find time to complete, despite the pressures and demands of our regular program of university work.

In this chapter we write about our most recent CABAE collaboration, in which we set the focus of our inquiry to an area of increasing concern to us, and a question increasingly being asked by teachers: how do we understand and describe our practice of climate change education? We collaborate to interrogate how we are adapting our practices and why, and inevitably turn to exploring how we might do climate change education more effectively. This is research with rippling effects. As we inform our own practice, we are also working out ways to communicate and share the practice of climate change education for the preservice teachers we teach, modelling dispositions, innovations, and practices to inform climate change education within their own classrooms.

Embedded in systems

Climate change education is not encompassed by, or the same as, environmental education, education for sustainability or sustainability education (Saylan & Blumstein, 2011; Jickling et al., 2018; Shellenberger & Nordhaus, 2004). Climate change education is different–necessarily (Monroe et al., 2019), which doesn't imply that the others haven't worked (Saylan & Blumstein, 2011; OECD, 2009). Climate change education refers specifically to the key content and concepts that are central to climate change crises: the entanglements of climate science and the social systems that have generated unprecedented and significant human impact on all Earth systems (OECD, 2019).

Humans have profoundly changed ecosystems (Lewis & Maslin, 2015) since the Industrial Revolution of the 1850s and in exponential and devastating ways since then (IPPC, 2021). We already have evidence of ecosystem failure globally, alongside the onset of the 6th mass extinction (Dirzo et al., 2014). Understanding human impact as the amalgamation of all our actions and choices, as individuals and societies, is central to reimagining ways forward (The Story of Stuff, 2007; The Story of Change, 2012; The Story of Solutions, 2013). We recognise that each choice we make carries a consequence and that the result is often the environment subsidising our comfort. For example, we choose to travel by plane as it transports us in less time, or we choose food products from Italy because they are familiar, or named in the recipe, rather than using those grown or manufactured locally. Learning to read the ingredients, value local products, and to make transport and living choices that reduce our resource use and recycle materials in meaningful ways is becoming imperative. Reimagining how we can live affluent lives with some sense of material circularity (Circular Economy, 2022) is central to our future wellbeing.

When considering pedagogical strategies to unpack and explore climate change and the impact of humans, educators look for ways to clarify the science while also highlighting the social choices, pressures and actions that undergird human impact on Earth's systems. For example, socio-scientific issues (Hsu, et al., in press) are pedagogic strategies that science educators have been developing to engage students in scientific reasoning with the social impact analysis often in the context of sustainability. Stakeholder analysis, where the perspectives of the various stakeholders are analysed and considered, is a valuable strategy (Raphael & White, 2021). Similarly, arts-based strategies, such as the ones employed in this research, usefully engage reflective practice while focusing on the representational connections between practice, theory, and action (Hannigan et al., 2016).

Collaborative Arts-based Autoethnography (CABAE)

Having engaged in collaborative self-study in 2021, which focused and refined on our practices from 2014, the three authors realised a shift to collaborative autoethnography better suited our research. Ethnographic insights had emerged from our practice-based research including our own cultural positionings, the culture within education institutions where our work was based, and the social enactment of issues like climate change. We are inspired by the work of Ergas & Ritter (2021, p. 18) who recognise there are other aspects of self to explore in more autoethnographic ways than self-study. We took hold of their suggestions to look at:

(a) "Ontology: How we define self affects who self is in relation to". This might include "body/mind, narrative/core self, thoughts/emotions, true/false self, me/I), which can then affect the self-other relationship".

(b) "Epistemology: A relationship to ourselves implies the need for exploring novel ways of knowing ourselves". This supports our use of arts-based inquiry as well as narrative inquiry.

(c) "Ethical commitments: We understand the wish to improve teaching practice; however, we need to acknowledge the politics of who determines what counts as improving practice and what doesn't't".

(d) "Methodology and authority of self".

Our CABAE approach doesn't imply a turn away from self-study of teacher practice as we were aware of the overlaps and intersections between collaborative autoethnography and self-study. As Hamilton et al. (2008) put it, "When a self-study involves social or cultural issues, it could fit the definition of auto-ethnography" (p. 25). We also acknowledge that self-study brings a focus on teaching practice (Loughran, 2007) that is implicit in our CABAE research. In early 2021 we developed, refined, and practiced CABAE (see Hannigan et al., in press) and continue applying and refining our practice.

Arts-based inquiry was key to our research because a shared art-making inquiry offered us individual contemplation and sharing within an interdisciplinary, supportive but critical group. Over the years of researching together and in other projects (see Hannigan & Raphael, 2020; Raphael et al., 2016), we had found art-based inquiry highly effective for revealing information about our professional selves, social selves, place-based, cultural and practice issues. Arts-based inquiry approaches were useful for (1) formulating a provocation from topics or issues which had already been expressed by the group or in the data we were generating, then (2) responding to the provocation creatively, and (3) sharing and discussing the artefacts or processes generated. Step 3 enabled deep sharing of issues of which we were often not conscious. Therefore, provocations based on our issues of teaching provided a valuable process for unearthing issues relevant to our personal and professional lives.

Drawing upon the arts in our research approach seemed particularly relevant for an inquiry focused on climate change education. As Braidotti (2013) suggests, to move beyond anthropocentrism, we need "resources of the imagination, as well as the tools of critical intelligence...to be enlisted for the task" (p. 82). Cahnmann-Taylor (2018) elaborates further when she explains we "find in art the tools to arouse others and ourselves to think more critically about new forms of becoming in the human, organic, and inorganic world" (p. 256). There is a sense of urgency about climate change education, and it can be daunting, so inquiring into the practice collaboratively with trusted colleagues, using the generative practice of arts-based inquiry, is heartening and reassuring.

The application of CABAE to climate change education

We began our collaboration with a discussion about some of the main educational challenges we face in these Anthropocentric times. We agreed that a responsibility to teach climate change weighed heavily on us as teacher educators, and while we felt we had evolving and emerging ideas, we also recognised tensions and felt inadequacies. Taking a CABAE approach allowed us to focus our attention on our practice of climate change education. We asked ourselves the question, "how do we understand and describe our practice of climate change education?" We resolved to reflect on this by way of creating an artwork (in any art form or medium) in response to the question. We set a date a few weeks later to bring our artworks to share with each other.

Our research took place over the summer of 2022 and being in different geographical locations and still cautious amid the global pandemic, we met via Zoom. Each of us took a turn to show our artworks, invite a response, and then talk about what we had discovered in the process of making and reflecting on the artwork. While doing this, we took notes of common themes and ideas that emerged from our discussion together. We found that the process of reflecting and creating these artworks was a kind of active meditation and sharing them with one another led to questions and critical conversations.

Following these conversations, we each wrote an autoethnographic account reflecting the processes through which we addressed the question, made our artwork, and considered suggestions for our developing pedagogy of climate change education. We also reflected on how each other's works challenged and therefore strengthened our own pedagogical approach. Finally, we collaboratively analysed them for emerging ideas and themes of climate change education, noting the common threads and new understandings ensuing, confirming the benefit of multiple perspectives. Our artworks and written accounts are presented in the next section, followed by the emerging themes and conclusions.

Jo's CABAE story – 'Basket-nest'

As I set about the task of reflecting on teaching climate change, I realised there were many questions that influenced my pedagogy. There is the question of who I teach, and for what aims and purpose. There is the question of what I teach, including the mandated curriculum, and the question of knowledge in a rapidly changing world. There is the question of where I teach, including my desire for place-based learning and connection to nature, real-world spaces beyond a sterile classroom, or transforming the classroom by bringing in nature. There is the question of when I teach, particularly what is most relevant for diverse groups of students at that time. There is also the question of how I teach, including teaching about climate change in a sustainable way.

I teach preservice teachers how to teach the arts. These preservice teachers might be on the way to becoming specialists in the arts, or generalist teachers who will teach the arts along with other curriculum areas. As a teacher/artist/activist, when teaching the arts, I believe it is important to engage with big ideas like climate change that really matter. I understand that the arts have always been a way that humans have explored and made meaning of their world. The arts require aesthetic engagement, honouring experiencing of the world through our senses. Artists engaging with nature can be prompted to create in response to beauty and wonder. Similarly, artists confronted by environmental devastation can be driven to create art out of care, concern, despair, or anger. It can be a short step between artist and activist. Rather than being narrowly focused on teaching content, I recognise the way the arts promote these felt connections, build empathy (for human and non-human life), and can encourage an attitude of caring necessary to prompt action for sustainability.

Reflecting on my teaching contexts and aims along with the strands of who, what, when and where I teach, led me to the desire to create an artwork that explored this complexity, weaving these many aspects of pedagogical consideration together. My artwork needed to be produced sustainably, and to be inspired by and reflect some of the beauty of the natural world. However, I felt strongly that it should also reflect the realities and tensions of living sustainably in the 21st century. I decided that I would attempt to draw upon the resources in my own backyard to closely consider their quality and to weave them together. I walked around the garden collecting different grasses, vines, and supple branches, careful to take only what I needed. I began to weave what I thought would be a basket.

Making basket-nest

Holding our research question in my mind as I weave, I realise that the process of making art is a meditation on pedagogy or thinking through action. As thoughts come into my mind, I write them on a page in my journal. Each idea about the process of making the artwork, could equally relate to the process of teaching climate change. I write:

It's daunting at first.

It must be done.

Just make a start.

The hardest part is beginning.

I need to work with what I have.

If it doesn't work, try a different approach.

I need to build a strong base.

I work in the open air, under the watchful gaze and intermittent song of two magpies. 'What are you doing?', they ask. 'Are you making a nest?'

'Yes, I imagined it was a basket, but maybe it is a nest. Maybe it's a basket-nest.'

'Keep going!' they encourage through song, as I weave my basket-nest towards understanding my pedagogy of teaching climate change. Basket-nest is both a product and a process (Figure 5.1).

Figure 5.1: Basket-nest weaving

Process

Process is important, as making a basket-nest is a series of problems to be solved. Weaving puts me in the sweet spot, being both creative and challenging enough that I quickly enter a state of 'flow'.

It's hard to stop once you start.

It's a good problem, one worth the struggle.

Better not to have fixed expectations, surprises are always possible.

There is beauty in imperfection.

I'm going back and forward over the same ground, learning something new every time.

Over, under, over, under, weaving makes the structure strong.

Birds do this too.

The magpies watch, '*She's new to this, but she's catching on.*' I am aware I could learn so much from others who have mastered this skill of basket/nest making before me.

Over and under, pull it tight, make it stronger.

It's the tension that holds the basket together.

Figure 5.2: Basket-nest detail

Materials

I'm working mostly with natural materials drawn sustainably from nature and spending time closely exploring the qualities of the plant materials, the colours, the smells, the textures and shapes, the strength, pliability, or brittleness. I am aware that some of these things that I have collected to make my basket-nest are still living as I hold them in my hands. The choice of materials is important for basket-nest and my teaching:

Diversity (of materials) adds strength

Some materials just aren't useful

How far can I bend this before it breaks?

I wanted my artwork to reflect the realities and tensions of living sustainably and acknowledge the existence of synthetic materials that we are so accustomed to using. Although it felt like contamination, I chose to weave soft plastics from packaging waste into my basket-nest. I wondered if I could maintain a sense of beauty or harmony even as I wove in this waste, like a magpie incorporating human-made material into its nest (Figure 5.2). I note the plastic is opposing, resisting, asserting itself, dominating and being uncooperative. In my notes the plastic is personified, as if an adversary. In the end a contrasting band of plastics is woven into Basket-nest, visible when you look inside or turn it around. I reflect:

You have to look at a problem from all angles.

How can we use resources in a sustainable way?

Metaphor for teaching climate change

The final aspect of the creation of Basket-nest is how it works on a metaphorical level. Firstly, the materials, and the contrast between a diversity of natural plant materials and the synthetic coexisting in the artwork. Secondly, the process of weaving together, working with the tensions, and the complex range of considerations to make something stronger. Finally, the product, a basket has capacity, it contains, collects, holds, and carries, and a nest accommodates, protects, and nurtures new life. As well as a reflection on a pedagogical approach to teaching climate change, Basket-nest is about engaging with the beauty of nature, acknowledging that we need to use resources sustainably, nurturing an attitude of caring for the environment. Finally, it is the active making of an open basket-nest potentially filled with ideas for growing a pedagogy of climate change, and an offer of hope and possibility.

Peta's CABAE story – 'Weaving systemic entanglements'

I had been thinking about climate change education a lot–much of my research was coalescing around issues facing education and its need to adapt to the many global crises we are currently experiencing and generating. Adapting practices at the systemic level is challenging even without the threat of ecosystem and social failure. With it, it can feel emasculating and immobilising.

I cherish the opportunities to work with colleagues in CABAE. The process is always generative and replenishing as it is creative and productive. I was eager to apply our collective thinking to climate change education as I was confident that the impact of the outcome would be significant and useful.

The first step of this task was to conceptualise the representation. I chose the infinity symbol as the flowing intersecting circles represent the interrelationship of Earth systems entangled with human (social) systems. The symbol felt powerful in its representation of the kind of thinking that we need to understand and enact new onto-epistemological ways of being in our changing world–where we (humans) are the instigators of this change. The lack of control felt by me as a consumer is overwhelming, yet as an educator I can change my practice and influence the practice of those around me. As a teacher educator, I feel privileged to be able to support other educators to reconceptualise their practice in these challenging times. Often drawing connections and giving permission for change is all teachers need–that small prompt or encouragement and some useful resources to apply.

Once I had the symbol, I needed to consider how to represent it. I wanted to generate a model. I needed to get my 'hands on' this task to achieve the innate sense of needing to act in climate change education. Planning the model took time as I weighed the representational force of several items and processes. I wanted to include themes of living, belonging, and change. I considered growing a plant in a pre-determined form or making a water course where the flow mapped the symbol, but these ideas didn't deliver the impact.

I recently moved to a new area and in my many walks around these new streets, discovered that the wilder areas were encroached and dominated by invasive plant species. Biodiversity loss, resulting in the 6th mass extinction, is a silent horror often left unspoken or unnoticed. Invasive species and habitat destruction are central to these crises (as is the changing climate and intensified weather events).

I had the choice of many species, I could have selected thorny blackberries (*Rubus fruitcosus*) but I didn't want the resulting injuries. I could have chosen ivy (*Hedera helix*) as there was a lot of it around me, but I opted for a species new to me (representing my new location). This species had a common name that is unpalatable (Wandering Dew), so I made a commitment to find and use the species name–*Tradescantia*. I decided that I needed to weave invasive species into my representation. And then 'weaving' them in became literal.

Figure 5.3: The space created by my harvesting of Trad

I was undertaking a bit of gardening right before I needed to start work on this project and found a patch of "Trad" that was quite large (almost taking over). Harvesting this plant for my project was an important task as it also freed my garden from this invasive species (in that patch–for now). As I harvested (Figure 5.3), I reflected on my new learning and how framing this species as invasive resulted in my strong desire to remove it. It is actually a very nice and easy to grow plant–something a gardener might enjoy, when it's in the right place. However, the way *Trad* thrives as it takes over an area seemed to mirror the way that humans impact many ecosystems when they take root. A useful, yet uncomfortable metaphor.

Creating this model took very little time, once I had addressed the challenges of how and what. I gathered the invasive plant species and the wire for the frame and sat comfortably at my desk learning how to create while also discussing with my colleagues. We were meeting to discuss our project and as I learned how to generate my artwork, we engaged in discussion about their artworks…and the similarities were striking.

I created a wire frame with small-gauge wire coiled into the infinity symbol. The space between the wire coils was to become the new home for strands of *Trad*. I learned about weaving and stopped to consider the aesthetics and to capture the process with photographs. Weaving was therapeutic, mesmerising, and pleasing. The growing, thickening, form evoked reflections about the gradual change an ecosystem can experience as *Trad* takes over and inhibits endemic species from thriving (Figure 5.4).

Figure 5.4: Weaving the *Trad*

As I compared the process of model making, I drew out the representations.

- The wire frame was like the curriculum as it gave structure–but needed interpretation in some places as the space was too great for the framing to be explicit. I had to massage the wire to maintain the form that I wanted.

- The *Trad* often broke which reminded me of the fragility of humans, yet the sheer volume of our numbers and determination is a domineering force. In fact, that is how *Trad* takes over an area–it grows so vigorously that it out competes other species for nutrients, space, and light.

- Because I am me, I had to consider the need for neatness. I tend to be a bit excessive in this way. I wanted the *Trad* vine to be straight and the leaves to branch out on jaunty angles, yet the process of weaving and the *Trad* growing form didn't conform. This juxtaposing of my desire against the growing form and process caused me to reflect on the way that education starts with desire to do the best and yet needs to respond to the context in every way. Intentions are key in education as planning and delivery often diverge.

- The resulting structure of woven *Trad* stems reflects the complexity of climate change. Complexity is often removed from our teaching content for fear that our students will not comprehend. Yet, in my experience, young people understand the complexity and interconnections in embodied ways, if not academic…and that is what is necessary.

- The process of weaving is like teaching. As a teacher I weave my students' interests together with strategies and resources and to generate learning opportunities that enable us all to move forward (learn)–in predetermined directions (regarding the type and content of knowledge and skills and following the curriculum).

- Circularity and systemic thinking are particular kinds of thinking. The infinity symbol was useful in its insistence that systems interact, and the reflection of the ongoing and iterative nature is central and foundational. This reminded me that in teaching about climate change the climate science is important, but so is understanding our governance systems and how we organise ourselves as societies–and how we relate to the ecosystems we are immersed in.

The final step was to hang my work in an outside space so that I could watch it weather and transform over the next several weeks (Figure 5.5). I chose to see the impending death of the *Trad* (representative of humans) as metaphoric and not literal. Could humans learn new ways to interconnect with ecosystems instead of dominating them? Could human societies learn to value ecosystems in their complexity and intrinsic value not just for how they make our lives more comfortable? There are many useful questions here.

Figure 5.5: The weathering/wethering/whethering of *Trad* entangled with our Earth/social systems

Shelley's CABAE story

I focused on one art course that I teach each year to university students. They are working towards developing their own children's picture/story book which will be used as an arts-based educational resource to teach art to children. Across five years teaching this unit, many students have chosen to base their emerging picture/story books (and accompanying teaching resources) on environmentally themed or climate change education. One possible reason for this is that one of the topics of my lessons is based on their experiences "being in and with nature". They then explore this reflective experience and discuss this in small groups, which triggers conversations about their learnings from other classes (e.g., environmental science) and the different ways they might teach climate change education in transdisciplinary ways. We also explore many examples of children's story books including those based on environmental education and sustainability, and explore artists' practices, processes, and artefacts. Being an art class, more than half the time of each lesson is allocated to art-making workshops using sustainable materials which contribute to their emerging children's story book projects (drawings, collage, digital imagery, photographs etc.).

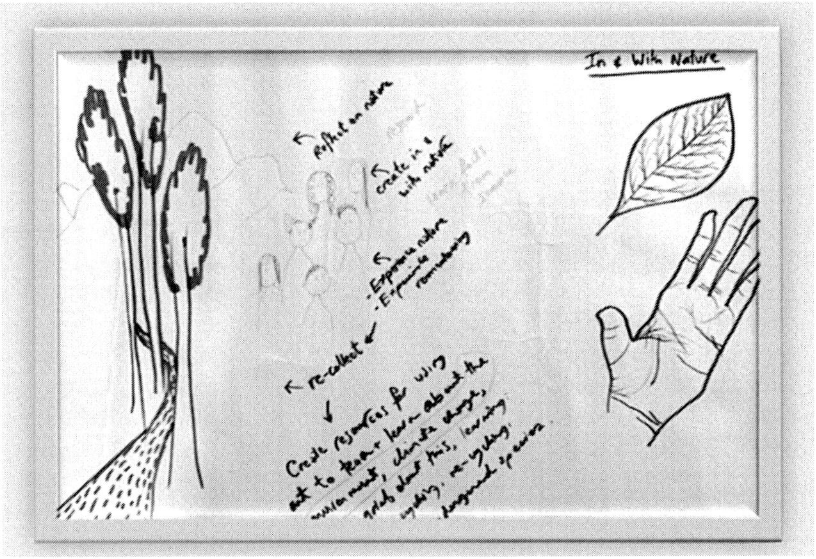

Figure 5.6: Shelley's representations

Figure 5.6 captured this aspect of my teaching. Contemplating its meaning made me think of the diminishing positive experiences that students in the future might be able to have in and with nature. This concerns me as it has been a strong initial impetus to the experiential learning of these art classes in lieu of the fact that we often don't have the time or space to venture out into the wilderness during class time to have these experiences.

Sharing my artwork with Jo and Peta (Figures 5.1-5.6), I was reminded about the exhaustive list of issues that we try to consider as arts educators: appropriate materials, eco-grief concerns, time and space issues, relevant processes to engage and more. It is due to this exhaustive list, that I have learned over the years to use this student-centred approach where I support students to reflect on their own experiences and discuss their <u>own</u> issues (environmental, social, cultural etc.) while they develop their art. This enables me to listen and teach them suitable processes and techniques relevant to the materials they are choosing to use. This is a method of teaching in, with and through art, in a relational way with their experiences, environment and place. Bentz (2020) articulates this approach when teaching art education and climate change together as: "Integrating climate change in arts courses, expressing (climate change issues) with art, learning (about climate change and our views about this) through art" (p. 1596, my bracketed inclusions). Bentz (2020) suggests these "can prepare and empower young people to address climate change themselves and add to the current discourse in a way natural sciences alone are unable to achieve" (p. 1596).

I expressed my concerns (from my art therapist's perspective) to Peta and Jo, about caring for the mental wellbeing of young people when teaching climate change issues. Peta advocated the need to teach Climate Change Education (CCE) as opposed to other terms we might use in the arts such as "sustainable" or "environmental". I contemplated this and unconvinced went off to find models from art therapy where climate change issues were explored. I was inspired by environmental art therapy where clients engage with nature in metaphoric ways. I realised I was seeking a pedagogical approach where the students could engage in relational ways with nature but also in ways that they could reflect on as meaningful, learning experiences. Siddons Heginworth (2016) shares how such approaches allow us to "feel our part in the great wounding of the earth, not just as perpetrators but as victims, for we are only doing outwardly what has been done to ourselves" (2016, p. 25). This felt like a better shift in our approach to climate change from worrying and feeling helpless, to more honestly realising humans' role, the Anthropocene, and realising we are nature therefore walking with nature as partners to make positive change.

I took myself into nature with all these thoughts and to engage in an embodied way with my natural environment. Kayaking through the Rocklands Reservoir (an artificial lake dammed for recreation and to manage water storage and allocation), I navigated dead trees from this flooded reservoir. Despite this Anthropocentric knowledge, I found signs of rejuvenation in the dried-up lake edges, pleased with the way nature finds ways to thrive, despite human interference. I observed a new gum tree opportunistically growing in the now permanent lower water level that climate change and water allocations had created and was fascinated by the homes that dead trees on the lake provide for birds, spiders, and insects (see Figure 5.7).

Figure 5.7: Rocklands Reservoir

Sharing Figure 5.7 developed a conversation about the difference between being in the wilderness and being in a place that humans have interfered with. The three of us pondered humans' habits of conquering wild places and wondered if there were any left that we could even call 'wild'. I considered ways in which Indigenous peoples of the world have lived more in harmony with nature and how environmental artists and art therapists (Boon, 2016; Siddons Heginworth, 2016; Sweeney, 2013) are finding creative ways to create and reflect in and with nature as ways of learning while also caring for ourselves as part of this intertwining of nature.

I came to realise that a shift had occurred for me through this project with Peta and Jo. The shift was from a fear of working through the depressing thoughts/realities of climate change with my students who are in turn going to teach children, to finding ways to incorporate more time with nature in my art classes and even raise CCE issues in class and explore these in relational and embodied ways with students.

However, I was still troubled by teaching CCE in my art classes because of the limited time we have allocated to teach art in universities and schools. There are so many other topics and aspects of the artworld to teach in this limited time, it felt a shame to take up too much of our art education time with CCE. I tried to capture the relationality of being in and with nature through art but also the education systems I work in when creating Figure 5.8.

Figure 5.8: Shelley's final art piece

Looking at this final art piece I was struck with how I seemed to weave the relational experience of art, nature, environment, embodied experience with the many competencies and administrative tasks that take my time away from teaching the way I want to (and the way research suggests I do). This is captured by the regular diamond forms (with the organic-ness of nature represented by the brown and blue forms behind). I realise I want more time for my students to engage with nature authentically–ideally on field trips as Gilbert et al. (2018) document. I come to realise that my fear of raising tough issues of CCE with students that I had at the start of this project (and that still lingers a little) is the lack of time and facilities we have to deal with some of the trauma and other issues that students are currently experiencing (compounded by COVID-19-related trauma). I worry that I might send them off after class with more burden or stress, when I really want my art class experiences to be joyful and even 'time out'. Perhaps, I wondered, this is a tension with transdisciplinary education.

Whilst this inquiry has made me more aware that there is very little true wilderness left for students to reflect on for their art projects, and therefore my strategy to get them to reflect on such experiences is troubled. Delving into the environmental art therapist's approaches and learning about more relational and embodied ways of creating art in and with nature, is a positive way forward for teaching art and climate change education together.

Finding common threads and weaving a pedagogy of Climate Change Education

As we shared our CABAE art works and stories and analysed them, looking for common threads, we noticed the recurrence of the motif of weaving. This was evident in Jo's basket-nest, Peta's infinity, and the interplay of shape and colour representing systems and nature in Shelley's layered final artwork. We recognised that we were all grappling with the challenges and tensions of teaching climate change within a system that can pose fears, expectations, and barriers that we must weave our teaching practice in and around. In each case, we acknowledge how we work creatively within these structures, and work productively with these tensions, while making space for learning in more local free flowing, organic, and learner-centred ways. We found the practical and embodied action of weaving heightened the process of theoretical weaving. We noticed our weaving practices reflect Deleuze & Gauttari's (1987) theories of striated (territorialised) and smooth (deterritorialised) spaces. We acknowledge the need to work within the striated/territorialised space of school, curriculum, and assessment.

However, our pedagogical approaches value deterritorialising, or breaking down the human-centric focus, creating a smooth space of becoming in harmony with all living things. We recognise the need for structure, particularly having climate change as part of a mandated curriculum, but flexibility in pedagogical approaches is required. As Deleuze and Guattari (1987) suggest, "progress is made by and in striated space, but all becoming occurs in smooth space" (p. 537). For us, facts, information, or knowledge about climate change are not enough; an attitude of caring brought on by posthuman aesthetic engagement is needed to prompt action.

We have each acknowledged the importance of the aesthetic encounter and affect in a pedagogy of climate change education. We each acknowledge the importance of prioritising a connection with the natural world as contributing to a pedagogy that creates the conditions for affective learning and a felt response. This is evident in the way in which our artworks involve a creative working with place, our natural environment, and the living things within it. We have been inspired and energised for the task, finding a sense of wonder through connecting with nature and avoiding reinscribing the binaries of human and non-human life. We share a belief that a pedagogy of climate change education is enhanced by authentic connection to and experience of the natural world, and eco-centric ways of being in the world. Just as we are engaging with arts in this process of inquiry (CABAE), we are also mindful of the potential of the arts for engaging bodies and hearts with attention to beauty and wonder for affective learning in a pedagogy of climate change. The process of deeply considering representational connections across the Earth systems and educational practices is not dissimilar to quilt or collage approaches to arts-based research (O'Toolc & Beckett, 2013), but more complex than mere rhizomes–perhaps the mycorrhizal network of the interconnection of trees passing nutrients to each other for survival is a more apt metaphor.

Weaving is a productive activity, taking up raw materials and creating a strong fabric. The process reflects our proactivity in taking up the challenge of climate change education. For Jo, the weaving process reflects her pedagogical process of continuous problem solving, working with what is at hand, an assemblage (Deleuze & Guattari, 1987) of learners, conditions, brought attitudes and tensions. For Peta, working the *Trad* into wire frames embodied the pedagogical activity of supporting her pre-service (and in-service) teachers to foreground climate change in their education practices exploring the entanglements of science and society. For Shelley, layering the shapes and textures highlighted the complexity and tensions when regulatory and natural systems overlap. We are endeavouring to produce a strong fabric, combining and creating resources for classroom teachers to further weave into their own contexts and adapt for their students. The weaving and working with weft and warp, implies bending, flexing, and shaping; this is exactly how teachers come to climate change education as they make it accessible, palatable, and learnable for their students. We are weaving together our uncertain futures through learning experiences that reveal and frighten and provoke and constrain.

We each acknowledged the CABAE process enabled us to share our concerns and motivated us to find solutions to our own personal climate change anxieties and fears for students' experiences of climate change that may manifest as trauma and eco-anxiety. When teachers practice self-reflection about this kind of anxiety, they are better able to help their students navigate their anxieties (Pihkala, 2020; Raphael, 2021). Metaphors helped us make meaning: Jo's basket-nest as an object that holds and nurtures suggests a gesture of hope and opportunity; Peta's infinity is a symbol of continuity; and Shelley's layered print brings contrasting elements into a kind of balance. Our commitment to and action in climate change education is our way of taking positive action in the face of the climate change crisis. We realised that as educators we can develop our practice and influence the practices of those around us as we weave climate change education into the different courses we teach. As teacher educators, we have the privilege of being able to support other educators to develop a pedagogy of climate change education. Through this research, we have emboldened each other to model transdisciplinary practice and show that we are also learning. In our teacher education practice, we facilitate connections, provide useful resources, and important encouragement for teachers to enact their own pedagogy of climate change education. We bolster each other by accepting the challenge and by being proactive, embarking on a path of "radical hope" (Hickman, 2021) in education for climate change.

Conclusion

Our collaborative arts-based autoethnographic inquiry allowed for the generation of greater insights than if we were to reflect alone. In doing so, we confronted some of the issues we face for climate change education, and we experienced a sense of solidarity in working on this together. This chapter maps out how auto-ethnography and creative work can clarify and articulate approaches within teaching practice that are at times difficult to conceive of or represent, especially when it comes to challenging and often distressing situations such as the urgency of climate impacts. The research is born of the collaborations we have shared. It was surprising to learn of the accidental similarities in the creative responses that then fed into each of our own approaches as educators, such as Jo's and Peta's physical weaving and Shelley's metaphoric weaving (Figure 5.8). This in turn generates a useful metaphor for ethico-onto-epistemology, where our interrelationships have helped us respond to methods in educating about a real, and what can feel like an existential, crisis. Through creating and sharing our artworks over time, reflecting and discussing, we physically, literally, and figuratively weave new ways of knowing through collaborative interactions and art practice.

Our CABAE inquiry started with a provocation which enabled us each the freedom to explore our pedagogy through any art form or medium and acknowledge that we each have our own interests and materials to explore, places in which to create, and for some, re-visiting past works that are relevant or have new understandings. We became aware that the process was as important as the product. This is a reminder of our diverse student needs and pedagogies that require not only structure and scaffolding but also time, space, and permission for them to freely explore their own approaches.

We learned what is important for a pedagogy of climate change. We need support to teach climate change including school environments, allocation of time in the curriculum for each subject area, cross-disciplinary meetings to understand who is covering what and when, and to check that our scientific perspectives and knowledges are up to date. We need teaching resources, and the ability to use these critically and flexibly according to pedagogical approaches that stir body, mind, and heart. We identified a shared concern for wellbeing of students when facing climate change education, and the need to carefully consider how we balance a sense of urgency and despair with a sense of possibility and empowerment for students.

These personal and ethnographic elements have emerged from our practice-focused research clarifying again that we must work together in interdisciplinary and transdisciplinary ways to tackle big issues like climate change, as we are doing in this research–learning from each other. This inspires each of us to go forward and teach with more conscious awareness of our relationship with climate change. For instance, the damages to the Earth that have resulted in climate change are what Siddons Heginworth (2016) describes as a "cultural neglect of the feminine, both outwardly and with nature and inwardly with the feeling self (that) has left us all deforested, polluted and ailing" (p. 25). Their approach is to view selves also as victims of "a culture that promotes and idolises the active masculine and neglects and shames the feeling feminine" (p. 25). Collaborative autoethnography allows for these ethnographic insights reminding us that all things are interconnected and we must work in these interconnected ways that involve attention to feeling and emotions if we want to work towards change.

We are all teachers of climate change, in some way–although this may not be realised. Reminding each other of the human impacts that devastate our environments or rallying youth to act in ways that encourage governments to listen and act. We have roles to play and the warp and the weft to push against and lean on as we weave our way to more effective pedagogy for climate change education.

References

Bentz, J. (2020) Learning about climate change in, with and through art. *Climatic Change* 162, 1595-1612. https://doi.org/10.1007/s10584-020-02804-4

Boon, L. (2016). The wild inside: Offering children natural materials and an eco-psychological understanding of self within art therapy. In (Eds.) I. Siddons Heginworth & G. Nash. *Environmental arts therapy: The wild frontiers of the heart.* (pp. 47-60). Routledge.

Braidotti, R. (2013). *The posthuman*. Cambridge: Polity.

Cahnmann-Taylor, M. (2018). Four guiding principles for arts-based research. In M. Cahnmann-Taylor & R. Siegesmund (Eds.), *Arts-based research in education: Foundations in practice*. Taylor & Francis.

Chang, H., Ngunjiri, F., & Hernandez, K. A. C. (2016). *Collaborative autoethnography*. Routledge.

Circular Economy. (2022). Wikipedia. Retrieved from https://en.wikipedia.org/wiki/Circular_economy

Deleuze, G. & Guattari, F. (1987). *A thousand plateaus: Capitalism and schizophrenia*. London and New York: Continuum.

Díaz, S., Settele, J., Brondizio, E. S., Ngo, H. T., Agard, J., Arneth, A., ... & Zayas, C. N. (2019). Pervasive human-driven decline of life on Earth points to the need for transformative change. *Science, 366*(6471).

Dirzo, R., Young, H. S., Galetti, M., Ceballos, G., Isaac, N. J., & Collen, B. (2014). Defaunation in the Anthropocene. *Science, 345*(6195), 401-406.

Ergas, O. & Ritter, J. K. (2021). Expanding the place of self in self-study through an autoethnography of discontents. *Studying Teacher Education, 17*(1), 4-21. https://doi.org/10.1080/17425964.2020.1836486

Gilbert, B., Cox, A., & Osborne, E. (2018). *Arts programming for the Anthropocene: Art in community and environment*. Routledge.

Hamilton, M. L., Smith, L., & Worthington, K. (2008). Fitting the Methodology with the Research: An exploration of narrative, self-study and auto-ethnography. *Studying Teacher Education*, 4(1), 17-28. DOI: 10.1080/17425960801976321

Hamilton, M. L. & Pinnegar, S. (2015). Knowing, becoming, doing as teacher educators: Identity, intimate scholarship, *Inquiry* (Vol. 26). Bingley: Emerald Publishing.

Hannigan, S. & Raphael, R. (2020). "Drawing out" understandings through arts-based inquiry in teacher education. *Qualitative Research Journal,* 1-14.

Hannigan, S., White, P.J., Raphael, J. (in press). Evolving the practice-self through autoethnographic methodology. In (eds) E. Anteliz, P. Danaher, D. Mulligan. *The Routledge International Handbook of Autoethnography in Educational Research*. Routledge.

Hannigan, S., Raphael, J., White, P., Bragg, L., & Cripps Clark, J. (2016). Collaborative reflective experiences and practice in education explored through self-study and arts-based research. *Creative Approaches to Research*, *9*(1), 84-110.

Hickman, C. (2021). In ten years to 1.5C: How climate anxiety is affecting young people around the world. *The Conversation Weekly Podcast.* Retrieved from https://theconversation.com/ten-years-to-1-5-c- how-climate-anxiety-is-affecting-young-people-around-the-world-podcast-171566

Hsu, Y-S, Tytler, R., & White, P.J. (in press). Innovative Approaches to Socio-Scientific Issues and Sustainability Education – Linking Research to Practice. Springer.

IPCC, (2021). The Physical Science Basis. Contribution of Working Group I to the Sixth Assessment Report of the Intergovernmental Panel on Climate Change [Masson-Delmotte, V., P. Zhai, A. Pirani, S.L. Connors, C. Péan, S. Berger, N. Caud, Y. Chen, L. Goldfarb, M.I. Gomis, M. Huang, K. Leitzell, E. Lonnoy, J.B.R. Matthews, T.K. Maycock, T. Waterfield, O. Yelekçi, R. Yu, and B. Zhou (Eds.)]. Cambridge University Press.

Jickling, B., Blenkinsop, S., Timmerman, N., & Sitka-Sage, M. D. D. (Eds.) (2018). *Wild pedagogies: Touchstones for re-negotiating education and the environment in the Anthropocene*. Palgrave-Pivot.

Lewis, S. L. & Maslin, M. A. (2015). Defining the Anthropocene. Perspectives, *Nature, 519*. DOI:10.1038/nature14258

Lomax, P. (1991). *Managing better schools and colleges: The action research way*. Bristol, UK: Longdunn Press.

Loughran, J. J. (2007). Learning through self-study: The influence of purpose, participants and context. In J. J. Loughran, M. L. Hamilton, V. K. LaBoskey, & T. Russell (Eds.), *International handbook of self-study of teaching and teacher education practices* (pp. 151-192). Dordrecht, The Netherlands: Springer.

Monroe, M. C., Plate, R. R., Oxarart, A., Bowers, A., & Chaves, W. A. (2019). Identifying effective climate change education strategies: A systematic review of the research. *Environmental Education Research, 25*(6), 791-812.

Mulvihill, T. M. & Martin, L. E. (2021). Global teacher education: Challenges and possibilities, *The Teacher Educator*, *56*(1), 1-3. DOI: 10.1080/08878730.2021.1844947

OECD. (2009). *Green at fifteen?: How 15-year-olds perform in environmental science and geoscience in PISA 2006*. Organisation for Economic Co-operation and Development.

OECD. (2019). OECD future of education and skills 2030 concept note student agency for 2030. Retrieved from https://www.oecd.org/education/2030-project/teaching-and-learning/learning/student-agency/Student_Agency_for_2030_concept_note.pdf

O'Toole, J. & Beckett, D. (2013). *Educational research: Creative thinking and Doing*. Oxford University Press.

Palmer, M., White, P. & Wooltorton, S. (2018). Embodying our future through nurturing conversations: The change is in the doing. *Journal of Environmental Education. 49*(4), 309-317.

Pihkala, P. (2020). Eco-Anxiety and environmental education. *Sustainability*, *12*(23), 10149. MDPI AG. http://dx.doi.org/10.3390/su122310149

Raphael, J. (2021) Eco-anxiety and drama education. *Australian Drama Education Magazine*, No.17, Drama Australia.

Raphael, J., Hannigan, S., & White, P. (2016). Drawing out understandings of collaborative self-study in teacher education. In (Eds.) D. Garbett & A. Ovens. *Enacting self-study as methodology for professional inquiry.* Chapter 12. Retrieved from www.castle-conference.com

Raphael, J. & White, P.J. (2021). Transdisciplinarity: Science and Drama Education Developing Teachers for the Future, Ch. 9 in White, P. J, Raphael, J., & van Cuylenburg, K. (Eds.). *Science and Drama: Contemporary and creative approaches to teaching and learning.* Springer.

Samaras, A. P. (2011). *Self-study teacher research: Improving your practice through collaborative inquiry.* Thousand Oaks, CA: Sage Publications Inc.

Saylan, C. & Blumstein, D. T. (2011). *The failure of environmental education [and how we can fix it].* Berkley, CA: University of California Press.

Siddons Heginworth, I. (2016). Turning. In I. Siddons Heginworth & G. Nash (Eds.), *Environmental arts therapy: The wild frontiers of the heart* (pp. 9-26). Routledge.

Shellenberger, M. & Nordhaus, T. (2004). *The death of environmentalism: Global warming politics in a post-environmental world.* Retrieved from http://thebreakthrough.org/PDF/Death_of_Environmentalism.pdf

Sweeney, T. (2013). *Eco art therapy: Creative activities that let earth speak.* Self-published by Theresa Sweeney. (ISBN13: 9780615901473)

Tidwell, D. L., Heston, M. L., & L. M. Fitzgerald. (2009). *Research methods for the self-study of practice.* Springer.

Tillmann-Healy, L. (2003). Friendship as method. *Qualitative Inquiry, 9*(5), 729-749.

The Story of Stuff (2007), The Story of Change (2012), The Story of Solutions (2013), Leonard, A., Fox, L. & Sachs, J. (Writers). Fox, L. (Director). Free Range Studios (Producer). Retrieved from https://www.storyofstuff.org/movies/

White, P. J., Ferguson, J. P., O'Connor Smith N., & O'Shea Carré, H. (2021). School strikers enacting politics for climate justice: Daring to think differently about education. *Australian Journal of Environmental Education* (pp. 1-14). http://doi.org/10.1017/aee.2021.24

White, P. J., Raphael, J., Hannigan, S., & Cripps Clark, J. (2020). Entangling our thinking and practice: A model for collaboration in teacher education. *Australian Journal of Teacher Education.* 45(8). https://doi.org/10.14221/ajte.2020v45n8.6

Acknowledgements

In the spirit of reconciliation, we acknowledge the Traditional Custodians of the various lands on which we gather and work. We recognise their continued connection to the land, air and waters and acknowledge that they never ceded sovereignty. We respect all First Nations people and acknowledge their Elders and Ancestors.

Jo Raphael (BEd, MEd, PhD, SFHEA) - Deakin University, School of Education, 221 Burwood Highway, Burwood, VIC 3125, Australia jo.raphael@deakin.edu.au
Jo Raphael is Senior Lecturer in arts education at Deakin University. She is actively involved in teaching and researching in community and education settings within Australia and internationally. Her areas of research and publication include applied drama and theatre, inclusive education, and teacher education. In her practice she applies drama as a pedagogy for learning in areas across the curriculum including science, humanities, and education for sustainability. Jo's research draws on arts-based and participatory action research methods that are inclusive of participants as co-researchers and collaborative arts-based autoethnography as part of a commitment to developing a supportive community of practice amongst teacher education colleagues.
https://orcid.org/0000-0001-5220-314X

Shelley Hannigan (BFA, MEd, GradDipEd, Phd) - Deakin University, School of Education, Pigdons Road, Waurn Ponds, VIC 3216, Australia shelley.hannigan@deakin.edu.au
Shelley Hannigan is a senior lecturer in art education. Her specialist area is visual and media art. She is a practicing artist and has a background in working as an art therapist in hospitals as well as teaching art in a range of education contexts. Found objects, collage, mixed media and textiles have been introduced to her practice since then as she also explores theories of materiality, embodiment, and place as both artist and academic. She researches in place, Identity, practice, arts health, STEAM, and arts discipline areas.
https://orcid.org/0000-0002-3232-5761

Peta J. White (BSc (Hons), MRSysMan, PhD, SFHEA) - Deakin University, School of Education, 221 Burwood Highway, Burwood, VIC 3125, Australia peta.white@deakin.edu.au
Peta White is a senior lecturer in science and environmental education at Deakin University. She educated in classrooms, coordinated programs, supported curriculum reform, and prepared teachers in several jurisdictions across Canada and Australia. Peta continues her commitment to initial teacher education leading courses, units, programs, and in-service teacher education through research-informed professional learning programs. Peta's current research follows three narratives: science and biology education; sustainability, environmental, and climate change education; and collaborative/activist methodology and research.
https://orcid.org/0000-0002-0225-5934

CHAPTER 6

WORKING WITH TEACHERS IN REFORM INITIATIVES

RUSSELL TYTLER AND PETA J. WHITE

Abstract

As interpreters and producers of contemporary research, teacher educators have a key role as advocates and agents of educational system reform. Design-based research (DBR) is increasingly used as a critical research methodology aimed at reform, a key aspect of which involves working with teachers in a collaborative process involving the production, trialing and refining of innovations in curriculum and pedagogy. This chapter explores the relative roles of the researcher and teacher given that they bring potentially differing perspectives and responsibilities to a DBR process intended to stimulate teacher and system change. We draw on our experience of an interdisciplinary primary school mathematics and science project involving challenges to traditional expectations, disciplinary structures and epistemological and pedagogical beliefs. We describe how the DBR process involved teachers with different degrees of alignment with the innovation, and the ways that teachers' knowledge and commitments intersected with and were differentiated from those of the research team. We argue that these differences are generative for DBR as a driver of the reform processes important for an adaptive education system.

Keywords: Science and mathematics education, representation construction, design-based research, co-construction with teachers

Introduction: Design-based research

Design-based research (DBR) has gained considerable traction as a methodology that is pragmatic, flexible and focused on solutions to problems of practice. Also referred to as design experiments, or design research, DBR has become a methodology of choice for the design and validation of interventions, being an answer to classical experimental methodologies which compare 'experimental' with 'control' conditions but suffer from a range of problems:

A. Innovative interventions will often include multiple elements so that it becomes difficult in a simple comparison to be clear on what aspects are making a difference.

B. The complexity of classrooms makes it difficult for experimental methods to capture the multiple elements that inevitably affect outcomes, including variable contexts, or teacher fidelity to the intended innovation.

C. There are inevitably issues with simply characterising the features of any 'control' condition given the complex nature of classrooms.

D. The innovation needs to be decided in its entirety beforehand so that it becomes difficult to adjust during an investigation if there are elements that are identified as non-optimal.

In contrast, DBR, or design experiments, offer the possibility of a flexible, iterative cycle of interventions aimed at reaching an optimal solution to a problem (Reeves, 2006). Further, design experiments offer the possibility of generating theory around complex educational practices rather than being the after-the-event validation of already developed theory in experimental designs. Cobb et al. (2003) emphasise this theory building aspect of design experiments, which they argue "ideally result in greater understanding of a *learning ecology*" (p. 9, italics in original), which they describe as including "the norms of participation that are established, the tools and related material means provided, and the practical means by which" learning is orchestrated, for instance through tasks used and discourse norms established.

Plomp (2013) describes the purposes of DBR as:

> … to design and develop an educational intervention (such as programs, teaching-learning strategies and materials, products and systems) as a solution to a complex educational problem as well as to advance our knowledge about the characteristics of these interventions and the processes to design and develop them, or alternatively to design and develop educational interventions (about for example, learning processes, learning

> environments and the like) with the purpose to develop or validate theories. (p. 15)

Cobb et al. (2003, p. 9) are very clear that design experiments aim to develop theory rather than 'merely' to empirically tune 'what works'. They make the point that such theories are 'humble' in targeting domain-specific learning processes, are closely linked to practice that puts them 'in harm's way', and that focus on why designs work and how they might be adapted to differing contexts. DBR can draw on a range of methodologies but has defining characteristics (Brown, 1992; Collins, 1992). These include being iterative in nature, based in authentic environments, influenced by research and also practice, assessing not only design outcomes but the design itself, generating theory, and being pragmatic. Shavelson et al. (2003) describe the strengths of DBR as:

> …testing theories in the crucible of practice; in working collegially with practitioners, co-constructing knowledge; in confronting everyday classroom, school, and community problems that influence teaching and learning and adapting instruction to these conditions; in recognising the limits of theory; and in capturing the specifics of practice and the potential advantages from interactively adapting and sharpening theory in its context. (p. 25)

Cobb et al. (2003, pp. 9-10) describe what they call 'crosscutting features' of design experiments that characterise this close relation between theory and practice: 1) their purpose is to develop theories about the process of learning and means to support learning; 2) they involve intervention and are typically test beds for innovation; 3) they are 'conjecture driven tests' that put theories 'in harm's way'; 4) they involve iterative design cycles of invention and revision; and 5) they have pragmatic roots in that the theory must do real work in practice.

Figure 6.1 draws on Plomp (2006/2013, p. 17) to represent the essential features of DBR, which are described in some detail by Reeves (2006). First, researchers and practitioners collaborate to identify and analyse a problem of interest. Second, prototype solutions are developed, informed by existing design principles. Third, iterative cycles of testing and refinement of solutions in practice are undertaken. Fourth, evaluation of design principles and solution implementations leads to a new iteration.

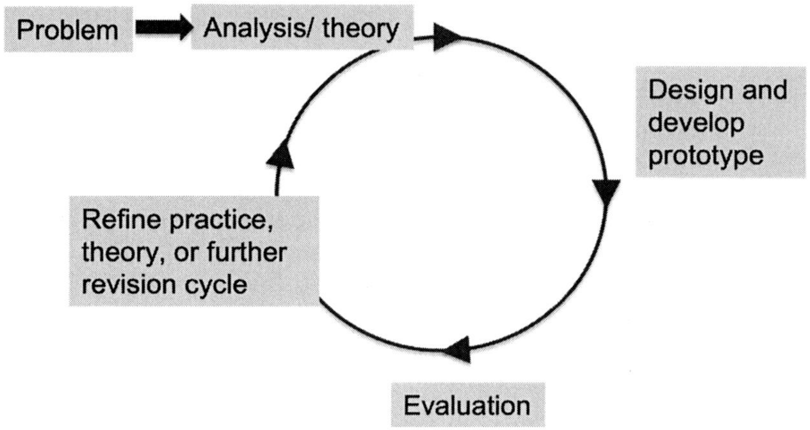

Figure 6.1: Design cycles in DBR (Based on Plomp, 2006/2013, p. 17).

Background to the current project

Since 2009, Science Education researchers at Deakin University have been developing and refining an approach to teaching and learning based on the semiotically informed views of the nature of scientific knowledge building of theorists such as Latour (1999) and Gooding (2004) who argue that scientific discovery practices involve the collaborative invention and refinement of multimodal representational systems. Science educators such as Lemke (1998) following the semiotics of Peirce (1998) and Lehrer and Schauble (2012) focusing on model-based reasoning, have pointed out the multimodal nature of learning in science classrooms, and the process of science learning as induction into the multimodal discursive practices of the discipline. From this perspective, representational tools are crucial resources for speculating, reasoning, contesting and justifying explanations, knowledge building, and communicating. Following these theorists, we have worked with teachers over a number of projects to develop a pedagogy that aims to bring science classroom practices closer to the epistemic practices within science itself, arguing the need to align epistemological processes, epistemic practices, and the role of material and symbolic tools with those of the discipline to support robust learning (Prain & Tytler, 2012). The pedagogy involved students generating, comparing, collaboratively assessing and revising, and coordinating representations in response to representational challenges.

In the first project we worked particularly closely with two teachers of Grade 5/6 students (aged 10-11 years) and two Grade 7/8 teachers to collaboratively develop, evaluate and iteratively refine learning sequences across a range of topics. The approach followed the DBR cycle. The research team developed draft learning sequences, workshopping with teachers and refining based on their input. These teachers worked collaboratively with each other and were receptive to the approach. They adapted it flexibly to their context, innovating with modification and addition of activities. Enactment of the sequences was monitored through video capture and transcription of selected sequences, ongoing conversations with teachers, student and teacher interviews, and analysis of student artefacts. From these analyses, the design principles were successively refined, based on teachers' ways of working and their own growing confidence with the approach. Through analysis of the teacher-student interactions and of student artefacts, interviews, and post-sequence assessments, the project advanced understandings of the nature of student learning and principles underpinning the pedagogy, as well as generating practical teaching and learning sequences. Thus, the research fulfilled the expectation of DBR in generating both practice solutions and theoretical insights into learning and pedagogy based on 'high theory' semiotic perspectives. The pedagogical principles were generated by repeat viewings of the data, interviews with teachers, and focus groups with the group of teachers. The 'representation construction approach' we developed from that project (Tytler et al., 2013) involves the key features:

1. The sequencing of representational challenges involving students generating representations to actively explore and make claims about phenomena, involving clarifying the representational resources underpinning key concepts, establishing a representational need, and aligning student generated with canonical representations.

2. Explicitly discussing representations including their selective purpose, establishing group agreement on generative representations, discussing representational form and function and the adequacy of particular representations.

3. The promotion of meaningful learning through establishing a perceptual context, for instance by exploring material phenomena, and privileging student engagement and agency.

4. Assessing 'as learning' through constant monitoring and providing feedback on student representations.

Two further projects followed this original study, extending the approach to more topics and year levels, and adapting it to a blended learning setting. As we pursued the approach in these projects, we became more aware of its key features and how to adapt it flexibly to contexts. In the project we discuss in this chapter (Interdisciplinary Mathematics and Science Learning; IMS), we extended the approach to investigate an interdisciplinary mathematics and science pedagogy for primary schools. The interdisciplinary approach had been pioneered by Lehrer and Schauble (Lehrer, 2009; Lehrer & Schauble, 2012) but in IMS we further extended it to include more topics, a wider range of year levels and an approach designed to investigate the possibility of scaling to system level. Each year over 3 years we worked with five cohorts of students, each with between one and six classes of 25-30 students, and between 13 and 16 teachers. We followed the cohorts of students longitudinally to track their progress in representational competence, which meant that each year we started with a fresh group of teachers. Having this number of teachers involved over the three years of the project allowed us to interrogate teachers' differential roles within the DBR methodology. That is the focus of this chapter.

Method

As outlined above, the IMS project followed a Design-based Research (DBR) methodology in which we collaborated each year with teachers to design, refine, implement and evaluate lesson sequences focused on mathematics and science learning in different topics. The principles underpinning the interdisciplinary approach were:

- Classroom activity is grounded in exploration of phenomena within engaging contexts that raise questions that can be investigated through both science and mathematics;
- The mathematics and science concepts and practices are mutually reinforcing;
- The interdisciplinary tasks lead to fresh learning in each subject, and not simply application of known procedures and concepts; and
- The reasoning and learning in each subject reflects disciplinary knowledge-building practices, involving the invention and guided refinement of representational systems.

Over the three years we developed, with teachers, a range of learning sequences in each topic, representing different interactions between mathematical and science concepts. Table 6.1 represents a subset of these sequences from the first two years, showing the science and mathematics concepts and practices that were the sequence foci. These sequences are available on the project website https://imslearning.org. The sequences are structured around the pedagogical stages revealed in the research, and feature teacher resource notes, activities and suggested questions and prompts, examples of the range of student responses, advice to teachers about the conceptual foci to be interrogated and emphasised. The structures and advice are based on research findings concerning the nature of student learning, and key features in the pedagogy that emerged from the research analyses. Thus, the outputs of the project to date illustrate this close relationship between theory and practice that is the hallmark of DBR.

For each cohort we worked closely with one case-study teacher over the year, teaching between 16 and 21 of our case study students. At the beginning of each year, we held workshops for newly recruited teachers, meetings to review and plan next steps during the sequence implementation, and review workshops at the end of the year. Each learning sequence was initially planned by the researchers and through these workshops and meetings was refined in consultation with the teachers. Detailed lesson guidelines and resources arising from the initial planning process were provided for each sequence, along with equipment and resources. Each sequence comprised between three and eight weekly lessons of one to two hours duration, shaped to accommodate contextual constraints. Each cohort experienced three to four sequences over the school year.

Table 6.1: Learning sequences developed in the first year of the project (adapted from Tytler et al., 2021, p. 1933)

Grade	Topic	Science Content	Mathematics Content
1, 4	Astronomy	Shadows, sun movement, explaining day and night: modelling, earth and space perspectives.	Angle as rotation and length of shadow (formal and informal measures), graphing shadow length, time sequencing, perspective taking.
1, 4	Ecology	Living things, diversity, distribution and adaptive features related to habitat.	Variation, data modelling of living things in sample plots, tables and graphs, space and mapping, measurement, area, coordinates, directionality, sampling, using a scale.

1, 4	Motion	Representing speed, distance and time relations, constant speed, acceleration.	Embodied representation of relation between distance, time and speed, length measures, modelling variation, graphing speed, distance, time, distance-speed-time relations for motion down a ramp.
2	Chemistry	Dissolving and mixing, physical change, particle ideas, chemical reactions, change to materials in cooking.	Representing time sequences of mixing and dissolving under different conditions, timing cooking measurement for cooking recipes: standard and informal units, fractions, proportion.
2,4	Fast Plant Growth	Plant growth, structure and function, growth needs and patterns, plant life cycles including germination, flower structures, fertilisation processes.	Measurement of plant height, leaf size, shape and number, tracking growth over time, tables, line graphs, units (cm and mm- formal units), time intervals, using a scale.
2	Force and Motion: Helicopters	Flight and air flow, gravity, representing forces, modelling and design, variable control.	Measure of helicopter parameters, time, data variation, data modelling, number line.
4/5/6	Light	Vision and properties of light: Vision, reflection and image creation, refraction.	Angle, rotation, reflection, directionality, symmetry.
5/6	Measurement: Body Height	Height measure and variation, differences between populations.	Measurement (m and cm), data modelling, variation, measures of central tendency and of variation, comparison through graphs, categorising/organising data, sampling.
6	Magic Microorganisms	Structure and function of microorganisms: bacteria, fungi. Investigative skills.	Multiplication, ratio, area, measurement.

Data were collected from consenting students, including photographs of written/drawn/constructed artefacts, classroom observations of teachers, video capture of each case study teacher's interactions with students during each lesson and interviews or focus groups with case study students and teachers. Video capture involved an iPad camera mounted on a SWIVL to follow the teacher, who wore a blue-tooth microphone, and another fixed camera was used to capture group conversations. We used a micro-ethnographic analysis of key episodes in the pedagogical cycle that presented themselves to the team as indicative examples of support for student learning, viewed through the Peircean semiotic lens on which the research is based. The analysis methodology is described in more detail in Ferguson et al. (2018) and Tytler et al. (2021). In the current chapter, we are concerned with the differential roles of the teachers in the collaborative DBR cycle. First, we illustrate the process of theory building arising from analysis of case study teachers in the first year of the project.

The involvement and roles of the teachers

One of the challenges in the project was the need to introduce new teachers to the principles and approaches of the project each year, as our longitudinal cohorts of students moved through the grades over three years. The approach involved challenges for teachers in terms of a) their epistemological orientations, in that the approach asks of teachers that they prepared and challenged students to generate representations and work with them to evaluate and refine these, rather than teach content directly as examples, and b) their pedagogical skills, in asking teachers to flexibly respond to students' representations and ideas and build towards a consensus (Hubber et al., 2010). These are approaches and skills with which teachers need to develop familiarity. Teachers in our project often expressed surprise and satisfaction at the depth of engagement with learning resulting from the approach. The other challenge for these generalist primary teachers was the focus on science or mathematics knowledge and practices that in some cases went beyond their expertise. In this, the research team worked to support teachers to gain familiarity with the target concepts and specific practices such as sampling strategies for living things, ways of modelling data, or cultivating microorganisms. In the project, teachers' roles included:

1. Participation in beginning of year workshops to introduce the approach;

2. Commentary on sequence plans, in year level teacher meetings;

3. Interpretation and enactment of sequence and ongoing discussion about next steps to support learning; and

4. Debriefing in cohort meetings, and engagement in interviews or focus group reviews.

The schools were recruited at the beginning of the project and agreed to have at least one teacher engaged with the research in each of the three years. In fact, in each school, partly because the project's reputation became established, the entire cohort of teachers at each year level worked with us to plan and implement the sequence. From these three to six teachers, one teacher in each cohort agreed to become our case study teacher. This involved working more closely with the research team in setting up to be videotaped and having the team's research assistant in the room to support as needed. This sometimes meant in situ discussions about interpreting students' work, or further challenges and questions that might be appropriate.

With this large group of teachers there was variation in their confidence with mathematics and science ideas and in teachers' flexibility in opening space for students to invent representations, in their pedagogical beliefs, and in the classroom cultures they developed. The uptake of the pedagogy therefore varied. We were, however, fortunate that in many cases our case study teachers proved committed and generative in taking up the approach, and from these teachers we gained valuable insights in the operation of the pedagogical approach in practice, and the way teachers can effectively shape students' learning through collaborative representational work.

Above we have described how teachers in the project were involved in co-design processes that led to the iterative development and revision of the learning sequences. However, teachers also played an active role in refining our theoretical perspectives about learning and teaching through representational work, as they flexibly and in some cases creatively enacted the sequences. To illustrate the process by which this occurred, we first refer to the key features of the representation construction approach, involving representational challenges, explicit discussion of representation, meaning-making grounded in material processes, and ongoing monitoring of students' ideas. With this, however, we recognised there was wide variation in the nature of representational challenges and how these were prepared for, depending on topic, and also some lack of specificity about the process by which the teacher orchestrated agreement about productive representational work. We wished, therefore, to research in more detail the nature of teachers' craft in designing and enacting these sequences. We were also informed by a growing body of work on students inventing and refining representations (e.g., Lehrer et al., 2006; Manz et al., 2020; Lehrer & Schauble, 2020).

During the first year of the project our two Grade 1 case study teachers proved to be flexible and creative in the way they adopted and adapted the approach. We found, in analysis of the video records of their classrooms of a Grade 1 ecology sequence, the second sequence they engaged with, evidence of significant innovation in their approaches and flexible support of learning through individual and whole class discussions. We therefore as a team analysed the video record of various parts of the sequence to uncover the different ways that they were supporting students' representational invention and refinement to build consensus. The analysis resulted in the development of a pedagogical model with four phases, shown in Figure 6.2. The model was further validated against teachers' pedagogical moves in a range of learning sequences and incorporated into the sequence design reporting. The development of the model and explication of the phases has been reported in two papers (Tytler & Prain, 2022; Tytler et al., in press).

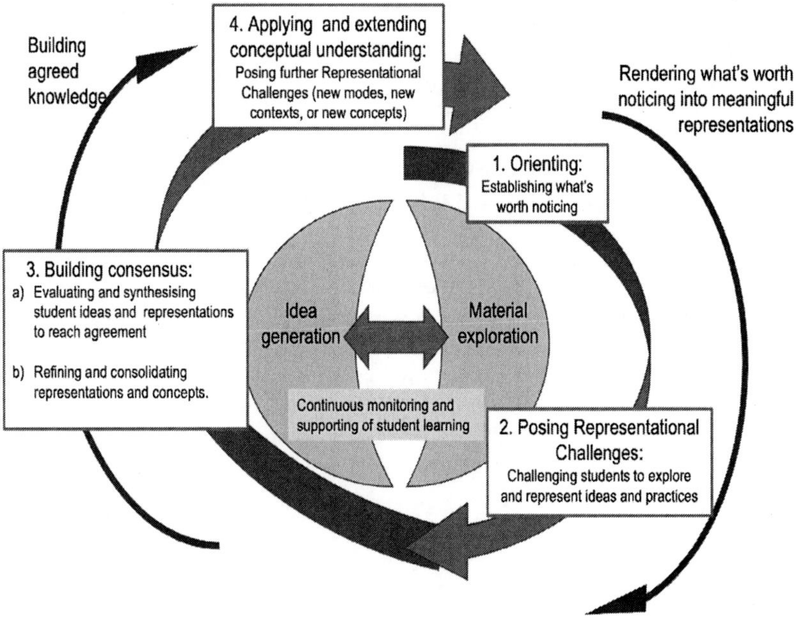

Figure 6.2: The IMS pedagogical model.

In constructing the model, we came to identify a more complex set of pedagogical moves associated with establishing a need that drove investigation and preparing students for the representational challenges that were a key aspect of the sequence.

The ecology sequence involved students being guided to explore the types and numbers of living things in the schoolground, achieved through different groups exploring living things in different sample plots representing distinct habitats. To engage Grade 1 students' interest and sense of purpose for the activity, we discussed with teachers the need to establish what was meant by 'living thing' and open up discussions about how the task of locating and representing living things would proceed. For the representational challenge they were encouraged to tally, draw and eventually graph the distribution of different organisms as would occur in a scientific study. Teachers varied in the details of how they did this, but we focused on our two case study teachers to identify the particular interactions they initiated, and the logic of these within the wider purposes.

In the analysis of Colin's moves, for instance, we identified a number of stages within this first phase that provided an overarching logic which we named 'orienting'. Prior to the major investigation of sample plots, Colin initiated a number of activities in which he:

- Established what we mean by 'living thing' through guided discussion, including probing challenging cases such as 'fire' or 'picked fruit';

- Organised a preliminary exploration to alert students to what living things might be in the schoolground, where, and why ('what do you see here?', 'why are they here and not there?', 'might we look over there in the trees, and not just on the ground?');

- Challenged students to represent areas in the schoolground to explore and used students' work to discuss the features of a clear map;

- Asked the class to predict what they might find, where, and represent this. He then created a list on the board.

As part of the identification of 'orientation' as a particular phase, we identified a set of purposes underpinning it that included but went beyond characterisations of 'engagement' and 'establishing prior knowledge' (as with the '5Es': Bybee et al., 2006) to an active building of representational competence (mapping), conceptual tools (living thing), epistemic practices (as with identifying not only where one might find living things, but looking for reasons why different organisms might be in different places) and values of curiosity and targeted questioning. We came to see all these purposes as contributing to an overarching purpose of establishing what is worth noticing, and how we might structure this noticing.

A further insight came from a close micro-ethnographic analysis of teacher-student interactions, for instance in Colin's exchange with Samantha who was drawing a caterpillar which she thought might be found in the schoolground where there are 'lots of leaves'. Colin asks why she has put trees, grass and leaves in the background:

Samantha:	Because–caterpillars live there.
Colin:	Okay, fantastic–keep going. Show me… I like how you have this bit here, caterpillars are living because they move and eat (*reading student sentence–following with finger*). See if you can think, show me a way–how you could add to this picture to maybe give a bit more information about how they move and eat.
Samantha:	Oh–maybe I could show it eating a leaf.
Colin:	Ok–yes, maybe you could show it eating a leaf.

We argue in our analysis (detailed in Tytler et al., 2022) that Colin bridges between Samantha's everyday sense of caterpillars and their diet to guide her attention to the drawing as a sign that 'establishes new relations between the caterpillar, its needs, and the habitat in which it is found'.

Theoretical outcomes from analyses of teachers' practice

Through the micro-ethnographic analysis of the two case study teachers in the project's first year, including Colin's pedagogy illustrated by the examples above, we achieved four outcomes across the practice-theory spectrum:

a. A clarification of the stages of the pedagogical practice underpinning our representation construction approach, and their purposes;

b. Articulation of the relationship between students' imaginative projections strategically framed by the teachers' prompting, and the explicit authoritative 'direct instruction' moves of the teacher;

c. Attention to the nature of the interactions, from our guiding semiotic perspective, that foregrounded the need to support students' linking of meaning across signs, in the case above between their everyday sense and the more formal sign systems of science.

d. The production of refined learning sequences that are published on our project website, drawing on the insights we gained from these DBR processes, along with examples of student productions to illustrate the range of representational inventions.

For the first outcome, a theoretical advance which is strongly tied to practice, we have identified a range of key features of each of the pedagogical stages, including:

- The close relationship between material practices and representations in investigating numbers of living things in each plot (represented in the centre of the model);
- In the representational challenge stage, the active intervention of the teacher in circulating amongst groups and individuals to challenge, affirm and model;
- In the first part of the building consensus stage, the variety of ways teachers engage students in sharing, evaluating and refining representations;
- The strategies through which teachers established cxplicit agreement in the refining and consolidating stage of building consensus;
- The variety of ways in which representational systems were extended, for instance in restructuring data in the ecology unit to model data on individual organism counts, or the orchestration of different representational accounts in the class display.

The second outcome is a theoretical advance that cast light on the debate between inquiry and direct instruction in showing the way teachers offered strong guidance, sometimes implicit, sometimes explicit, in supporting students' representational work and associated conceptual advancement. We argued on this basis the advantages of a strongly guided pedagogy set within an inquiry context that supported students' engagement with mathematics and science epistemic practices, particularly featuring the imaginative production of representations that mirror fundamental disciplinary practices (Tytler & Prain, 2022a).

The third outcome is a theoretical advance that we have since extended through further analyses, of different sequences (Tytler & Prain, 2022b), regarding the critical importance of transduction–the coordination of meaning across modes–that is creatively supported by both case study teachers in ways that draw attention to the challenges for student learning in coordinating meaning across the material experience and multiple representations in different modes including diagrams, tallies, tables, graphs, talk and annotations (Tytler et al., 2022).

The fourth outcome is a practice outcome; the production of refined learning sequences whose shape is informed by the particular teacher co-design processes described above, for each of the sequences and also through the development of general principles such as with the pedagogy described in Figure 6.1., based on analyses of particular teachers in particular sequences and validated by extension across sequences. This outcome, the refinement of an approach, can be recognised as the explicit, avowed aim of the DBR process, but as we can see it is not the only outcome. Through the process we have generated theoretical outcomes that have been 'put in harm's way' in being tested and found to do useful work in interpreting and reshaping these sequences.

We can add to these outcomes a further set generated by co-design interactions with the group of collaborating teachers. These include the tangible contributions to the learning sequence designs made by teachers in prior workshops and planning, informal discussions during the sequences, and post sequence reviews. We also learnt from these interactions, and by field notes and observations of teachers generally, a range of challenges faced by teachers in negotiating the approaches, including challenges concerning:

- Teachers' epistemological beliefs concerning the appropriate focus of learning in each subject and the role of the teacher supporting learning. This included some teachers' reluctance to pursue mathematical practices beyond the grade level normally designated;

- Disciplinary knowledge where teachers sought clarification from the research team, or occasionally when students' science ideas and insights caught teachers by surprise, or were unrecognised; and

- Pedagogical beliefs and practices when teachers were unused to organising classroom discussions that involved building on students' contributions, or opening up opportunities for student representation invention without pre-figuring their productions.

We have reported on these challenges in other publications (Tytler et al., 2021; Tytler et al., in press). The insights generated have informed the refinement of the layout of sequences and supporting ideas within these resources.

Discussion: How should we think about the relationship between researchers and teachers, and the operation and outcomes of the DBR cycle?

In the analysis above, we have identified the close relationship in the DBR (design experiment) methodology between practice and theory outcomes, and a range of roles for teacher participants in the research, depending on their buy-in to the project, their take-up of its pedagogical underpinnings, and the different roles assigned by the research team. In the previous volume in this series, we have explored the complex relations between practice and theories in DBR (Tytler et al., 2021), and the role of teachers in a DBR project in senior science (Unsworth et al., 2021). The current chapter explores these themes further in terms of the layered nature of teacher participation in DBR, and relations between the practice aims of the research and theory development at different levels. Figure 6.3 articulates the respective roles of teachers in the DBR process, either in explicitly helping the researchers frame and evaluate the sequences, or implicitly through observation and analyses of their practice and student responses, during teaching of the sequences. In the first phases of the cycle, teachers' contributions are generally explicit, and all teachers to some extent are involved in providing these co-design contributions. In the final stages of the cycle, the evaluation and analysis of practice, the role of the teachers is more implicit, offering insights as described above for the case study teachers such as Colin, or insights into challenges from the teacher cohort generally.

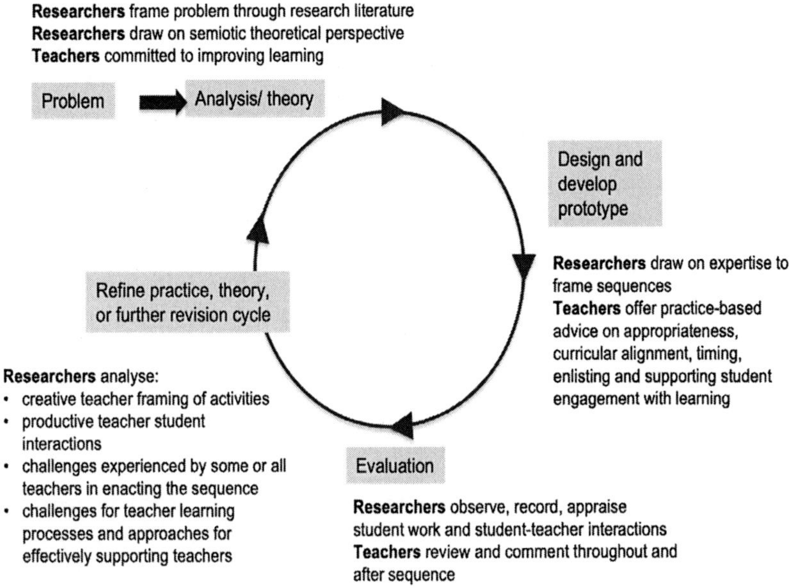

Figure 6.3: Respective roles of teachers and researchers in the DBR cycle.

Table 6.2: Roles of DBR researcher and teacher participants.

Aspect of research	Researchers	All teachers	Committed case study teachers
Theoretical perspectives and overall design	√		
Judgments of practicality of learning sequence		√	√
Creative adaptation of teaching and learning processes			√
Evaluation of learning quality	√	√	√
Judgement of challenges for teachers	√		
Redesign processes	√		

Table 6.2 breaks down these roles more explicitly, identifying the distinctive roles of teachers generally, and those of the committed case study teachers who pushed the practice-related and theoretical insights of the team through their interpretation and extension of the approach, allowing closer analysis of the theory-in-practice.

We should note, however, that the coincidence of volunteering teachers with committed and adaptable teachers is to an extent a matter of luck. We were fortunate with the two case study teachers who volunteered in the first year of the project, in that they quickly adapted to and refined the pedagogy in inventive ways. This also occurred with some non-case study teachers, but it was also true that all case study teachers did not necessarily demonstrate these epistemological and pedagogical alignments with the project.

Conclusion

The principle of co-design within a DBR methodology can be powerful in investigating approaches to teaching and learning based on fresh theoretical insights. The research cycle then becomes a testing ground for both the development of new practices, and the application and refinement of theories. In this chapter, we have drawn on our experience of a major initiative investigating an interdisciplinary approach to mathematics and science teaching based on pragmatist semiotic principles. We have focused particularly on the roles of different teachers in the co-design process, at different stages in the cycle. We argue, on the basis of the research, the value of conceptualising teacher participation in terms of 1) the general cohort of teachers involved in providing practice-based judgments on details of the planned sequences and further evaluative judgments on the practicalities of the approach and the student learnings that resulted, and 2) a select set of teachers–in this case volunteering case study teachers–who proved to be particularly insightful and adaptive in extending the pedagogy in ways that allowed both practice-focused and theoretical insights to emerge.

We have also argued distinctive roles for the research team, who generated the broad shape of the sequences on the basis of theoretical perspectives and prior research, and teachers who needed to be introduced to the perspective and approach but who brought a range of practice-focused qualities to the co-design contract. This relationship will vary depending on the nature and intent of the initiative. In other cases, for instance, the teacher-researcher relationships may be more equally distributed. Nevertheless, we argue that in co-design DBR processes, it can be productive to conceptualise the distinct roles that researchers and teachers can play.

Finally, on the basis of the varied outcomes described in the chapter, we have illustrated what Cobb and colleagues (2003) referred to as a distinctive type of theory associated with design experiments, and the fundamental linking of pursuit of 'what works' with theory development in DBR.

References

Brown, A. L. (1992). Design experiments: Theoretical and methodological challenges in creating complex interventions in classroom settings. *The Journal of the Learning Sciences*, *2*(2), 141-178.

Bybee, R. W., Taylor, J. A., Gardner, A., van Scotter, P., Powell, J. C., Westbrook, A., & Landes, N. (2006). The BSCS 5E instructional model: Origins and effectiveness. *Colorado Springs, Co: BSCS*, *5*, 88-98.

Cobb, P., Confrey, J., Di Sessa, A., Lehrer, R., & Schauble, L. (2003). Design experiments in educational research. *Educational Researcher*, *32*(1), 9-13.

Collins, A. (1992). Towards a design science in education. In E. Scanclon & T. O'Shea (Eds.) *New directions in educational technology*. Berlin: Springer Verlag.

Ferguson, J., Aranda, G., Tytler, R., & Gorur, R. (2018). Video research – Purposeful selection from rich data sets. In Xu, L., Aranda, G., & Clarke, D. (Eds.) *Video-based research in education – Cross-disciplinary perspectives,* (pp. 124-139). Oxon, UK: Routledge. https://doi.org/10.4324/9781315109213

Gooding, D. (2004). Visualization, inference and explanation in the sciences. In G. Malcolm (Ed.), *Studies in Multidisciplinarity* (pp. 1-25), Vol 2. Elsevier.

Hubber, P, Tytler, R., & Haslam, F. (2010). Teaching and learning about force with a representational focus: Pedagogy and teacher change. *Research in Science Education*, *40*(1), 5-28.

Latour, B. (1999). Pandora's hope: essays on the reality of science studies. Cambridge, MA: Harvard University Press.

Lehrer, R. (2009). Designing to develop disciplinary dispositions: Modeling natural systems. *American Psychologist, 64*(8), 759-771.

Lehrer, R. & Schauble, L. (2012). Seeding evolutionary thinking by engaging children in modeling its foundations. *Science Education, 96*(4), 701-724.

Lehrer, R. & Schauble, L. (2020). Stepping carefully: Thinking through the potential pitfalls of integrated STEM. *Journal for STEM Education Research*, *4*, 1-26. https://doi.org/10.1007/s41979-020-00042-y

Lehrer, R., Konold, C., & Kim, M. J. (2006). Constructing data, modeling chance in the middle school. Paper presented at the annual meeting of the *American Educational Research Association*, San Francisco, CA.

Lemke, J. (1998). Multiplying meaning: Visual and verbal semiotics in scientific text. In J. R. Martin & R. Veel (Eds.), *Reading science* (pp. 87-113). London, UK: Routledge.

Manz, E., Lehrer, R., & Schauble, L. (2020). Rethinking the classroom science investigation. *Journal of Research in Science Teaching*, *57*(7), 1-27.

Peirce, C. S. (1998). *The essential Peirce: Selected philosophical writings*, vol. 2. Bloomington, IND: Indiana University Press.

Plomp, T. (2013). Educational design research: An introduction. In T. Plomp & N. Nieveen (Eds.), *An introduction to educational design research* (pp. 9-36). Enschede, The Netherlands: Netherlands Institute for Curriculum Development (SLO).

Prain, V. & Tytler, R. (2012). Learning through constructing representations in science: A framework of representational construction affordances. *International Journal of Science Education*, *34*(17), 2751-2773.

Reeves, T. C. (2006). Design research from a technology perspective. In J. Van den Akker, K. Gravemeijer, S., McKenney, & N. Nieveen (Eds.), *Educational design research* (pp. 52-66). London, UK: Routledge.

Shavelson, R. J., Phillips, D. C., Towne, L., & Feuer, M. J. (2003). On the science of education design studies. *Educational Researcher, 32*(1), 25-28. DOI:10.3102/0013189X032001025

Tytler, R. & Prain, V. (2022b). Supporting student transduction of meanings across modes in primary school astronomy. *Frontiers in Communication.* DOI:10.3389/fcomm.2022.863591

Tytler, R. & Prain, V. (2022a). Interdisciplinary Mathematics and Science – a guided inquiry approach to enhance student learning. *Teaching Science, 68*(1), 31-43.

Tytler, R., Mulligan, J., Prain, V., White, P., Xu, L, Kirk, M., Nielsen, C., & Speldewinde, C. (2021). An interdisciplinary approach to primary school mathematics and science learning. *International Journal of Science Education, 43*(12), 1926-1949. https://doi.org/10.1080/09500693.2021.1946727

Tytler, R., Mulligan, J., White, P., & Kirk, M. (in press). Promoting effective interactions between mathematics and science: Challenges of learning through interdisciplinarity. In Y. Li, Z. Zeng, & N. Song (Eds.),

Changes and innovations in disciplinary and interdisciplinary education in STEM: An international perspective. Springer.

Tytler, R., Prain, V., Hubber, P., & Waldrip, B. (Eds.) (2013). *Constructing representations to learn in science*. Rotterdam, The Netherlands: Sense Publishers.

Tytler, R., Prain, V., Kirk, M., Mulligan, J., Nielsen, C., Speldewinde, C., White, P., & Xu, L. (2022). Characterising a representation construction pedagogy for integrating science and mathematics in the primary school. *International Journal of Science and Mathematics Education,* DOI:10.1007/s10763-022-10284-4

Unsworth, L., Tytler, R., Fenwick, L., Humphrey, S., Chandler, P., Herrington, M., & Pham, L. (2021). *Multimodal literacy in school science: Transdisciplinary perspectives on theory, research and pedagogy*. Routledge.

Russell Tytler (MSc, MEd, PhD) - Deakin University, School of Education, 221 Burwood Highway, Burwood, 3125, VIC, Australia. russell.tytler@deakin.edu.au
Russell Tytler is Alfred Deakin Professor and Chair in Science Education at Deakin University. He is SMSI project lead. He researches and writes on student engagement with science and mathematics, teacher and school change, school-community partnerships, and STEM curriculum policy and practice. https://orcid.org/0000-0003-0161-7240

Peta J. White (BSc (Hons), MRSysMan, PhD, SFHEA) - Deakin University, School of Education, 221 Burwood Highway, Burwood, 3125, Victoria, Australia. peta.white@deakin.edu.au
Peta White is a senior lecturer in science and environmental education at Deakin University. She educated in classrooms, coordinated programs, supported curriculum reform, and prepared teachers in several jurisdictions across Canada and Australia. Peta continues her commitment to initial teacher education leading courses, units, and programs and in-service teacher education through research-informed professional learning programs. Peta's current research follows three narratives: science and biology education; sustainability, environmental, and climate change education; and collaborative/activist methodology and research.
https://orcid.org/0000-0002-0225-5934

CHAPTER 7

HONOURING THE ETHOS OF CO-DESIGN METHODOLOGY IN CONTEMPORARY SOCIO-SCIENTIFIC TEACHING RESEARCH

SEAMUS DELANEY AND MADELEINE SCHULTZ

Abstract

Research co-designed with participants is considered challenging by all stakeholders involved and can be made more so when the focus of the research is on complex educational problems. This chapter describes an ongoing professional learning program, which is supporting senior secondary chemistry teachers to integrate socio-scientific issues such as climate change and sustainable development into their chemistry teaching. The design of the professional learning program is itself undergoing an iterative design cycle, informed by peer reflection between the researchers, artefacts collected from involved students and teachers, and semi-structured interviews with teachers. This chapter provides selected researcher peer reflection and feedback from one group of teachers, implementing an intervention specific to an example of chemistry education research, the Mahaffy Tetrahedron, which models a teaching approach to humanising chemistry education. Through this example and the context of how the researchers presented it initially to the teachers, the methodological considerations of co-designed research-practice partnerships are investigated. We consider issues that arise as researchers guide teachers to incorporate innovative approaches into their teaching.

Keywords: Co-design research, socio-scientific issues, professional learning, sustainability, chemistry

Introduction

The past decades have seen an increase in extreme weather events including heatwaves, droughts and floods, which have been directly linked to climate change (Pörtner et al., 2022; Seneviratne et al., 2012). In addition, there is now more anthropogenic mass than biomass on earth, and twice as much plastic as there is animal matter (Elhacham et al., 2020). Our whole society is experiencing these environmental challenges, including teachers and students.

To prepare them to become conscious and politically active citizens who are open to science that addresses these challenges, today's students need to be equipped with tools to make sense of their world and the rapid and complex changes that they will face. This has been recognised by governments worldwide, and curricula have been modified over the past several years to include socio-scientific issues (SSIs) and sustainability; for example, within the Next Generation Science Standards (https://www.nextgenscience.org/) and the Australian National Curriculum (https://www.australiancurriculum.edu.au/f-10-curriculum/cross-curriculum-priorities/sustainability/). A recent Royal Society of Chemistry report surveying both young people (11-18 years) and education professionals in the United Kingdom (UK) and Ireland found that the vast majority of educators perceive there is not enough climate change or sustainability focus in the current curriculum (RSC, 2021). In parallel, students surveyed overwhelmingly stated that they want climate change and sustainability to be the main priorities of what they learn in the chemistry classroom. However, a recent global survey of national curriculum documents from 48 countries in 2020, published in the *Learn for our Planet* report (UNESCO, 2021) found that less than half of national curriculum documents made mention of climate change explicitly. This misalignment between curriculum policy and priorities was recently expressed as follows:

> We cannot fail to recognize the untenable disconnect between our current learning objective and the types of chemical understandings and ways of thinking that our students need to analyze critically and productively to help address the global challenges we face. (Talanquer et al., 2020, p. 2697)

As found in the UK and also observed in Australia (Australian Education for Sustainability Alliance, 2014), teachers are looking for opportunities to incorporate these topics in their science teaching, for which they require professional development.

A large body of literature has explored features of teacher professional development that make it likely that teachers will adopt innovations and that these will impact student learning (Darling-Hammond et al., 2017; Penuel et al., 2007; Sims et al., 2021) although the 'consensus' from those studies has been questioned (Sims & Fletcher-Wood, 2021). Self-determination has been shown to be associated with motivation to participate in professional development and change teaching practices (Gorozidis & Papaioannou, 2014).

Research-practice partnerships (RPPs) are a vehicle through which teachers and researchers can work together to implement research findings within teaching practice (Coburn & Penuel, 2016). They can result in educational improvement and transformation through spanning the boundaries of research and practice (Farrell et al., 2022).

Design-based research is an approach to educational research that focuses on practical applications and increasing the relevance of research through interventions (Plomp & Nieveen, 2013; van den Akker et al., 2006). Proponents of the approach argue that it is best suited to develop a solution to a problem in a specific context (van den Akker et al., 2006). Thus, design-based research has been identified by Tytler et al. (2021) and others as an optimal methodology for a development study, where teachers engage and participate in authentic environments (professional learning programs; their own classrooms), and their relationship with the subject content is complex, such as teaching SSIs in a chemistry classroom.

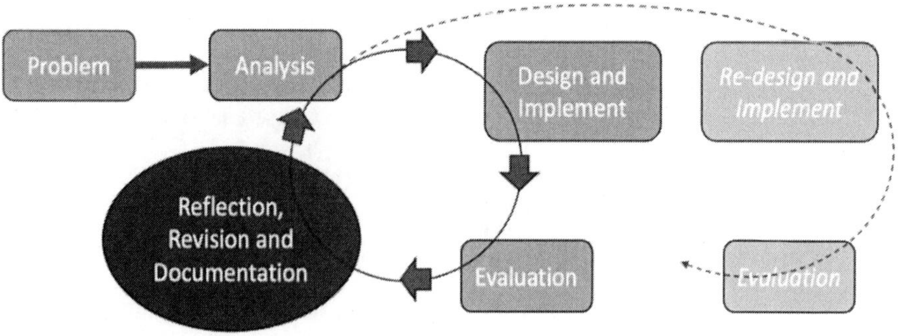

Figure 7.1: Iterations of design-based research cycles. Based on Plomp and Nieveen (2013, p. 17) and adapted from Tytler et al. (2021, p. 55)

Figure 7.1 outlines the cyclic nature of design-based research: analysis of problem, design and implement, evaluate and reflect, revise and re-implement (Plomp & Nieveen, 2013). This iterative approach ensures that subsequent cycles of the professional learning program are informed by the previous steps of the research and its implementation is optimised to the benefit of both the researchers and subsequent participants.

Co-design-based research

One approach to developing practical innovations in the classroom within RPPs is through co-design methodology. Co-design has been defined as:

> …a highly-facilitated, team-based process in which teachers, researchers, and developers work together in defined roles to design an educational innovation, realize the design in one or more prototypes, and evaluate each prototype's significance for addressing a concrete educational need. (Roschelle et al., 2006, p. 606)

Co-design is a dynamic process that can evolve during a project. Common tensions in co-design projects include different definitions of success and workplace norms, and lack of time, particularly on the part of teachers. The former may lead to disengagement by teachers if they perceive that researchers are focused on writing research publications, while the latter is an issue in any project involving teachers. However, research suggests that managing these tensions can increase the agency of the teachers and their ownership of the outputs of co-design projects (Roschelle et al., 2006).

Nonetheless, co-design is an effective approach that gives teachers ownership of programs (Severance et al., 2016). Participation in co-design projects has been shown to increase the leadership capacity of teachers (Scornavacco et al., 2021). Preconditions for reliable change through co-design are on-going RPPs, promoting collective capacity and a focus on both curriculum and infrastructure (Penuel, 2019).

Application to SSI

Based on a number of projects in different science classrooms, Sadler and colleagues developed a SSI Teaching and Learning Framework (Sadler et al., 2017), which gives a useful way to structure activities, based on the NGSS dimensions. Three examples of co-design to address SSIs in classrooms using the SSI T&L Framework have been reported (Friedrichsen et al., 2020). In each case, teachers were successfully able to focus on SSI in their teaching.

In related work, a co-design project around Responsible Research and Innovation in primary classrooms was recently found to lead to empowerment of both teachers and researchers (Kyza & Agesilaou, 2022).

We have used a co-design methodology to work with teachers to run chemistry education research projects in their classrooms. The year-long professional learning program, which is summarised in the methodological design section below, has just completed its second year, and thus second cycle, of its implementation. The impetus for writing this chapter did not come from the urge to explore the benefits and affordances of the co-design research methodology *per se*. Rather, as both researchers and organisers of the professional learning program we were keen to unpack and explore an interesting but at the time baffling choice made by a group of teachers that occurred in the second year of the project, and to position this example as a methodological consideration when working with teachers as *co-designers* of design-based research. We explore this through a narrative peer self-reflection (both co-authors were part of the research design team) and evaluation of the iterative educational (co)-design research model (Plomp & Nieveen, 2013; Tytler et al., 2021) used to design the professional learning project. To further this investigation, we also present a small selection of quotes from teachers who were part of this specific group, collected through semi-structured interviews. Finally, we make some recommendations about the use of co-design that are applicable across contexts.

Methodological design of the co-design professional learning program

As part of a year-long professional learning program, groups of senior chemistry teachers were invited to participate in co-designed, group-based research projects, which they would implement and evaluate in their own classrooms. Data collected on the co-design research projects included de-identified student work samples and teacher reflections collected through an online platform, as well as a series of semi-structured interviews conducted halfway through the program, shortly after the program and after 12 months.

Institutional ethical approval for the project was received from Deakin University (HAE-19-206), the Victorian Department of Education and Training (2019_004196) and Melbourne Archdiocese Catholic Schools (0949). The professional learning program was financially supported by the Chemistry Education Association, which enabled the teachers to attend the professional learning program at no personal expense. Table 7.1 summarises the first two years of the project, which were heavily impacted by COVID-19 restrictions causing school closures.

The overall goal for the professional learning communities was to involve teachers in chemistry education research projects of their choosing. Teachers in each group came from different schools and school sectors (Government, Catholic, Independent) and participation in the program provided an opportunity for the teachers to become part of the chemistry education teaching community in Victoria.

Teachers self-selected their groups, coming together on the first day of the program based on a topic (as shown in Table 7.1) that they wished to collectively investigate in their co-designed research. In the first year of the program, four full days of face-to-face interaction were planned across the year; two at the start, to introduce participants to chemistry education research and allow them to discuss together and choose topics, one mid-year and a wrap up day at the end. During the year, meetings with individual groups were planned. Unfortunately, the advent of COVID-19 shortly after the two face-to-face days in 2020 meant that all subsequent meetings were online, and much of the classroom implementation was also conducted online. In 2021, the year-long program was shortened to just two full days of face-to-face interaction, one at the start and one at the end, but further COVID-19 lockdowns again shifted all planned meetings after the first full face-to-face day to an online format. Each teacher had a unique circumstance enabling some to carry out some activities in their classrooms, while others interacted with their students only online.

Figure 7.2 outlines the year-long program as it was intended to be implemented in its second cycle (2021), and how it will be implemented in its third cycle (pending an absence of further lockdowns).

Table 7.1: Summary of participants and projects in chemistry professional learning communities (SSI relevant topics chosen by particular groups are bolded)

Year	No. teachers	No. groups	Topics chosen by each group	Program Format	Implementation format (with students)
2020	12	3	• **Systems maps** • Reading comprehension (**incl. SSI relevant texts**) • Interpreting exam questions	• 2 days F2F (Feb) • ~Monthly online meetings with indiv groups • ½ day online workshop at end of year	• Mostly online during remote learning (due to COVID lockdowns)
2021	18	3	• **Mahaffy tetrahedron** • Experimental design literacy • Linking to prior concepts	• 1 day F2F (May) • ~Monthly online meetings with indiv groups • 2 hour workshop at end of year	• Mostly online during remote learning (due to COVID lockdowns)

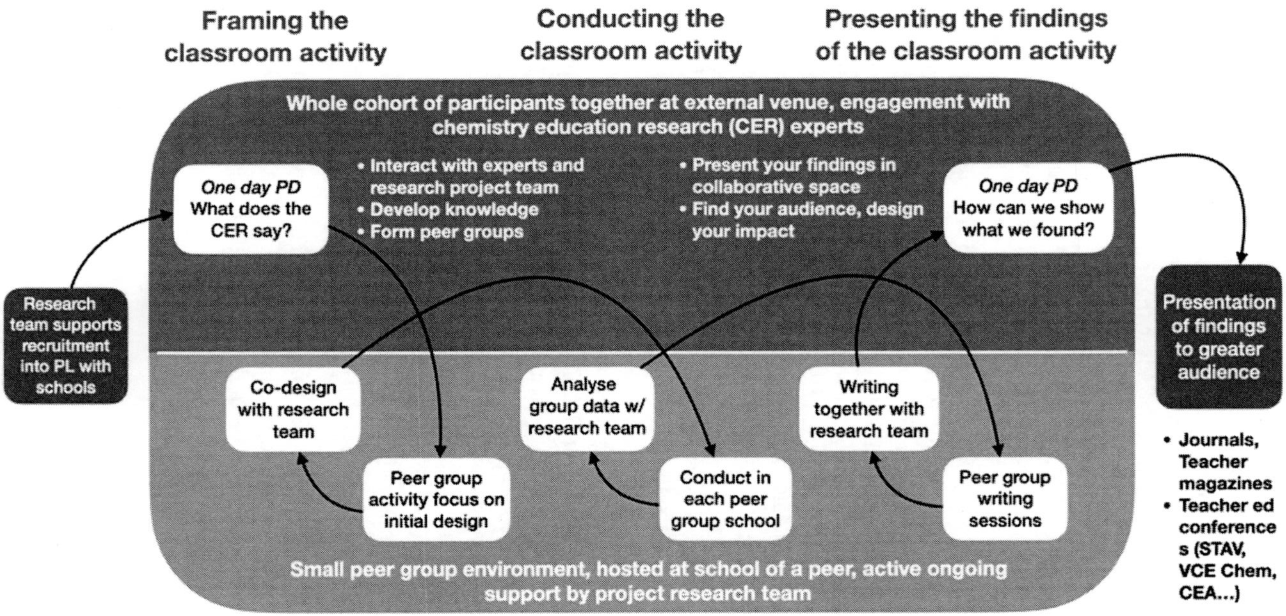

Figure 7.2: Outline of year-long professional learning program, demonstrating which parts of the program are either conducted with all participants together (top), or working in individual groups supported by the researchers (bottom)

Figure 7.2 demonstrates how the researchers intend to model co-designed research, framed around implementing, evaluating and communicating a ‘classroom activity’. Some aspects of the program were delivered to the whole cohort of teachers together (above the line), while others were with each group of teachers separately (below the line), in order to focus on each group’s research topic.

Here iterative design-based research cycles (as shown in Figure 7.1) were modelled. As researchers, we constantly revised and reflected on how each group was implementing their school-based research, and revised our approach to the next interaction with each group. Figure 7.2 also demonstrates how we interacted with the teachers, our co-designers, *before* and *after* the year-long program. Beforehand, as part of the recruitment process, researchers communicated with schools and school leadership about the program and what involvement of a teacher would entail. After the program, groups of teachers were supported to find an audience for their research findings. To date, four groups have presented at teacher education conferences, and two peer-reviewed journal articles are in preparation.

On the first day of the professional learning program, examples of chemistry education research on selected ‘hot topics’ were presented to the teachers (Cooper & Stowe, 2018). As co-designed research that would be implemented in their own schools and classrooms, we emphasised to the teachers that it was important for them to choose a topic or approach that they wanted to implement with their students. This was one way we could minimise the difference in how the researchers and the participants perceived the outcomes of the research (Roschelle et al., 2006). For example, one ‘hot topic’ focused on domain-specific literacy, in other words to support students to ‘read like a chemist’ (Shanahan et al., 2011). Two teacher groups co-designed and implemented literacy-focused classroom activities.

The remainder of this chapter will focus on reflecting on the methodological considerations of co-designed research focused on SSI teaching, though it should be noted that as shown in Table 7.1, not all groups chose SSI-relevant chemistry education research topics.

Contextualising the co-design of a SSI-relevant professional learning program

As outlined above, the teachers who were present at the first day of each year's program self-selected into groups based on their interest to co-design an activity around the innovative chemistry education research they had been presented and interacted with through workshop activities. An example of one *PowerPoint* slide presented to the teachers in attendance is shown in Figure 7.3. The slide is presented exactly as shown, and included the relevant passage from Talanquer et al. quoted above (Talanquer et al., 2020, p. 2697).

The figure in the slide draws upon many years of chemistry education research. Now four decades ago, Johnstone neatly encapsulated the difficulties that students encounter when learning chemistry as being due to the different types of representations used, in the so-called Johnstone's triangle of macroscopic, microscopic and symbolic (Johnstone, 1991). Fifteen years later, Mahaffy extended the triangle to a tetrahedron by adding the human element (Mahaffy, 2006). Others have drawn upon this seminal research into 'humanising' the chemistry curriculum in subsequent years (Burmeister et al., 2012; Sjöström & Talanquer, 2014).

More recently, this approach to chemistry education research has greatly expanded, including a focus on incorporating systems thinking into chemistry education (Mahaffy et al., 2018; Mahaffy, Matlin et al., 2019). Systems thinking has been identified by members of the chemistry education researcher community as a means to *integrate* global sustainable development challenges into chemistry teaching and learning, and a special issue in the Journal of Chemistry Education represented a milestone in this research direction, with multiple examples of action research and theoretical frameworks presented (Mahaffy, Ho et al., 2019; Murphy et al., 2019; Orgill et al., 2019).

A number of these articles were used to inform and guide interactive workshop activities with the teachers attending the first day of the professional learning program. For example, teachers at the workshop drew 'systems maps' to consider the interconnected environmental, economic and societal aspects of a chemical process (Eaton et al., 2019). This in turn inspired one teacher group in 2020 to research and evaluate systems maps created by their students.

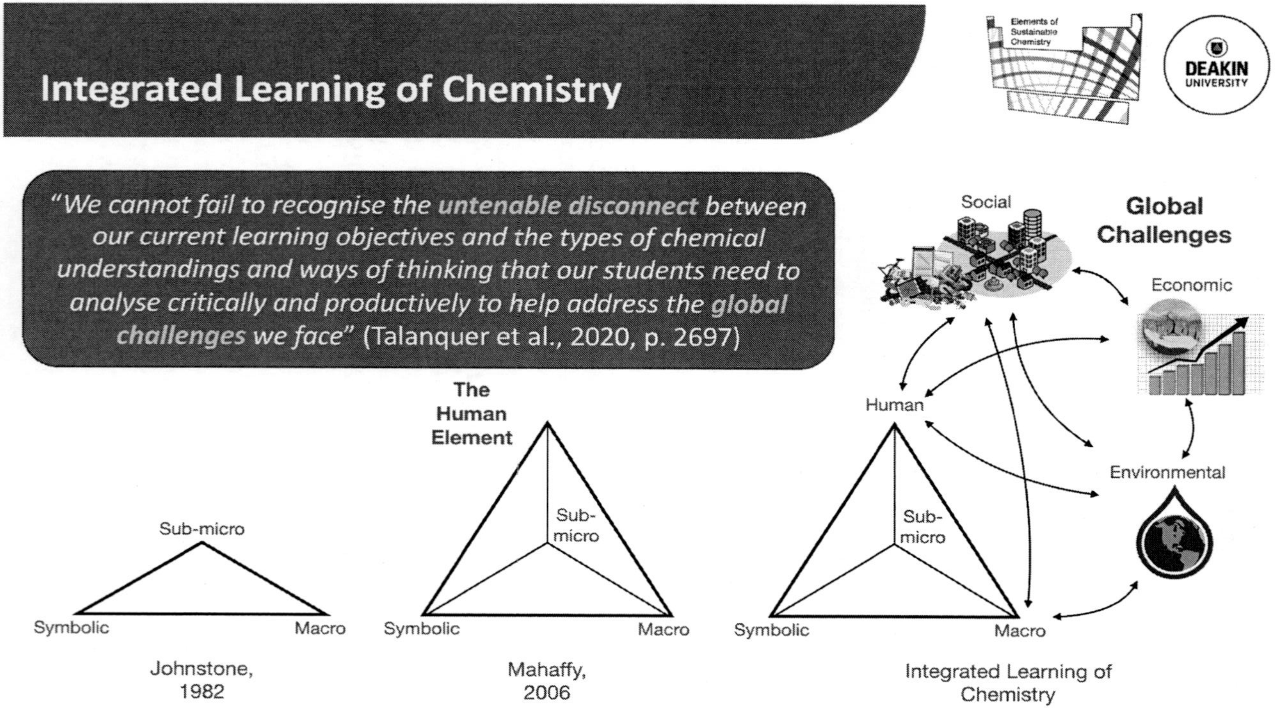

Figure 7.3: Sequential development of approaches to understanding chemistry, as presented to the 2021 cohort of teachers participating in the professional learning program. Adapted from Mahaffy (2006) and Talanquer et al. (2020)

The 'baffling choice' alluded to in the introduction above occurred at this stage. This group of six teachers, seemingly inspired by the research presented on embedding sustainability and global challenges into chemistry teaching and learning, came together based on the workshop slide shown in Figure 7.3. An observation of the teachers that day and from semi-structured interviews conducted later in the year confirmed that a particular representation inspired them collectively. However, it was not the diagram on the far right of Figure 7.3, which represented visually the focus of the *latest* and *most innovative* chemistry education research they were presented. Rather it was the *middle* diagram, the Mahaffy tetrahedron (Mahaffy, 2006). No doubt this would be an excellent piece of SSI-relevant research for the teachers to embed into their teaching practice. But it was fifteen years old and did not therefore meet *our* research intentions ('our' here being the research team) for the teachers, as co-designers, to incorporate a *recent* example of chemistry education research.

As educators delivering the professional learning program but also the researchers, we were therefore torn as to how to proceed as the *other* co-designers of this group-based teacher action research. As educators with greater expertise in this topic field, should we step in and subtly redirect the teachers to incorporate a more substantial and recent focus on socio-scientific issues in chemistry teaching and learning? As researchers, how much of the *ethos* of co-design should be honoured? We had to make an on-the spot decision on how to proceed, because each teacher group had only one afternoon to develop their research intervention before returning to their own schools. Therefore, perhaps to the detriment to our own research objectives, we decided that the desire for teachers to feel they have ownership on the direction of their group's research project outweighed other considerations, and so we encouraged them to proceed in designing their classroom activity around the Mahaffy tetrahedron.

Unpacking this example in the context of the research's larger intentions was then of particular interest for the research team when analysing and subsequentially re-designing the professional learning program for subsequent years. The following section presents an operationalisation of the iterative co-design cycle, motivated by (Tytler et al., 2021) which itself was inspired by (Plomp & Nieveen, 2013) and (Reeves, 2006). As a methodological approach, the operationalisation is being used to validate and question different aspects of our professional learning program. Selected quotes from semi-structured interviews from teachers are presented here to position the validation of the design and subsequent re-design of the co-design cycle.

Documentation and revision of the co-design process cycle

Figure 7.4 represents the iterative design process cycle as it was implemented in this professional learning program. At each stage, we represent (i) how our co-designers are involved, and how as researchers we are contributing to our research question, namely how can we make meaningful change to a chemistry teacher's practice regarding SSIs. The representation deliberately draws on a similar figure in Tytler et al. (2021, p. 55), which contends that each stage of the process cycle is able to "inform/validate/raise questions about presumptions about context… and will enable insights into and judgements about theories at every level" (p. 54). Whilst the context in Tytler et al. (2021) was design-based research, we posit in this chapter that the same contention can be made for *co*-design-based research.

Beginning with our *analysis* of the *problem*, as researchers we identified a professional learning program as a logical means to change teacher practice. Each member of our research team had expertise leading and designing professional learning for secondary school teachers. Our analysis of the problem mirrored the findings of the Royal Society of Chemistry *Green Shoots* report (RSC, 2021), where surveyed chemistry teachers recognised an urgent need to incorporate more climate change and sustainable development focused content into the classroom, a view overwhelmingly held by their students, but did not feel ably supported to do so in their teaching.

We then set out *designing* and *developing* the first iteration of the co-designed professional learning program. As chemistry education researchers ourselves, we identified examples of recent literature that could be easily explained to the teachers, and included examples of the innovative practices being trialled or evaluated in chemistry classrooms. In the case of the SSI-specific examples, we focused on innovative teaching and learning activities that engaged students with systems thinking or with 'humanising' the chemistry curriculum by connecting students with authentic real-world contexts.

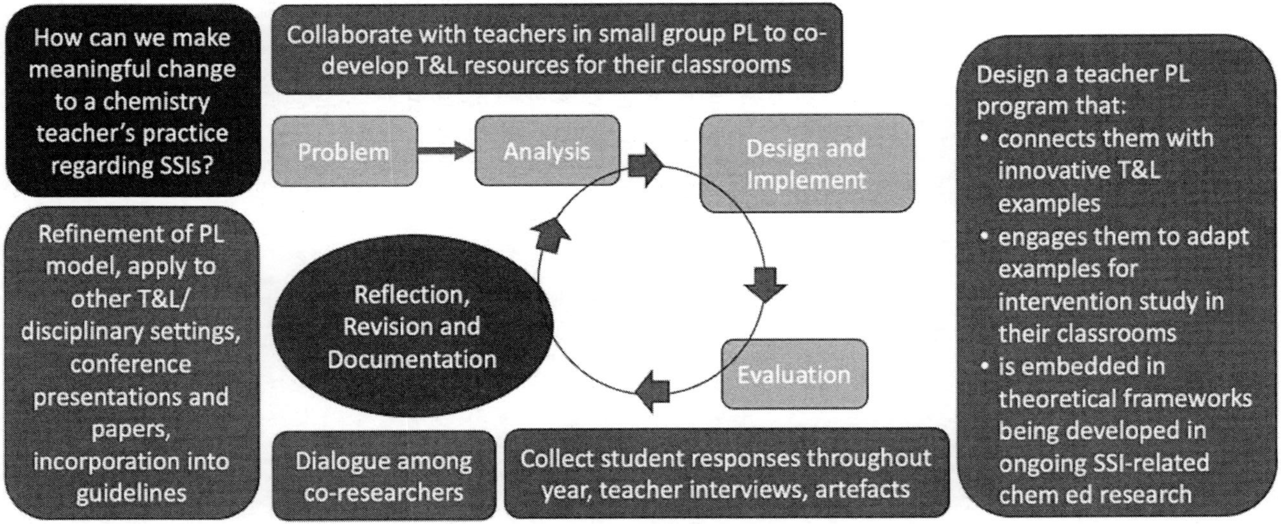

Figure 7.4: Operationalisation of the iterative co-design approach used in the professional learning program

We wanted the research-practice divide (Chi, 2021; Rycroft-Smith & Stylianides, 2022) to not appear insurmountable to our teachers, who as co-designers would be asked to consider literature examples when designing their own *research-informed* classroom activity. However, when *evaluating* our professional learning program, we observed that teachers did struggle with the research-practice divide, specifically with examples that felt more 'education discipline' and less 'chemistry discipline'. As one teacher described:

> *I guess it* (the first PL day) *was a bit overwhelming. I wasn't really so... you know, how education research was... I guess with chemistry research, it's more tangible, right? You've got... it's more results oriented. We've got, you know, an experiment. And here is the results. Whereas this* (education research) *seemed a bit more qualitative to me. And I wasn't quite sure how that works. But I'm sort of getting an idea of how things are happening.* (mid-program interview; after implementing the classroom activity with their class)

We suggest that this teacher entered the professional learning program expecting to undertake a co-designed research activity that was more quantitative with a learning gain measured by a pre-post experimental study. We did allocate time on the first day to illustrate common educational research methods that teachers could practically expect to implement in their schools. We didn't want them to be disappointed, and as co-designers we wanted them to feel that their designed classroom activity would generate some student data. However, this data would not necessary be used to measure a tangible learning gain, but rather could elucidate student knowledge about a socio-scientific issue, so they could change their teaching practice to better support their students. One teacher, after completing the program, acknowledged the influence the program had had on their teaching:

> *I think that's when I realised that there was a huge gap (*in student knowledge about the SSI studied–biopolymers). *In the understanding of connecting these ideas, that's when I realised... actually, hang on a minute, perhaps I need to think about changing my approach to keep teaching. I think we need to have a story in almost across every chapter.* (post-program interview)

Another issue that became apparent was that each classroom activity would be implemented by different teachers at different schools, so this could not be a *controlled* experiment in any case. There would be no control sample (such as a second chemistry class in the same school who didn't undertake the activity), and each school would implement the activity differently. As one teacher noted:

> *I guess at the beginning, I would have expected we'd all kind of do the same one, to make it more comparable... And I guess that's something I've learned as well about education research... that everyone's, you know, you've got to take into account everyone's school situation and what they (the school) expect of you and what they're telling you to do as well.* (post-program interview)

Reflecting on and *documenting* the implementation of the program so far, we observed that our co-designer teachers appreciated interacting with each other, as well as the gentle reminders and multiple check-ins throughout the year:

> *I think it's always a good idea to talk to some other chemistry teachers, and sort of being gently pushed to try something new. Because a lot of the time we want to, but you know, X, Y and Z. And if we don't have the added pressure, a lot of us aren't going to take up anything new.* (post-program interview)

> *It was nice that we all sort of had a focus. We all were sort of on this journey. And it was nice to discuss and to brainstorm, you know, this project that we were doing how, you know, sort of flesh out, and it was nice to have you guys there helping us.* (mid-program interview)

The first teacher here in particular articulates a common barrier for teachers to be co-designers of education research, namely the myriad other calls on their time. Participating in research, or conducting, documenting and evaluating research in their class is challenging for teachers in the best circumstances, so it is worth noting again that the first two years of this program occurred during COVID-19 lockdowns and remote schooling.

Another challenge for the teachers was that, though they were co-designers of their group's classroom activity, they were also participants in *our* research evaluation of this program, which meant that as Deakin researchers we had to manage the ethical aspects. Methodologically but also ethically, the teachers couldn't know which of their students was involved in the research. Students, after their parents/legal guardians had provided consent, were invited to participate by email or in-person by a non-participating teacher at the school, and student data (responses to worksheets, surveys) were collected by an online *Qualtrics* interface only accessible to the Deakin researchers. As multiple schools completed each classroom activity or survey, we as researchers could then collate and de-identify the student samples and share responses with the teachers, so they could not determine if any particular response was from one of their own students. This was made all the more challenging when for large periods of 2020 and 2021, previously approved projects were suspended for government and Catholic schools due to ongoing COVID-19 restrictions in schools. This meant that much student and teacher data could not be included in our research evaluation, and we became 'only' educators providing a professional learning program.

Finally, considering how we have *revised* our program so far, most of our changes have been to deliver the program predominantly online. Putting this dilemma to the side, as researchers we would say our program has changed; we have improved how the teachers are initially introduced to, and how they subsequentially interact with, chemistry education research. Teachers, as co-designers, needed some tangible experiences that they could draw upon, such as interactive workshop experiences that involved them *undertaking* some of the presented research as part of their professional learning; for example, making a systems map of a process of making ammonia with other teachers (Eaton et al., 2019) or categorising chemical vocabulary on a page from a year 11 chemistry textbook (Shanahan et al., 2011). Also, we regularly found ourselves summarising key outcomes from a research article that could inform on their co-design activity, rather than expecting the teachers to read the article themselves. Methodologically, this impressed upon us that the teachers were participants of a professional learning program *first*, and co-designers of a research project *second*. Therefore, using the iterative design process cycle, we have revised our program to provide more support to teachers to implement the co-designed activity.

Finally, we have some insight regarding design research from a *curriculum* perspective (McKenney et al., 2006). The senior secondary chemistry curriculum that all of these teachers used in 2020 and 2021 did *not* include social, ethical, economic or explicitly environmental aspects of global challenges. For instance, students might study water quality in terms of dissolved oxygen or particulate matter, but then not study the *chemistry* that led to this situation, such as nitrate-based fertiliser runoff from a nearby farm or the *chemistry* that is contributing to innovative solutions to clean drinking water in low-income countries or in rural areas which have increasingly scarce water supply. A revised study design from 2023 will improve on this deficiency in Victoria, and teachers need to be prepared to deliver this type of content.

We contend that one reason that our group of teachers chose a 15-year-old study over more recent example shown was that they could envisage a related classroom activity for the Mahaffy tetrahedron (Figure 7.3) for their co-design research that could be incorporated into the curriculum they *implement* in their classroom. In other words, from a curriculum perspective, they could more easily identify how they could engage students with *what* is the real-world context of the chemistry content in the curriculum (for example, copper is used because it is electrically conductive, aluminium is used to make planes because it is lightweight). However, they could not connect sufficiently with recent teaching innovation examples demonstrating the interconnected nature of chemistry, and *how* chemistry is addressing sustainable development challenges.

Conclusions and implications for research practice

Co-design is an effective approach to empower teachers to make changes to their practice, supported by researchers. We have demonstrated here, through our peer reflection and teacher semi-structured interviews from the first two iterations of our professional learning program, that an iterative design process cycle can also be utilised to reflect, revise and document on a co-designed research project. Utilising this research framework has influenced how we have revised the SSI-relevant professional learning program in subsequent iterations.

However, it is imperative that as researchers and educators we recognise that our co-designers are still teachers first, who are contending with a raft of external and internal pressures that determine what and how their ‘co’ in co-design can be. Clearly, extended lockdowns and remote teaching across 2020 and 2021 has impacted how participating teachers could implement and evaluate evidence collected from students regarding their co-designed SSI-relevant classroom activity. But other issues are also significant, not least how closely related the teachers perceived the innovative teaching examples they were presented to the curriculum they teach. Without further changes to the intended curriculum of senior secondary chemistry, any teachers’ actions towards incorporating an integrated learning of chemistry, particularly for climate change and sustainable development, may come from only their own personal imperatives to engage and motivate their students.

References

Australian Education for Sustainability Alliance. (2014). *Education for sustainability and the Australian Curriculum Project: Final report for research phases 1 to 3*. Retrieved from Melbourne, Australia. Retrieved from http://www.aaee.org.au/wp-content/uploads/2017/08/AAEE-Education-for-Sustainability-and-the-Australian-Curriculum-Project-Final-Report-2015.pdf

Burmeister, M., Rauch, F., & Eilks, I. (2012). Education for Sustainable Development (ESD) and chemistry education. *Chemistry Education Research and Practice, 13*(2), 59-68. doi: 10.1039/C1RP90060A

Chi, M. T. H. (2021). Translating a theory of active learning: An attempt to close the research-practice gap in education. *Topics in Cognitive Science, 13*(3), 441-463. doi: 10.1111/tops.12539

Coburn, C. E. & Penuel, W. R. (2016). Research–practice partnerships in education: Outcomes, dynamics, and open questions. *Educational Researcher, 45*(1), 48-54. doi: 10.3102/0013189x16631750

Cooper, M. M. & Stowe, R. L. (2018). Chemistry education research - From personal empiricism to evidence, theory, and informed practice. *Chemical Reviews, 118*, 6053-6087. doi: 10.1021/acs.chemrev.8b00020

Darling-Hammond, L., Hyler, M. E., & Gardner, M. (2017). *Effective teacher professional development*. Retrieved from: https://learningpolicyinstitute.org/sites/default/files/product-files/Effective_Teacher_Professional_Development_REPORT.pdf

Eaton, A. C., Dclaney, S., & Schultz, M. (2019). Situating sustainable development within secondary chemistry education via systems thinking: A depth study pproach. *Journal of Chemical Education, 96*, 2968-2974. doi: 10.1021/acs.jchemed.9b00266

Elhacham, E., Ben-Uri, L., Grozovski, J., Bar-On, Y. M., & Milo, R. (2020). Global human-made mass exceeds all living biomass. *Nature, 588*(7838), 442-444. doi: 10.1038/s41586-020-3010-5

Farrell, C. C., Penuel, W. R., Allen, A., Anderson, E. R., Bohannon, A. X., Coburn, C. E., & Brown, S. L. (2022). Learning at the boundaries of research and practice: A framework for understanding research–practice partnerships. *Educational Researcher* (Advance online publication). doi: 10.3102/0013189x211069073

Friedrichsen, P. J., Sadler, T. D., & Zangori, L. (2020). Supporting teachers in the design and enactment of socio-scientific issue-based teaching in the USA. In M. Evagorou, J. A. Nielsen, & J. Dillon (Eds.), *Science teacher education for responsible citizenship: Towards a pedagogy for relevance through socioscientific issues* (pp. 85-99). Cham, Switzerland: Springer International Publishing.

Gorozidis, G. & Papaioannou, A. G. (2014). Teachers' motivation to participate in training and to implement innovations. *Teaching and Teacher Education, 39*, 1-11. doi: 10.1016/j.tate.2013.12.001

Johnstone, A. H. (1991). Why is science difficult to learn? Things are seldom what they seem. *Journal of Computer Assisted Learning, 7*, 75-83. doi: 10.1111/j.1365-2729.1991.tb00230.x

Kyza, E. A. & Agesilaou, A. (2022). Investigating the processes of teacher and researcher empowerment and learning in co-design settings. *Cognition and Instruction, 40*(1), 100-125. doi: 10.1080/07370008.2021.2010213

Mahaffy, P. G. (2006). Moving chemistry education into 3D: A tetrahedral metaphor for understanding chemistry. Union Carbide Award for Chemical Education. *Journal of Chemical Education, 83*(1), 49. doi: 10.1021/ed083p49

Mahaffy, P. G., Ho, F. M., Haack, J. A., & Brush, E. J. (2019). Can chemistry be a central science without systems thinking? *Journal of Chemical Education, 96*(12), 2679-2681. doi: 10.1021/acs.jchemed.9b00991

Mahaffy, P. G., Krief, A., Hopf, H., Mehta, G., & Matlin, S. A. (2018). Reorienting chemistry education through systems thinking. *Nature Reviews: Chemistry, 2*, 1-3. Doi: 10.1038/s41570.018.0126

Mahaffy, P. G., Matlin, S. A., Holme, T. A., & MacKellar, J. (2019). Systems thinking for education about the molecular basis of sustainability. *Nature Sustainability, 2*(5), 362-370. doi: 10.1038/s41893-019-0285-3

McKenney, S., Nieveen, N., & van den Akker, J. (2006). Design research from a curriculum perspective. In J. van den Akker, K. Gravemeijer, S. McKenney & N. Nieveen (Eds.), *Educational design research* (pp. 67-90). London, UK: Routledge.

Murphy, K. C., Dilip, M., Quattrucci, J. G., Mitroka, S. M., & Andreatta, J. R. (2019). Sustainable consumer choices: An outreach program exploring the environmental impact of our consumer choices using a systems thinking model and laboratory activities. *Journal of Chemical Education, 96*(12), 2993-2999. doi: 10.1021/acs.jchemed.9b00400

Orgill, M., York, S., & MacKellar, J. (2019). Introduction to systems thinking for the chemistry education community. *Journal of Chemical Education, 96*, 2720-2729. doi: 10.1021/acs.jchemed.9b00169

Penuel, W. R. (2019). Co-design as infrastructuring with attention to power: Building collective capacity for equitable teaching and learning through design-based implementation research. In J. Pieters, J. Voogt & N. Pareja Roblin (Eds.), *Collaborative curriculum design for sustainable innovation and teacher learning* (pp. 387-401). Cham, Switzerland: Springer International Publishing.

Penuel, W. R., Fishman, B. J., Yamaguchi, R., & Gallagher, L. P. (2007). What makes professional development effective? Strategies that foster curriculum implementation. *American Educational Research Journal, 44*(4), 921-958. doi: 10.3102/0002831207308221

Plomp, T. & Nieveen, N. (Eds.). (2013). *Educational design research*. Enschede, The Netherlands: Netherlands Institute for Curriculum Development.

Pörtner, H.-O., Roberts, D. C., Tignor, M., Poloczanska, E. S., Mintenbeck, K., Alegría, A., . . . Rama, B. (Eds.). (2022). *Climate change 2022: Impacts, adaptation, and vulnerability. Contribution of Working Group II to the Sixth Assessment Report of the Intergovernmental Panel on Climate Change*. Cambridge, UK: Cambridge University Press.

Reeves, T. (2006). Design research from a technology perspective. In J. van den Akker, K. Gravemeijer, S. McKenney & N. Nieveen (Eds.), *Educational Design Research* (pp. 52-66). London, UK: Routledge.

Roschelle, J., Penuel, W. R., & Shechtman, N. (2006). *Co-design of innovations with teachers: Definition and dynamics*, Paper presented at the The International Conference of the Learning Sciences, Indiana University, USA.

Royal Society of Chemistry (RSC). (2021). *Green shoots: A sustainable chemistry curriculum for a sustainable planet*. Retrieved from: https://www.rsc.org/new-perspectives/sustainability/a-sustainable-chemistry-curriculum/

Rycroft-Smith, L. & Stylianides, A. J. (2022). What makes a good educational research summary? A comparative judgement study of mathematics teachers' and mathematics education researchers' views. *Review of Education, 10*(1), e3338. doi: 10.1002/rev3.3338

Sadler, T. D., Foulk, J. A., & Friedrichsen, P. J. (2017). Evolution of a model for socioscientific issue teaching and learning. *International Journal of Education in Mathematics, Science and Technology, 5*(2), 75-87. doi: 10.18404/ijemst.55999

Scornavacco, K., Kelly, M. R., & Boardman, A. (2021). Leading with curricular co-design: An exploration of teacher leadership through the co-design process. *Journal of Research on Leadership Education, Advance online publication*. doi: 10.1177/19427751211017595

Seneviratne, S. I., Nicholls, N., Easterling, D., Goodess, C. M., Kanae, S., Kossin, J., . . . Zhang, X. (2012). Changes in climate extremes and their impacts on the natural physical environment. In C. B. Field, V. Barros, T. F. Stocker, D. Qin, D. J. Dokken, K. L. Ebi, M. D. Mastrandrea, K. J. Mach, G.-K. Plattner, S. K. Allen, M. Tignor, & P. M. Midgley (Eds.), *A Special Report of Working Groups I and II of the Intergovernmental Panel on Climate Change (IPCC)* (pp. 109-230). Cambridge, UK and New York, NY: Cambridge University Press.

Severance, S., Penuel, W. R., Sumner, T., & Leary, H. (2016). Organizing for teacher agency in curricular co-design. *Journal of the Learning Sciences, 25*(4), 531-564. doi: 10.1080/10508406.2016.1207541

Shanahan, C., Shanahan, T., & Misischia, C. (2011). Analysis of expert readers in three disciplines: History, mathematics, and chemistry. *Journal of Literacy Research, 43*(4), 393-429. doi: 10.1177/1086296x11424071

Sims, S. & Fletcher-Wood, H. (2021). Identifying the characteristics of effective teacher professional development: A critical review. *School Effectiveness and School Improvement, 32*(1), 47-63. doi: 10.1080/09243453.2020.1772841

Sims, S., Fletcher-Wood, H., O'Mara-Eves, A., Cottingham, S., Stansfield, C., Van Herwegen, J., & Anders, J. (2021). *What are the characteristics of teacher professional development that increase pupil achievement? A systematic review and meta-analysis*. Retrieved from: https://educationendowmentfoundation.org.uk/education-evidence/evidence-reviews/teacher-professional-development-characteristics

Sjöström, J. & Talanquer, V. (2014). Humanizing chemistry education: From simple contextualization to multifaceted problematization. *Journal of Chemical Education, 91*(8), 1125-1131. doi: 10.1021/ed5000718

Talanquer, V., Bucat, R., Tasker, R., & Mahaffy, P. G. (2020). Lessons from a pandemic: Educating for complexity, change, uncertainty, vulnerability, and resilience. *Journal of Chemical Education, 97*, 2696-2700. doi: 10.1021/acs.jchemed.0c00627

Tytler, R., Hobbs, L., Brown, J., White, P. J., Campbell, C., Cripps Clark, J., . . . Widjaja, W. (2021). Theory and practice relations in design-based research: Designing professional learning with teachers teaching mathematics and science out-of-field. In P. J. White, R. Tytler, J. Ferguson & J. Cripps Clark (Eds.), *Methodological approaches to STEM education research* (Vol. 2). Cambridge, UK: Cambridge Scholars Publishing.

UNESCO. (2021). *Learn for our planet: A global review of how environmental issues are integrated in education*. Retrieved from: https://unesdoc.unesco.org/ark:/48223/pf0000377362
van den Akker, J., Gravemeijer, K., McKenney, S., & Nieveen, N. (Eds.). (2006). *Educational design research*. London, UK: Routledge.

Acknowledgements

We thank the teachers who participated in our program for their huge effort during a stressful time. We are grateful to Lisa Chiavaroli (project officer) as well as the Chemistry Education Association for their financial support in facilitating the professional learning program for participating teachers.

Seamus Delaney (BSc(Hons), MTeach(Sec), MEdPol(Int), PhD, SFHEA) - Deakin University, School of Education, 221 Burwood Highway, Burwood, 3125, Victoria, Australia. s.delaney@deakin.edu.au
Seamus Delaney is a chemistry and science education senior lecturer at Deakin University. His research interests are teacher professionalisation, chemistry learning approaches and science education in informal/out-of-school contexts. He works with schools and teachers on addressing global challenges such as sustainable development in the classroom.
https://orcid.org/0000-0003-3271-1686

Madeleine Schultz (BSc(Hons), PhD, MEd, FRACI, SFHEA) - Deakin University, School of Life and Environmental Sciences, 75 Pigdons Rd, Waurn Ponds, 3216, Victoria, Australia. madeleine.schultz@dcakin.edu.au
Madeleine Schultz is an honorary fellow in the School of Life and Environmental Sciences. She has taught undergraduate chemistry at universities in Australia and overseas since 2002. Her interests in chemistry education research include concept development, assessment and incorporation of sustainability into chemistry curricula.
https://orcid.org/0000-0001-7967-5147

CHAPTER 8

COLLABORATING ACROSS DISCIPLINARY LINES: GENERATING A FRAMEWORK FOR OUT-OF-FIELD TEACHERS IN THE SCIENCES, ENGLISH, MATHEMATICS, DESIGN TECHNOLOGIES AND THE HUMANITIES

LINDA HOBBS, LUCINDA MCKNIGHT, JARED CARPENDALE, SUSAN CALDIS, CORAL CAMPBELL, SEAMUS DELANEY, CHRIS SPELDEWINDE, LEISSA KELLY AND COLLEEN VALE

Abstract

This chapter highlights the value of doing interdisciplinary research in relation to teaching out-of-field, a phenomenon which itself requires teachers to assimilate different disciplines into coherent practice. As a team of experts in a range of diverse disciplines and subject areas, we set out to develop a framework for thinking about what a teacher needs to know, do and be, when crossing subject boundaries. This chapter focuses on the interdisciplinary collaborative inquiry process we undertook to develop this framework. We explain our focus, research process and the role of signature pedagogies in providing a conceptual frame for working together across disciplines. We recount our methodological steps, illustrate the complexity of our interdisciplinary dialogue, and suggest how our collaboration has led to new understandings of teaching out-of-field, to inform teacher professional development. We emerge from this study with a heightened empathy for out-of-field teachers who face the realities of working across disciplines.

Keywords: Teaching out-of-field, definitions, literature review, interdisciplinarity, collaboration

Introduction

Interdisciplinary research and collaboration are being encouraged at all levels of research and practice to promote sharing of ideas and enhanced innovation. Collaborative research enables "researching 'with' rather than research 'on'" (Given, 2012, p. 2), whereas interdisciplinary research advances understandings through juxtaposition of what constitutes disciplinary knowledge and practice (Williams & Roth, 2019). This research involves disciplinary experts/teacher educators/researchers (the authors) from different disciplines generating a framework of subject-specific expertise for out-of-field teachers that is translatable across disciplines.

Teaching out-of-field, that is, teaching specialisations or year levels without the necessary background (Hobbs, 2013; Ingersoll, 1998; Weldon 2016), results in a teacher being placed in a subject or discipline that is unfamiliar to them. A secondary teacher has training in some subjects but not others; however, when asked to teach out-of-field they need to learn new subject matter (e.g., knowledge, skills, attitudes, stories) and new teaching strategies specific to that content. In addition, Hobbs (2020) identified that science teachers teaching out-of-field also learned which aspects from their in-field subject matter and pedagogies could be translated to their out-of-field teaching, and vice versa. These teachers also discovered an appreciation for what other teachers do and where they can gain support, as well as how to make the content relevant, how students behaved and learned across different subjects, and how to remain a passionate teacher despite teaching out-of-field.

This chapter focuses on the interdisciplinary collaborative inquiry process we undertook to arrive at a framework that elucidates what a teacher needs to know, do and be, when crossing subject boundaries. In the same way that teachers move into unfamiliar territory when teaching out-of-field, we also needed to empathise and traverse unfamiliar territory to understand similarities and differences with respect to what it means to be a teacher and teacher educator in specialist areas other than our own.

The research team consists of school classroom teachers who are now teacher educators/researchers/experts from the fields of Science (Linda), Chemistry (Seamus), Physics (Jared), Design Technology (Coral), English (Lucinda), Geography (Susan) and Mathematics (Colleen). We have various experiences of the out-of-field phenomenon. We have all taught out-of-field, either at school or university, and we bring this nuanced life experience to the research process, even as we officially 'represent' the discipline and related subjects of our first degrees. Lucinda and Susan have contributed to the work of their professional associations in understanding and advocating for the needs of out-of-field teachers. In relation to research, Lucinda is a newcomer to specifically researching out-of-field teaching, while all others have published in this space in relation to: out-of-field teacher identity, beliefs and learning (Linda and Coral); defining and mapping out-of-field teaching (Linda, Coral and Seamus); supporting the growth in out-of-field in-service or pre-service teacher knowledge and practice (Jared and Susan); and designing and delivering professional learning specifically for out-of-field teachers (Colleen, Seamus and Coral). These various experiences, expertise, and commitments inform what we bring to the study.

Using the lens of Shulman's (2005) signature pedagogies and Akkerman and Bakker's (2011) notion of boundary crossings, the researchers drew on what is currently known (through their lived experience, practice and the literature) to elaborate on and differentiate subject ways of knowing (knowledge and expertise), doing (practice), and being (identity). This collaborative research is experiential, with the researchers' crossing boundaries into other disciplinary fields and experiencing out-of-field-ness as they negotiate salient elements for disciplinary/subject comparison.

The research focus

The issue of teaching out-of-field centres on the disciplinary divide especially at the secondary school level. In this research, we are problematising what specialisation means. According to Hobbs and Törner (2019):

> To specialise means to become immersed and expert in a defined and bounded body of knowledge and skills such that there is coherence, connectedness and flexibility to what is known and what one can do…For the subject specialist teacher, a history of learning, knowing and using the specific disciplinary knowledge and skills through formal qualifications

> provides at least a background to their teaching; at most a filter that colours every pedagogical move. (p. 4)

Subjects have long been the major reference point in secondary schooling (Goodson, 1993), and to some extent "shared traditions, shared experiences, shared problems, values and idioms" (Schwab, 1964, p. 12) make the subject identifiable. For the out-of-field teacher, understanding the content and teaching strategies is accompanied by challenges of participation in the broader subject culture, that is, the tensions and varying discourses relating to contemporary and best practice; what it means to 'belong' to and identity with the group; and the support and collaboration needed and available (Crisan & Hobbs, 2019). Our collaboration has focused on these types of challenges that emerge within subject departments, more broadly within and across schools, and in the milieu of the broader subject, disciplinary and public discourses.

Teaching out-of-field has been conceptualised as a general lack of background in whatever is being taught, which could include teaching students with special needs or phases of schooling without the pedagogical training (e.g., Du Plessis et al., 2014; Sharplin, 2014). In our study we have focused on the disciplinary differences of teaching across different subjects. For the subject specialist, some parts of their practice can be translated or transferred from one subject to another; however other elements are subject-specific. Shulman's (1986, 1987) model of Pedagogical Content Knowledge (PCK) speaks to the relationship between teacher knowledge of teaching approaches, the content for teaching (including curriculum and its structure), and the students' learning needs (Carlson et al., 2019; Carpendale & Hume, 2022; Nixon & Luft, 2015). However, while PCK has been acknowledged as an important consideration for quality teaching and provides a useful way to understand teachers' specialised knowledge, the current conceptualisations of PCK do not explicitly examine the moral dimensions of the subject matter, nor matters of teacher identity to great depths (see Carlson et al., 2019). These considerations are at the heart of our interrogation of what it means to be and become a specialist teacher.

As teacher educators/researchers/experts from different disciplines, we are collaborating to generate a framework of subject-specific expertise for out-of-field teachers that is translatable across disciplines. The aim is to generate an enabling framework that acknowledges the complexities of teaching out-of-field while simultaneously finding a way to build bridges across disciplines. The framework will be designed to compare and contrast subject-specific expertise and nuance in a way that can support principals, heads of department, and teachers themselves as they plan for teaching out-of-field. One research question and two sub-questions underpin the inquiry:

What expertise is core to teaching a subject, including what is subject-specific and what is common or generic?

> *How does the experience of teaching out-of-field in one subject differ to teaching out-of-field in another?*
>
> *What subject teaching expertise do teachers teaching out-of-field need?*

The chapter is organised in a way that represents the processes of our interdisciplinary collaboration. Throughout the chapter we draw on excerpts of our conversation about our research methodology to illustrate how we co-constructed meaning. We begin by describing the theoretical resources that we found particularly useful in our deliberations: firstly Shulman's (2005) construct of signature pedagogies, which we used as an organisational framework for thinking about what it means to know, do and be a subject teacher; and secondly what we mean by discipline and subject in the context of subject teaching. We then describe and situate our inquiry in relation to current understandings of collaborative and interdisciplinary research. The methodological steps are then explained, followed by an illustration of interdisciplinary collaboration as we debate Schwab's (1964) understanding of the structure of disciplines. We finish with some reflections on the research process so far. This chapter does not introduce the framework that we developed, but it does describe and illustrate the process we have gone through to create it.

Framing our inquiry with signature pedagogies

This project explores the subject-specific nature of teaching out-of-field. In an attempt to move beyond the confines of PCK, we explored Shulman's more recent construct of signature pedagogies because of its acknowledgement of the 'profession' and the focus on salient and fundamental pedagogies.

Shulman (2005) introduced signature pedagogies as "the types of teaching that organise the fundamental ways in which future practitioners are educated for their new profession" (p. 52). This focus on professional learning acknowledges the connection between disciplinary practices and the ways people are apprenticed into the profession (Hennessey & Selvakumaran, 2018). Beyond education for the professions, the construct of signature pedagogies has become a useful framing of what is core to teaching and learning because of the way it focuses on the surface structure, deep structure, and implicit structures of pedagogy (pp. 54-55):

- Surface structure: "concrete, operational acts of teaching and learning, of showing and demonstrating, of questioning and answering, of interacting and withholding, of approaching and withdrawing"

- Deep structure: "a set of assumptions about how best to impart a certain body of knowledge and know-how"

- Implicit structure: "a moral dimension that comprises a set of beliefs about professional attitudes, values, and dispositions"

Shulman's seminal paper on signature pedagogies from 2005 has been influential in informing conceptualisations of core pedagogy for particular disciplines. Further, the language of signature pedagogies has crept into system-level (e.g., Victorian Department of Education and Training, 2019) and school-level (e.g., Hennessey & Selvakumaran, 2018) framing of quality and purposeful teaching. In our interdisciplinary collaboration we sought out differences and commonalities in the signature pedagogies of different subjects. Table 8.1 lists selected research relating to different subjects and contexts in school education (all reference Shulman 2005), organised chronologically.

Table 8.1: References to signature pedagogies in particular subjects and contexts for schools, organised chronologically

Authors	**Title**	**Discipline or practices**
Carter (2007)	Ways of knowing, doing, and writing in the disciplines	All/ English
Gurung, Chick, & Haynie (2009)	Exploring signature pedagogies: Approaches to teaching disciplinary habits of mind (Edited Book)	General / English / Biology / Geography / Physics / Maths
Komoto (2009)	Moving toward a signature pedagogy in geography	Geography

Darby (2010)	Characterising secondary school teacher imperatives as subject (signature) pedagogies in maths and science	Maths / Science
Abramovich et al. (2012)	Parallel structures of computer-assisted signature pedagogy: The case of integrated spreadsheets	Mathematics
Beck & Eno (2012)	Signature pedagogy: A literature review of social studies and technology research	Social sciences / Technology
Crippen & Archambault (2012)	Scaffolded inquiry-based instruction with technology: A signature pedagogy for STEM education	STEM
Larkin et al. (2012)	TPACK and pre-service teacher mathematics education: Defining a signature pedagogy for mathematics education using ICT and based on the metaphor 'Mathematics is a language'	Mathematics
Lovin et al. (2012)	Examining beliefs and practices of self and others: Pivotal points for change and growth for mathematics teacher educators	Mathematics
Passey (2012)	Educational technologies and mathematics: Signature pedagogies and learner impacts	Mathematics
Thomson, Hall, Jones, & Selft-Green (2012)	The signature pedagogies project: Final report. Culture, creativity and education.	Creativity
Quinnell et al. (2013)	It's not maths; it's science: Exploring thinking dispositions, learning thresholds and mindfulness in science learning	Science

Asunda, P. & Weitlauf. J (2018)	STEM habits of mind: Enhancing a PBL design challenge—integrated STEM instruction approach	STEM / Engineering
McLain (2018)	Emerging perspectives on the demonstration as a signature pedagogy in design and technology education	Design Technology
Seow et al. (2019)	Field-based inquiry as a signature pedagogy for geography in Singapore	Geography
Vincent-Lancrin, Stéphan et al. (2019)	Eleven signature pedagogies related to the fostering of creativity and critical thinking	Creativity, Critical thinking
Koeper et al. (2020)	Turning chemistry education on its head: Design, experience and evaluation of a learning-centred 'Modern Chemistry' subject	Chemistry
McLain (2021)	Towards a signature pedagogy for design and technology education: A literature review	Design Technology

As shown by the different titles in Table 8.1, in school education, signature pedagogies have become a useful construct to define: what is salient to the practice of a subject teacher (Gurung et al., 2009; McLain, 2021; Seow et al., 2019); core beliefs in relation to teaching and learning (Crippen & Archambault, 2012; Lovin et al., 2012; McLain, 2021); fundamental practices associated with learning, such as critical and creative thinking (Thomson et al., 2012; Vincent-Lancrin et al., 2019), technology use (Beck & Eno, 2012; Passey, 2012); and norms of learning and certain habits of mind, head and heart in particular contexts (e.g., Komoto, 2009). Signature pedagogies set out to teach the disciplinary ways of thinking (learning to know), doing (learning to do, practice), and being (learning to be). Thomson et al. (2012) recognise the epistemological, ontological, and axiological nature of signature pedagogies, which they believe becomes tacit knowledge of the practitioner as demonstrated in both how they orient themselves to questions and tasks, and what they say and do.

So, in thinking about what is core to teaching a subject and what is subject-specific or common, we needed to consider the experience of teaching the out-of-field subject, recognising that fundamental to teaching a subject is the disciplinary ways of knowing, doing and being. In doing so, through collaboration we are reflecting on and teasing out the subject teaching expertise inherent to teaching that subject.

> Coral: Some of the very early discussions as a group raised my awareness of different types of subject-specific understanding, so when Susan has talked or when others have talked, you sort of start to get an understanding of the nuances that exist between different subjects.

The expertise needed by teaching out-of-field includes both the actual teaching act but also how a teacher prepares themselves for teaching, and how they themselves become a learner of teaching the subject.

> Susan: Something common is that teachers engage with networks and communities of practice in engaging with expert others to find ways to overcome and manage it.

> Leissa: Something core and subject-specific might include threshold concepts that most teachers need to know to teach their discipline properly.

The process of identifying what is core to teaching a subject has enabled us to develop a framework that includes: the set of disciplinary knowledges and practices that a teacher needs to know; and how these knowledge and practices are translated into teaching through curriculum frameworks (knowing), modes of inquiry and practice (doing), and how one relates to the subject (being).

> Lucinda: So, if teachers are coming from say an English background into a science classroom, they're going to bring different attitudes or different ways of seeing and knowing.

Collaborative and interdisciplinary research

We use a three-part theoretical framing (see Figure 8.1) of interdisciplinary collaboration to generate this framework of subject-specific expertise. Collaborative and interdisciplinary research (Given, 2012; Mansilla & Gardner, 2003) offer ways of thinking about how to invite dialogue between different disciplinary experts.

> Lucinda: I think really, it's just about being able to achieve something together that couldn't be achieved individually. But you know that's my main

understanding and that's how I've used it in my interdisciplinary research, you know between medicine and education.

Further, the theory of collaborative advantage (Huxham, 2003) is used as a practice-oriented theory that focuses on how joint working across disciplines can enhance practical understanding. Naturally occurring data, such as through conversation, provide theoretical insights, as illustrated in this chapter.

Boundary crossing (Akkerman & Bakker, 2011) is used to recognise the educative nature of crossing boundaries through collaboration. Akkerman and Bakker define boundaries as "sociocultural differences leading to discontinuities in action and interaction" (p. 21) rather than any move between different practices. Hobbs (2013) iterates that "[a]ccording to this view, for out-of-field teachers, a boundary exists when the differences between the practices and perspectives required to teach the subject are 'discontinuous'" (p. 274). Boundary crossing enables interrogation of the nature of the boundary crossing itself and the mechanisms that can support that boundary crossing, that is, moving from a familiar in-field space to teaching out-of-field (Hobbs, 2013, 2014). Space here refers to sociocultural differences, but in this context of signature pedagogies, the social practices are also set within a physical space–the classroom–which changes depending on the subject: the science lab, the technology workshop, the geography field trip, the mathematics classroom, and drama/role play spaces for English. As a result, teachers need to adapt to different spatially enacted pedagogies when teaching out-of-field.

Linda: It's also a temporal experience for the out-of-field teacher where they move from one disciplinary space to another throughout a day, a week, a term. And actually, all teachers and students are constantly changing from one disciplinary way of knowing, doing and being, needing to appreciate the different ways of looking at the world and understanding it.

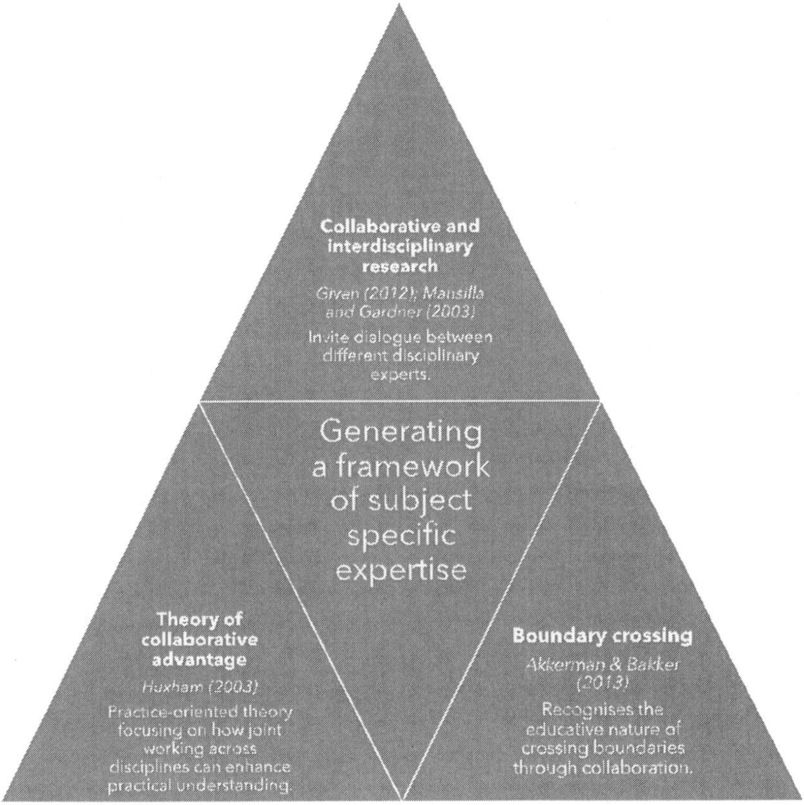

Figure 8.1: Theories informing interdisciplinary collaborative inquiry into teaching out-of-field

Lucinda: These theories are enabling us to focus on what arises as a result of us working together and sharing our own disciplines and then agreeing on what is useful for comparing. We could argue that it also reflects what the out-of-field teacher experiences; they're having to meld disciplines themselves, so it makes sense to be echoing that in the kind of research that we're doing.

Coral: In higher education we're being asked to collaborate in interdisciplinary ways with people from other faculties and clearly other disciplines because there's a recognised advantage in doing that in terms of the richness of the outcomes of the projects and the research that you undertake.

Linda: I think through a collaborative ideation process we've been able to cross boundaries and understand a little bit about the others. We appreciate the collaborative advantage as a result of meeting and coming together through that dialogue between us as different disciplinary experts.

Defining discipline and subject

As indicated, specialist teachers are assumed to understand and appreciate disciplinary knowledge and practices. One of the challenges for our interdisciplinary discussions was to understand what is meant by 'the disciplines' and then how this relates to what occurs in the classroom through teaching and learning.

Leissa: From an internet search, I found there wasn't a clear definition of what discipline was. *Discipline* tends to be referred to as a branch of academic study which usually took place in universities and consisted of theoretical knowledge, learning and practice (such as 'academic discipline' or 'field of study' in Wikipedia), whereas S*ubject* refers to a branch of knowledge studied generally in schools.

Williams and Roth (2019) refer to "'disciplinarity' in the professional sense" (p. 15) outside the world of academia in their efforts to explore the interplay of disciplines involved in interdisciplinary teaching. They define disciplinarity from a sociocultural activity theory perspective in two mutually reinforcing ways (p. 16):

a) a social phenomenon caused by increasing specialisation and differentiation of labour involving social, material, and discursive practices; and

b) a form of discourse making the specialisation 'thematic' (located in a coherent body of transmittable or 'teachable' knowledge).

According to Schwab (1964), it is through the limiting or distorting lenses of the bodies of knowledge of each discipline that we look at things, here aligning with Williams and Roth's (2019) second perspective. Disciplines are distinctive in terms of: how they are organised; the substantive conceptual structures used by each discipline; and the syntax of each discipline in terms of what its canons or evidence and proofs are and how well they can be applied (Schwab, 1964). According to Deng (2015), Schwab's characterisation of this as a theory of knowledge was useful for curriculum reform in the 1960s as it delineated not only the disciplinary knowledge needed for schools but also "the conceptual and methodological aspects of each individual type" (Deng, 2015, p. 776). Schwab's legacy lives on in contemporary schooling. For example, in the current Australian Curriculum (Australian Curriculum, Assessment and Reporting Authority (ACARA), n.d.), *Learning Areas* represent the developmental sequences of disciplinary knowledge, skills and understanding, which are used as the organising framework for schools, determining teacher and student timetables, resourcing and teacher professional learning.

Schwab's thinking of the structure of the disciplines and curriculum has resonances with Shulman's (2005) signature pedagogies as both consider the utility of knowledge. For Schwab (1973), the academic disciplines are considered a resource that can be put to use by students "in the development of intellectual powers and moral disciplines" (Deng, p. 776). "To know what structures underlie a given body of knowledge is to know what problems we face in imparting this knowledge" (Schwab, 1964, p. 13). In a similar way, Shulman's (2005) signature pedagogies capture the disciplinary practices, knowledge and moral dimensions that are needed for those entering the profession.

Methodological steps

The research questions above focus our attention on what we want to learn about the subject-specific nature of out-of-field teaching. How we determined to do this required another set of questions to guide our approach to collaboration and framework development. Therefore, four methodological questions guide our interdisciplinary collaborative inquiry, leading to the development of a framework to support out-of-field teacher learning:

1. How does looking across the disciplines deepen our understanding of teaching in an out-of-field context?

2. What value is there in having a framework that provides a comparative lens for teachers and researchers?

3. What might be included in such a framework?

4. What limitations are there with such a framework?

The below section illustrates how we have been teasing out what is common to teaching out-of-field across the disciplines, and how looking, comparing, and contrasting disciplinary ways of knowing, doing and being deepens our thinking in relation to teaching in an out-of-field context. The comparative approach also helps to develop our understanding of our own discipline because we are noticing nuances of our discipline that we may not have noticed before. It also enhances our ability to empathise with out-of-field teachers, especially for those of us who are new to researching the out-of-field teaching phenomenon.

Questions 2-4 continue to guide our inquiry. We have developed a framework based on discussions, but it has yet to be fully theorised, aligned with the literature and then tested with other experts in the field. The fourth question adds a self-reflexive process in recognition that there may be limits to what can be represented and achieved through a single framework. The guiding metaphor of reflexive boundary crossing (Akkerman & Bakker, 2011) reminds us to acknowledge that we often feel confounded in our conversations. This happens for example when those representing aligned disciplines use terms that others simply do not understand, for example, ‘modes of inquiry’ and ‘misconceptions,’ which are central to science and mathematics teachers from a constructivist view of learning, less so to the English teacher. What did seem to be common was the notion of pedagogical imperatives or barriers to learning progression. Sometimes we were left with more questions than answers. *What would this possibly mean in my discipline or subject?* we found ourselves asking.

We have four broad methodological steps, the first of which is the focus of this chapter:

1. **Develop a proposed framework:** mostly completed through conversation, debate and documenting subject-specific ideas against agreed elements of what it means to know, do, and be in order to be a teacher of the given subject. This began with general discussion of the scope and brainstorming of useful constructs, or ‘elements’. After a series of meetings, Jared translated and transferred these elements into a spreadsheet. Each expert/researcher then detailed what each of these elements looks like in their discipline. Salient elements that were useful for differentiating between the disciplines and subject teaching were isolated and organised into a diagram, which becomes the basis of the framework. The diagram represents what we considered to be salient elements of the discipline and subject teaching that an out-of-field teacher needs to learn. The salient elements thus far are: knowledge for subject teaching; how we inquire and generate knowledge; connections; and pedagogical imperatives.

2. **Align to the Literature:** underway. We are engaging with the literature for each salient element of the framework.

3. **Interview experts:** to be completed. Experts outside of our research group will be asked to comment on the generalised framework as a way of representing what an out-of-field teacher needs to learn.

4. **Refine the framework:** based on expert feedback and literature.

The illustration below is part of a conversation about how to communicate our interdisciplinary collaborative inquiry (Colleen was not present). A slide deck was developed by Linda and then presented to the research team to discuss and respond. An interesting exchange developed in relation to the meanings ascribed to a quote from Schwab (1964) about the structure of the disciplines.

Illustration of interdisciplinary collaboration

During our discussion about our methodology, Leissa introduced the following quote by Schwab to the group:

> *For to know what structures underlie a given body of knowledge is to know what problems we shall face in imparting this knowledge.* (Schwab, 1964, p. 13)

Below are excerpts of our interdisciplinary collaborative conversation relating to how these ways of thinking about disciplines and signature pedagogies has implications for out-of-field teachers in our varying disciplines. Two parts of the conversation are emphasised: *italicised* text denotes markers of collaborative co-construction; and **bold** text denotes reflections on the out-of-field teaching phenomenon.

> Susan: *That really resonates with what I've read* from Wyse et al (2018) from BERA[1], when they're talking about being able to recognise and understand what the distinctive core of your discipline is. And once you know what that is in terms of ways of knowing and ways of doing, then you look for that in the syllabus and curriculum structures and your pedagogical

[1] Wyse, Brown, Oliver and Poblete (2018) produced *The BERA Close-to-practice research Project* report, arguing for education to be considered a discipline on the basis that its origins are grounded in the academic, practical and integrated knowledge traditions.

practice. *I think with the experience of my participants*[2], that was something that really helped them extract that information. In the geography education context, **when they were confronted with teaching out-of-field, which sometimes may have still been within the HSIE[3] Key Learning Area or sometimes also quite outside of it, such as in the technologies.** For example, they thought about: what did I go through; and what did I know to be able to understand; what is the distinctive core of Geography; I know that it's this; **all right, now I need to kind of work through the same process and go through that with this out-of-field subject. It's about knowing the discipline within which you teach and what's distinctive, and then working through that to try and find it in the subject area you are not familiar with.**

Seamus: *I can give my two cents as well. In terms of the structure absolutely*, in science and particularly for Chemistry, the research for decades now focuses on how to structure Chemistry knowledge, so first figuring out how to structure what is the observable and what is the unobservable, how do we represent the invisible, dealing with sub microscopic levels of matter when we can't see it, using the language of Chemistry, the language of science to describe it. They're all things that the science teacher or the chemistry teacher has to deal with, and it is very structured knowledge, it's very particular and has to be done a certain way. **And so, then for an out-of-field teacher, they might have a handle on how to engage students into science, how to motivate students and do the practical activity, but they won't be able to dive into or use the language of chemistry to describe the things that they can't see, the unobservable, because they haven't dealt with the structure first.** *So, I think when you said that before I was really thinking* **the out-of-field teacher needs to get a lot more first on how chemistry knowledge is structured before they can start to think about how they can connect with that knowledge and teach.**

Lucinda: *Just to be a bit of a devil's advocate here* and *with my post-structuralist hat on and my critical discourse analysis hat* on, the idea of structures underlying a body of knowledge really kind of cements and reifies or ossifies it, and *the idea that we're going to impart this knowledge,* it's a particular epistemological approach to thinking about a discipline. *I just think we need to also keep in mind all the time* the notion that these structures are constantly changing and being contested. Just look at what's happening with the History curriculum in the media at the moment, and for us with phonics in the English curriculum, just the incredible amount of disruption and change and everything that's going on. *So, I don't know whether to just have a coda or something that acknowledges that or can we say something*

---

[2] See Caldis (2022), participants are pre-service geography teachers who participate in a reflection exercise.

[3] Human Society and Its Environment (HSIE)

about that. **Because a way that out-of-field teachers could really struggle would be if they had a very old fashioned or traditional or antiquated version of a subject based on their own learning when they were in school that they just tried to deliver or impart.** '*Impart', for me, is a really problematic word because that's not my epistemological understanding. I also think this is a good example of how we are each bringing our own disciplinary and theoretical framing, so within English I would be teaching post structuralist reader response as a way of analysing texts.*

Coral: *I was just thinking of what Seamus said earlier, and so my understanding of what Seamus said was slightly different to how Lucinda took it. Because when Seamus started talking about that structured knowledge, he was talking specifically about* things that kids can't see. And so, when we talk about some of the subatomic structures and things in chemistry, *for myself, I have a visual representation that's constructed* of probably ten different representations that I've engaged with over the years. And so that development of my own understanding of some of those things then translates to how we offer students the way to represent that for themselves. *So that's what I was seeing in what Seamus was saying so, not that the knowledge was structured in a particular way, I was seeing it as an open way of trying to place that in context* so that students could develop their own representations around that.

Linda: *Yeah, and I think you're right, there is an epistemological even ontological difference between what you're doing in English versus what we're doing in science.* So, in science there is a natural phenomenon there that needs to be understood and we have ways of thinking about that that have developed over a long period of time. Whereas English is, *I think,* much more around discourse with contestation of ideas and ways of communicating those, which is not about understanding a phenomenon in a particular way, or maybe you're more open as to how you view it.

Lucinda: *Yeah definitely.*

Seamus: *I think what I was saying was probably closer to what Coral was saying.* I'm thinking as well as the Physics class when they're asked to represent a force, students will be doing things and teachers will be doing things in front of them, they can't see it, but they all can see the effect of the force. And then they're asked to draw it and the teacher goes up on the board and starts writing things, and students are meant to somehow understand what's on the board and how that represents what they've just done in the practical activity. **Dealing with the out-of-field teachers this year in that course[4] was interesting in that they would want to know how to describe**

[4] Seamus taught in a Graduate Certificate of Secondary Science designed specifically for out-of-field teachers of science.

that, that's the bit that they wanted to know how to do. They felt they'd be a better Science teacher if they could describe that to their students. *I'm sort of in Lucinda's camp, I don't think that's what we should be doing,* **but I think that's often what the out-of-field teacher wants. They want the words, the language, the terminology, they see that as the knowledge that they gain from spending time in the discipline, because then they can describe what they can't see.**

Lucinda: *I agree, I absolutely agree obviously that they need the words and language and terminology and everything, but I just want to point out* that those things are always shifting and changing themselves for teachers in any discipline, because otherwise you're having a particular positivist view of what science is. Look at how scientific theories and things change over time and understandings of natural phenomena that have changed. *But what I was thinking about was,* **imagine a teacher coming from an English kind of discipline and having to teach in science, or coming from science and having to teach in English if you're in such different paradigms.**

Linda: *Absolutely, and when you look at that quote there*–to know what structures underly a given body of knowledge is to know what problems we shall face in imparting this knowledge–that's exactly what we've been talking about, **you're actually making decisions: what's my understanding of that body of knowledge at this particular time that I have to make decisions around how to then impart that knowledge.**

Lucinda: *I don't even want to impart it, I want to* encourage fresh responses from students, original creative responses. *I think this is one of the key differences between the Humanities (or the creative arts certainly), and the sciences, we want something different* so we're not imparting, if anything, we're I suppose eliciting but we're also inspiring.

Leissa: **If you're a teacher teaching out-of-field, do you really want to be in there inspiring or are you in survival mode trying to just make your way through?**

…

Coral: *But I also think that a science teacher has an interest in science and enjoys science so I don't think you could be* a science teacher if you didn't want to be doing science, teaching science, living science, being curious about the world, looking at all the new stuff, reading science journals all of that sort of stuff. So, you can't maintain that passion for a subject unless it's part of who you are.

Susan: *I agree, and I also think that* **the personal values and beliefs held about what it means to be a teacher, I think that's also quite important for those who are managing and grappling with out-of-field teaching**. *I*

> *speak from the experience of one of my participants,* Anna teaching technologies, agriculture, German and in the primary school when she's a specialist geography and history teacher and it was her immense strength of personal belief that it was her responsibility to provide these students with the best possible education that she could regardless of what she was facing. **That really got them to take stock and almost get a grip and find ways to manage this very difficult situation.** *But I do agree with Coral as well* that within your own subject domain, you have to be interested in it, and that's where we would say every geography teacher is a geographer who engages with the discipline beyond school contexts.

Reflection on the process

Reflection on this interaction illustrates the value of bringing together different expertise, backgrounds in the out-of-field phenomenon, and disciplinary perspectives when attempting to understand this cross-disciplinarity problem facing teachers. Interaction within this collaborative space involves: being stimulated to propose ideas, meanings and emphases for later consideration; responding to others; disputing and justifying; identifying differences and similarities across and within disciplines; multiple readings and interpretations of words and constructs; and attempting to comprehend the nuances of the other disciplines in comparison to our own.

The outcomes of this short interaction are several key learnings about the teaching out-of-field phenomenon, including the need to be aware of:

- Subject-related variation in how knowledge structure is conceived of and valued, and the difficulties this can pose for teachers in comprehending what is important for the subject;
- Variation in how to engage students with the knowledge and discourses within different subjects;
- Juxtaposition of the common aspects that teachers can bring into any subject, as well as the subject-related understandings of how the knowledge for teaching should/could be structured and the philosophical or paradigmatic underpinnings of the subject that determine the signature pedagogies;
- Contestations in curriculum and pedagogy based on political and disciplinary ideologies;
- The importance of teacher personal beliefs and values about the subject and their approach to being a learner in determining their response to learning teach the out-of-field subject;

- The mechanisms and supports that teachers can use to reflect and learn; and,
- The potential for teachers to rely on traditional teaching practices when teaching out-of-field that may not represent contemporary teaching approaches.

Conclusion as reflections on the research process so far

This interdisciplinary collaborative inquiry has been useful in finding those areas that are tacit, which we have taken for granted as being central to teaching and then realising that they are not common across subjects. These types of realisations can present as discontinuities for out-of-field teachers (Akkerman & Bakker, 2011; Hobbs, 2014) as they struggle to identify and overcome differences between familiar and unfamiliar ways of knowing, doing and being. The methodological challenge for us as we develop our framework has been determining what was salient across disciplines. If our ideas differed dramatically, we needed to decide on an overarching element that was inclusive of other disciplines or other subject ways of thinking about it.

In this chapter, we have reflexively demonstrated the challenges in attempting boundary crossing to understand each other's disciplines. The conversation above touches on desires, passions and commitments in our professional lives, as we seek to locate and orientate ourselves within both this interdisciplinary research group and our own disciplines and subjects. Seeking a shared language, overcoming misunderstandings, and carving out positions in our discussions, we mirror the experiences of out-of-field teachers in schools.

Alliances form, dissolve and re-form through dialogue, as does knowledge, and we are aware that many teachers do not have the luxury of time to have these vital conversations. It is our hope that the framework we develop can provide key support to teachers in finding ways to navigate unfamiliar subjects in both rich and timely ways. Teaching out-of-field is a phenomenon the contours of which are still being mapped, and which involves both discomfort and rewards. As Jared says:

> Going through this process enlightened the way that I was thinking about teaching. These conversations have made me appreciate different ways of thinking about concepts. One of the questions you posed at the very start, Linda, about knowing, doing and being made me really uncomfortable to start with, because 'being' is a little out of my research area. But *being out-*

of-field as a researcher doing out-of-field research has been quite interesting.

References

Abramovich, S., Easton, J., & Hayes, V. O. (2012). Parallel structures of computer-assisted signature pedagogy: The case of integrated spreadsheets. *Computers in the Schools, 29*(1-2), 174-190.

Akkerman, S. & Bakker, A. (2011). Boundary crossing and boundary objects. *Review of Educational Research, 81*(2), 132-169. doi:10.3102/0034654311404435

Asunda, P. A. & Weitlauf, J. (2018). STEM habits of mind: Enhancing a PBL design challenge-integrated STEM instruction approach. *Technology and Engineering Teacher, 78(*3), 34-38.

Australian Curriculum, Assessment and Reporting Authority (ACARA). (n.d.). Learning areas (Version 8.4). ACARA. Retrieved from https://www.australiancurriculum.edu.au/f-10-curriculum/learning-areas/

Beck, D. & Eno, J. (2012). Signature pedagogy: A literature review of social studies and technology research. *Computers in the Schools, 29*(1-2), 70-94.

Caldis, S. (2022). Transitioning into the profession with an out-of-field teaching load. In L. Hobbs & R. Porsch (Eds), *Out-of-field teaching across teaching disciplines and contexts* (pp. 261-284). Springer.

Carlson, J., Daehler, K. R., Alonzo, A. C., et al. (2019). The refined consensus model of pedagogical content knowledge in science education. In A. Hume, R. Cooper, & A. Borowski (Eds.), *Repositioning pedagogical content knowledge in teachers' knowledge of teaching science* (pp. 77-92). Springer. https://doi.org/10.1007/978-981-13-5898-2

Carpendale, J. & Hume, A. C. (2022). Collaborative content representation design to support out-of-field teachers' pedagogical content knowledge in science. In L. Hobbs & R. Porsch (Eds.), *Out-of-field teaching across teaching disciplines and contexts* (pp. 191-220). Springer. https://doi.org/10.1007/978-981-16-9328-1_10

Carter, M. (2007). Ways of knowing, doing, and writing in the disciplines. *College Composition and Communication, 58*(3), 385-418. Retrieved from https://www.jstor.org/stable/20456952

Crippen, K. J. & Archambault, L. (2012). Scaffolded inquiry-based instruction with technology: A signature pedagogy for STEM education. *Computers in the Schools, 29*(1-2), 157-173.

Crisan, C. & Hobbs, L. (2019). Subject specific demands of teaching: Implications for out-of-field teachers. In L. Hobbs & G. Törner (Eds.), *Examining the phenomenon of "teaching out-of-field": International perspectives on teaching as a non-specialist* (pp. 151-178). Dordrecht, The Netherlands: Springer.

Darby, L. (2010). Characterising secondary school teacher imperatives as subject (signature) pedagogies: A pedagogy of support in maths and a pedagogy of engagement in science. In S. Howard (Eds.), *AARE 2010 Conference Proceedings of the 2010 Australian Association for Research in Education Conference.* Retrieved from http://www.aare.edu.au/10pap/2499Darby.pdf ,

Deng, Z. (2015) Content, Joseph Schwab and German Didaktik. *Journal of Curriculum Studies, 47*(6), 773-786

Du Plessis, A. E., Gillies, R. M., & Carroll, A. (2014). Out-of-field teaching and professional development: A transnational investigation across Australia and South Africa. *International Journal of Educational Research, 66*, 90-102.

Given, L. M. (2012). Collaborative research. In *The SAGE encyclopedia of qualitative research methods*. Thousand Oaks, CA: SAGE Publications.

Goodson, I. (1993). *School subjects and curriculum change* (3rd ed.). Bristol, UK. The Falmer Press.

Gurung, R. A. R., Chick, N., & Haynie, A. (Eds.). (2009). *Exploring signature pedagogies: Approaches to teaching disciplinary habits of mind*. Stylus.

Hennessey, D. & Selvakumaran, Y. (2018). What is the role of signature pedagogies for subject-based expertise? Australian Curriculum Studies Association Symposium, Brisbane, August 30 2018. Retrieved from https://www.acsa.edu.au/pages/images/Signature%20pedagogies%20for%20subject%20based%20expertise.pdf

Hobbs, L. (2013). Teaching 'out-of-field' as a boundary-crossing event: Factors shaping teacher identity. *International Journal of Science and Mathematics Education, 11*(2), 271-297. http://dx.doi.org/10.1007/s10763-012-9333-4

Hobbs, L. (2014). Boundary crossings of out-of-field teachers: Locating learning possibilities amid disruption. In J. Langan-Fox & C. L. Cooper (Eds.), *Boundary-spanning in organizations: Network, influence, and conflict* (pp. 7-28). New York, NY: Routledge.

Hobbs, L. (2020). Learning to teach science out-of-field: A spatial-temporal experience. *Journal of Science Teacher Education, 31*(7), 725-745.

Hobbs, L. & Törner, G. (2019). Teaching out-of-field as a phenomenon and research problem. In L. Hobbs & G. Törner (Eds.), *Examining the phenomenon of "teaching out-of-field": International perspectives on teaching as a non-specialist* (pp. 3-20). Dordrecht, The Netherlands: Springer.

Huxham, C. (2003). Theorising collaboration practice. *Public Management Review, 5*(3), 401-423.

Ingersoll, R. M. (1998). The problem of out-of-field teaching. *Phi Delta Kappan, 79*(10), 773-776.

Koeper, I., Shapter, J., North, V., & Houston, D. (2020). Turning chemistry education on its head: Design, experience and evaluation of a learning-centred 'Modern Chemistry' subject. *Journal of University Teaching & Learning Practice, 17*(3), 1-17. https://doi.org/10.53761/1.17.3.13

Komoto, C. (2009). Moving toward a signature pedagogy in geography In R. A. R. Gurung, N. Chick & A. Haynie (Eds.), *Exploring signature pedagogies: Approaches to teaching disciplinary habits of mind* (pp. 121-137). Stylus.

Larkin, K., Jamieson-Proctor, R., & Finger, G. (2012). TPACK and pre-service teacher mathematics education: Defining a signature pedagogy for mathematics education using ICT and based on the metaphor "mathematics is a language". *Computers in the Schools, 29*(1-2), 207-226.

Lovin, L. H., Sanchez, W. B., Leatham, K. R., Chauvot, J. B., Kastberg, S. E., & Norton, A. H. (2012). Examining beliefs and practices of self and others: Pivotal points for change and growth for mathematics teacher educators. *Studying Teacher Education, 8*(1), 51-68.

Mansilla, V. B. & Gardner, H. (2003). Assessing interdisciplinary work at the frontier: An empirical exploration of 'symptoms of quality'. CNRS Society of Information Rethinking Interdisciplinarity Seminar, December 1.

McLain, M. (2018). Emerging perspectives on the demonstration as a signature pedagogy in design and technology education. *International Journal of Technology and Design Education, 28*(4), 985-1000.

McLain, M. (2021). Towards a signature pedagogy for design and technology education: A literature review. *International Journal of Technology and Design Education,* 1-20. https://doi.org/10.1007/s10798-021-09667-5

Nixon, R. S., & Luft, J. A. (2015). Teaching chemistry with a biology degree: Crosscutting concepts as boundary objects. In J. A. Luft & S. L. Dubious (Eds.), *Newly hired teachers of science: A better beginning* (pp. 75-86). Sense Publishers.

Passey, D. (2012). Educational technologies and mathematics: Signature pedagogies and learner impacts. *Computers in the Schools, 29*(1-2), 6-39.

Quinnell, R., Thompson, R., & LeBard, R. (2013). It's not maths; it's science: Exploring thinking dispositions, learning thresholds and mindfulness in science learning. *International Journal of Mathematical Education in Science and Technology, 44*(6), 808-816.

Schwab, J. J. (1964). The structure of the disciplines: Meaning and significance. In G. W. Ford & L. Pugno (Eds.), *The structure of knowledge and the curriculum* (pp. 1-30). Chicago, IL: Rand McNally.

Schwab, J. J. (1973). The practical 3: Translation into curriculum. *The School Review, 81*, 501-522.

Seow, T., Chang, J., & Neil Irvine, K. (2019). Field-based inquiry as a signature pedagogy for geography in Singapore. *Journal of Geography, 118*(6), 227-237.

Sharplin, E. D. (2014). Reconceptualising out-of-field teaching: Experiences of rural teachers in Western Australia. *Educational Research, 56*(1), 97-110.

Shulman, L. S. (1986). Those who understand: Knowledge growth in teaching. *Educational Researcher*, 15, 4-14.

Shulman, L. S. (1987). Knowledge and teaching: Foundations of the new reform. *Harvard Educational Review, 57*(1), 1-23. https://doi.org/https://doi.org/10.17763/haer.57.1.j463w79r56455411

Shulman, L. S. (2005). Signature pedagogies in the professions. *Daedalus, 134*(3), 52-59.

Siskin, L. S. (1994). *Realms of knowledge: Academic departments in secondary schools*. The Falmer Press.

Thomson, P., Hall, C., Jones, K., & Sefton Green, J. (2012). The signature pedagogies project: Final report. Culture, creativity, and education. Retrieved from http://www.creativetallis.com/uploads/2/2/8/7/2287089/signature_pedagogies_report_final_version_11.3.12.pdf

Victorian Department of Education and Training (2019). Victorian Teaching and Learning Model. Retrieved from https://www.education.vic.gov.au/school/teachers/teachingresources/practice/improve/Pages/Victorianteachingandlearningmodel.aspx

Vincent-Lancrin, S., González-Sancho, C., Bouckaert, M., Luca, F. d., Fernández-Barrerra, M., Jacotin, G., Urgel, J., & Vidal, Q. (2019). Eleven signature pedagogies related to the fostering of creativity and critical thinking. https://doi.org/10.1787/e316cb72-en

Weldon, P. (2016). Out-of-field teaching in secondary schools. Policy Insights, Issue 6. Melbourne: Australian Council for Educational Research.

Williams, D. & Roth, W-M (2019). Theoretical perspectives on interdisciplinary mathematics education. In B. Doig et al. (Eds.), *Interdisciplinary mathematics education, ICME-13 Monographs* (pp.13-34). Switzerland: Springer Nature. https://doi.org/10.1007/978-3-030-11066-6_3

Wyse, D., Brown, C., Oliver, S., & Poblete, X. (2018). The BERA Close-To-Practice Research Project. Research Report. BERA. Retrieved from https://www.llse.org.uk/uploads/datahub/4567ceb%5E03in-03/2018-12-10-BERA-Close-to-Practice_report_Nov2018.pdf

Acknowledgements

This research is funded by Deakin University, School of Education.

Linda Hobbs (BAppSci, GDE, MEd, PhD) - Deakin University, School of Education, 75 Pigdons Road, Waurn Ponds, VIC 3216, Australia. l.hobbs@deakin.edu.au
Linda Hobbs is a science and STEM educator. Her research focuses on teaching out-of-field, STEM and science education, and she works with schools and teachers in a range of capacities through professional development, student programs and research.
https://orcid.org/0000-0002-6898-8495

Lucinda McKnight (BA, MA, PhD) - Deakin University, School of Education, 221 Burwood Highway, Burwood, VIC 3125, Australia. l.mcknight@deakin.edu.au
Lucinda McKnight is an Australian Research Council (ARC) Discovery Early Career Researcher (DECRA) Fellow at Deakin University. Her research involves the study of curriculum and pedagogy for secondary English. She is an experienced preservice teacher educator and council member of the Victorian Association for the Teaching of English (VATE).
https://orcid.org/0000-0003-0997-6790

Seamus Delaney (BSci (Hons), MTeach(Sec), MEdPol(Int), PhD) - Deakin University, School of Education, 221 Burwood Highway, Burwood, VIC 3125, Australia. s.delaney@deakin.edu.au

Seamus Delaney is a chemistry and science education senior lecturer at Deakin University. His research interests are teacher professionalisation, chemistry learning approaches and science education in informal/out-of-school contexts. He works with schools and teachers on addressing global challenges such as sustainable development in the classroom.
https://orcid.org/0000-0003-3271-1686

Coral Campbell (BAppSci, BEd Ppost-grad), PhD) - Deakin University, School of Education, 75 Pigdons Road, Waurn Ponds, VIC 3216, Australia. coral.campbell@deakin.edu.au
Coral Campbell is a science/STEM educator and researcher. Her research encompasses science/ STEM education and out-of-field teaching across all levels. Coral is involved with industry-partners, DET, schools, teachers and educators through research, professional development, and student programs.
https://orcid.org/0000-0002-1373-9168

Jared Carpendale (BSc, BTchg, MEd(Hons), PGDip(SciEd), PhD) - Massey University, Institute of Education, Tennent Drive, 4472, Palmerston North, New Zealand. j.carpendale@massey.ac.nz
Jared Carpendale is a science, physics, chemistry, and STEM education senior lecturer at Massey University. His research interests are understanding teachers' pedagogical content knowledge for teaching science, effective professional learning and development for teachers, and supporting out-of-field teachers in science. He works predominately with pre-service and in-service secondary science teachers.
https://orcid.org/0000-0002-5904-4425

Susan Caldis (BEd, MEd, MRes, PhD) - Macquarie University, School of Education, Balaclava Road, Macquarie Park, NSW 2109, Australia. susan.caldis@mq.edu.au
Susan Caldis is a geography educator and recently appointed Lecturer. Her teaching and research focus on geography education, transition into the profession, transformation of pedagogical practice, and teaching out-of-field. Susan also completes extensive work with professional associations leading professional learning initiatives for geography teachers.
https://orcid.org/0000-0002-6659-8629

Colleen Vale (B.Ec(Hons), Dip.Ed.(Sec), M.Ed, PhD) - Monash University, School of Curriculum, Teaching and Inclusion, 19 Ancora Impara Ave, Clayton, VIC 3800, Australia. colleen.vale@monash.edu

Colleen Vale is professor of mathematics education in the Faculty of Education. Equity and social justice in mathematics education is a focus of her research and scholarly activity. She is renowned for her research in gender equity and digital technologies in mathematics education, out-of-field teaching in STEM and learning and teaching mathematical reasoning. She is currently working on projects about school mathematics leaders and adaptive expertise in STEM education.
https://orcid.org/0000-0002-9070-7071

Leissa Kelly (BEd, GDipEnv&HeritInt, MSci (pre), MEnvSci, PhD) - Deakin University, School of Education, 75 Pigdons Road, Waurn Ponds, VIC 3216, Australia. leissa.kelly@gmail.com
Leissa has a long-term involvement in community and professional education, and in science education practice and research. Her work has primarily focused on environmental education and STEM (including medical) education.
https://orcid.org/0000-0003-2046-5342

Chris Speldewinde (BA(Hons), GDT(Pri), Dip FP) - Deakin University, School of Education, 75 Pigdons Road, Waurn Ponds, VIC 3216, Australia. christopher.speldewinde@deakin.edu.au
Chris Speldewinde is a primary school teacher who is undertaking a doctorate at Deakin University that examines the Australian bush kinder pedagogy and STEM learning. Chris, along with Coral Campbell, are the preeminent scholars in Australian bush kinder research and have several academic and practitioner publications regarding bush kinders. Chris works as a research assistant on projects involving with multi-university research teams investigating issues in early childhood, primary and secondary school education. He also has interests in the implications of teaching out of field; policy and governance in education; and early childhood and primary school education.
https://orcid.org/0000-0001-9114-0911

CHAPTER 9

AN ETHICAL DILEMMA INSPIRING EXPLORATION INTO NETNOGRAPHY

EMILY ROCHETTE

Abstract

The COVID-19 pandemic has brought challenges to researchers and students completing short-term research subjects. Some may be inspired to turn to social media to conduct their research perceiving this to be a straightforward process to implement in ways that circumvent restrictions brought on by the global pandemic and avoid drawn-out processes of institutional review boards. This chapter seeks to tell the story of an ethical dilemma that sparked exploration into netnography, a methodological approach to "understand the cultural experiences that encompass and are reflected within the traces, practices, networks and systems of social media" (Kozinets, 2019, p. 14). By exploring the similarities and differences between netnography and other approaches applied in the online context, the chapter aims to engage readers in a conversation about overseeing research in the online space. The ethical practices enabling netnographic research to be conducted over social media through collection, participation and interpretation of online traces are introduced and explained. Literature critiquing netnography and highlighting its use in education research are woven throughout to offer readers further avenues to explore netnography in science, technology, engineering and mathematics (STEM) research.

Keywords: Netnography, online research, social media research, digital ethnography

An ethical and methodological dilemma

In the uncertain times of a global pandemic with face-to-face gatherings heavily restricted, researchers may face greater obstacles of participant recruitment and data collection. The pressure placed on post graduate students completing short-term research projects seems to be augmented. As such, these students may seek to conduct their research drawing on data from the Internet. This is something I have reflected on over the past two years through my service on the human research ethics committee at The University of Melbourne where I often find myself on an intellectual journey questioning and exploring the boundary between ethics and methodology. Here, I share some of my reflections that were inspired by an ethics application for which the student researcher proposed collecting and analysing data from social media.

Social media are websites and applications that enable users to create and share content. These include YouTube, Facebook, Instagram and Twitter to name a few. The use of social media is not a fixed, easy-to-define notion. Like LinkedIn, designed for professional networking, some social media platforms may be intended for a specific use, whereas others may enable users to fluidly traverse personal and professional boundaries. Consider a Facebook user who posts vacation photographs on a personal account to stay in touch with family abroad yet uses the same profile to participate in closed Facebook groups dedicated to teachers' professional learning. As such social media platforms and the possibilities for using them are in flux with the addition or removal of features and affordances. Kozinets (2019) explains:

> As social media changes, and people's adoption of it and adaptations to it alter, worlds of meaning also alter, and with them cultures. (p. 112)

In Australia, about 79.9% of the population are active social media users (Kemp, 2021). You probably have more than one social media account that can be accessed on your mobile phone, computer or even videogame console. How would you feel if you became aware that your social media data were being accessed and used by a researcher? Which of your identities might your social media posts reflect? Would these feelings change based on the social media account the researcher could access?

It might surprise you to learn that your agreement with most social media sites states that you retain ownership of the content you share, but you also grant the platform a non-exclusive, royalty-free, transferrable, sublicensable worldwide license to use your content (Kozinets, 2019, p. 219). This means that you own your content but you lose control over it. Although Kozinets (2019) reports that most people would be unwilling to supply their social media data and information for research investigations (p. 171), scholars have indicated that this is a complex space to negotiate (e.g., Golder et al., 2017; Salmons, 2018; Sugiura et al., 2017). For instance, Fiesler and Proferes (2018) explored 368 Twitter users' perceptions of research on the platform. Their findings indicated that attitudes toward researchers' use of tweets tended to be contextual depending on who the researcher is, what the research is about and how it is conducted and/or disseminated.

Linked to these procedural ethical regulations, I pondered the extent to which my comments on the student's application might impact the research design. Would I stifle this student's social media scholarship in ways that prevent understanding an important cultural phenomenon? Surely, there are unique cultural experiences that, when explored and explained through ethical and rigorous research designs, yield insightful findings about taboo or sensitive topics. Consider these examples: Tavakoli and Mura's (2015) exploration of Iranian women's tourist behaviour in the virtual world Second Life found that participants rejected gender-based stereotypes of Muslim women's bodily representations in Iran yet accepted stereotypical representations of femininity prevalent in Western media. Berdychevsky and Nimrod's (2015) study of seniors' online communities indicated that the online space offers older people an avenue for sharing sex-related information and support where face-to-face alternatives are limited. Both studies investigate online communities enabling users to explore aspects of their identity that may not be possible otherwise.

Closer to my own research interests, I wondered what might be learned about the lived experiences of teachers that remain undocumented by mainstream research endeavours. How could Pinterest enable researchers to understand the out-of-field teaching experience as educators search for and save ideas for use in their classrooms? How might initial teacher education providers harness the affordances of social media to maintain relationships with their alumni in ways that permit ongoing evaluation of their programs? To what extent could young people's posts inform the development of school programs aimed at improving the health and wellbeing of students? What insights could be gleaned about teachers' professional learning needs from closed Facebook groups? As such, I needed to do some of my own research to better understand the ethical and methodological boundary that social media researchers navigate.

Reflecting on real ethical considerations through fictional social media data

To take you on the journey traversing the ethical and methodological boundary of the student's ethics application, I invite you to participate in a role play with fictional social media data. Imagine that you are a researcher who is interested in individuals' perceptions of spiders. Your research passion lies in uncovering these perceptions to inform policy that protects our arachnid friends from harm and enables them to continue to control insect populations. Human perceptions of spiders have been well-documented, but you have noticed a gap in the literature with respect to understanding animals' perceptions of spiders. So, you turn to *Barker*, a social media application for animals, to explore and collect data from this unique cohort. Figure 9.1 presents some fictional data from *Barker*.

Figure 9.1: Fictional Social Media Data, adapted from White (2014, pp. 47-48)[1]

[1] These images were sourced from Creative Commons and marked with a CC BY 2.0 license.

Now, let us work through an exercise inspired by Kozinets (2019, p. 73) as you carefully reflect on this exchange. Consider the following questions through your researcher's lens:

1. What do you notice about the *Barker* feed as a type of data?
2. How might it be useful for research?
3. How does it make certain kinds of research more difficult?
4. How would you approach *Barker* as a site to conduct your research?
5. What cautions would you have?

Drawing on your research interest to study individual's perceptions of spiders perhaps you responded in the following ways: Posts on *Barker* would certainly give me access to animals' perceptions of spiders! Perhaps searches of #barnlife, #waroninsects and #rageagainsttheweb would enable me to initiate my research and better specify my conceptual foci. I am able to view who *liked* or *re-barked* the posts to gain access to more data. In addition, I would like to follow up with *Barker* users to interact with them either through interviews or the use of digital diaries, but there may be few animals willing to accept me into their communication space. I have to create an account to see some *Barker* posts and the terms and conditions specify that I can only create an account if I have between four and 10 legs! Is deception warranted? Hmmm... Four to 10 legs? Would some members of the animal community be overrepresented while others are excluded? Do I need informed consent from animals? How old are they? Is age calculated in dog years?

These questions were similar to those I had when reviewing the student's ethics application. As such, I needed to revisit the National Statement on Ethical Conduct in Human Research (The National Health and Medical Research Council (Australia) et al., 2007 (Updated 2018)) and consult the methodological literature to gain further insight into precautionary measures to be taken when researching cultural phenomena through social media. A colleague in my human research ethics committee recommended that I begin by exploring netnography (Kozinets, 2019) a methodology within which pragmatic, and systematised approaches are "intended to address many of the procedural, ethical and methodological issues specific to online research" (Nind et al., 2013, p. 654). The next section offers an introduction to netnography and is followed by a discussion unpacking some of the tools that netnographers employ to conduct research ethically and in ways aimed to avoid compromising the methodological approach.

Introducing netnography

Originating in the fields of market and consumer research, netnography is a qualitative research methodology that emerged in the late 1990s through the work of Kozinets (1998) and has continued to evolve since (Kozinets, 2010, 2015). Currently netnography is defined as "a way to study social media that maintains the complexities of its experiential and cultural qualities" (Kozinets, 2019, p. 4). Netnography is a form of digital anthropology, yet four elements distinguish it from other methodologies in terms of researchers':

1. Cultural focus to study cultures and communities that emerge through computer-mediated interactions;

2. Use of social media data, differentiating it from traditional ethnography;

3. Engagement, as opposed to distanced methods studying social media like text mining; and

4. Netnographic praxis, a set of 25 different practices netnographers may draw on for data collection, analysis, interpretation and presentation of findings, setting netnography apart from digital ethnography.

Netnography shares common positions and practices with ethnography such as focusing on the human experience and cultural understanding as well as being attuned to the context within which social systems of shared meaning are explored and informed by researchers' awareness of self and participants. However, the netnographic rules of engagement are distinct from those used by physically embedded, face-to-face ethnographers. Figure 9.2 illustrates the six procedural movements of netnography.

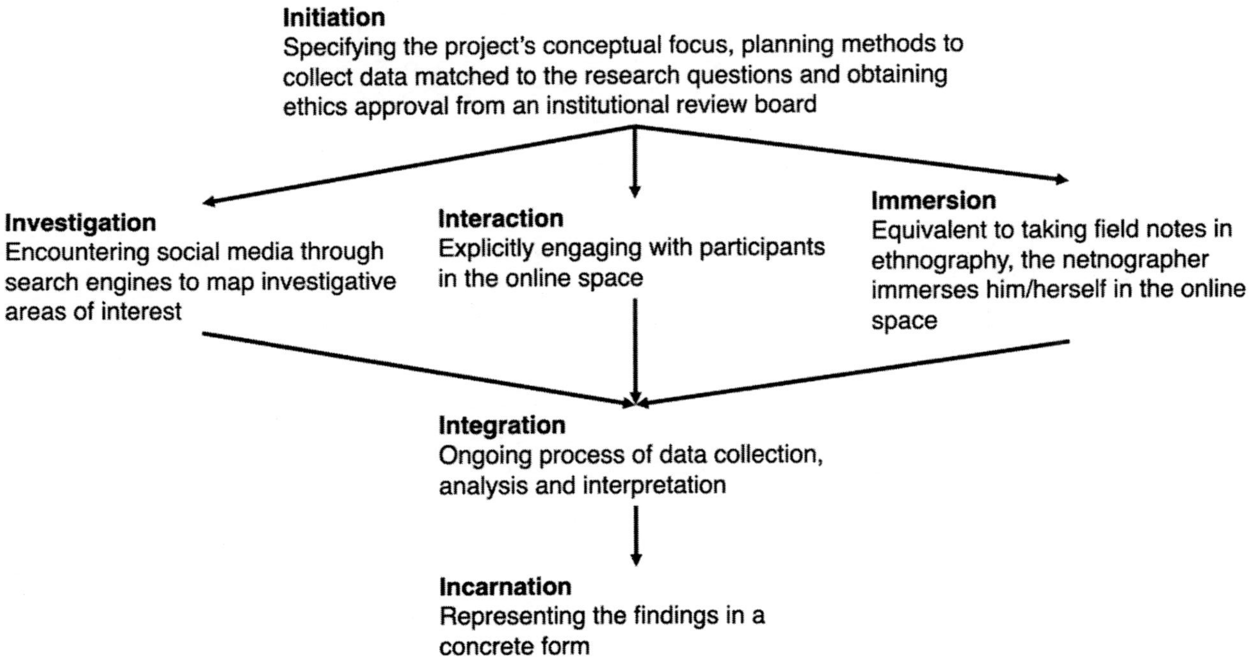

Figure 9.2: The Six Procedural Movements of Netnography, adapted from Kozinets (2019)

A methodological assumption underpinning netnography lies in the philosophy of Bernard Stiegler and technogenesis, the idea that human beings and technologies coevolve. Although the rates of change in contemporary society are swift, a trained netnographer has an ability to sense, record and analyse these cultural changes. After the researcher has specified the research focus, planned methods of appropriate data collection and obtained ethics approval from an institutional review board data collection begins through investigation, interaction and immersion.

Investigative data are created by unknown others; they can be searched for, scraped[2] and saved. Charlotte's sophisticated vocabulary presented in Figure 9.1, for instance, could become apparent through a search for #barnlife. Interactive data emerge through direct communication between the researcher and participants. This could include *liking*, *re-barking* or responding to Charlotte's post. Immersive data are equivalent to an ethnographer's field notes capturing analytical observations and annotating shared content. This might include a preliminary analysis of Charlotte and Wilbur's conversation suggesting that a pig's perception of beauty may enable a spider population to thrive in the barn.

Similar to ethnography and other qualitative methodologies (e.g., Charmaz, 2014) when researchers undertake netnographic studies the boundaries between data collection, interpretation and analysis are not explicit (Kozinets, 2019). The methodology offers researchers the freedom to draw on a variety of iterative approaches to collate, code and categorise data in ways that apply different interpretive approaches. Charlotte and Wilbur's data, for example, could be analysed through discourse analysis and triangulated with other members of the *Barker* community discussing similar topics. As the researcher does so, s/he may have to return to the platform, interact with a wider pool of users and immerse her/himself in the online space, then further analyse the data until sound conclusions can be drawn.

Finally, incarnation refers to communicating the findings of the research in a concrete form. How might the substance and affect of Charlotte and Wilbur's conversation be conveyed to readers without compromising their anonymity? Here, netnographers need to take care to offer enough context

---

[2] Scraping is a procedure by which computer software saves social media data from a website in a particular format such as a Microsoft Excel file.

about the analysis and interpretation of participants' lived experience while ensuring that ethical protocols are maintained.

Negotiating the ethical and methodological boundary

Networked individualism postulates that "people function more so as connected individuals and less as embedded group members" (Rainie & Wellman, 2012, p. 11) where the public and private spheres of life are far more intertwined than in the past. As such there are new realities and expectations about the transparency, availability and privacy of people's data and how they are treated (Kozinets, 2019, p. 123). This section invites readers to reflect on some of the ethical and methodological considerations with respect to researcher presence, negotiating consent and presenting social media data when communicating the research findings.

Balancing methodological rigour and researcher presence

Costello et al. (2017) draw researchers' attention to the imprecise interpretation of the term netnography leading to misperceptions and missed opportunities. They report that some so-called netnographic studies omit the once accepted need for human presence collecting interactive data and give way to non-participatory approaches. However, a researcher's participation in the online community can be difficult, at times, inappropriate and perceived to be antagonistic. When seeking to collect and analyse forum posts from patients who purchase medications online, Sugiura et al. (2017) reported negative reactions from members and moderators of online communities highlighting a flaw in the netnographic data collection framework.

More recently, Kozinets (2019) has clarified that netnographies do not have to include all three data collection operations. Investigative operations may be sufficient to collect data that respond to the research questions without the need for interaction or immersion. As such, Kozinets cautions researchers to be particularly mindful about disrupting the online community when the interactions are already providing rich data (p. 196). However, when passive approaches are taken because social media data are perceived to be easier and more cost effective to collect, netnographers miss opportunities to ensure the trustworthiness of their work. Further, without inviting users to challenge their research frameworks through ongoing co-creation of meaning, netnographers may depart from the qualitative rigour of the methodology (Costello et al., 2017).

Public versus private: The consent gap

The National Statement on Ethical Conduct in Human Research (The National Health and Medical Research Council (Australia) et al., 2007 (Updated 2018)) cautions that information available on the Internet can range from being fully in the public domain (e.g., newspapers, books or journal articles), to information that is public, but where individuals who have made it public may consider it to be private, to information that is fully private in character. *The consent gap* refers to the distinction between social media users' ascribed and actual beliefs about the need for permission to use their online content (Kozinets, 2019). Would the benefits of understanding individuals' perceptions of spiders through their social media data outweigh the users' beliefs that their posts should be private? For researchers seeking to utilise this data, the guiding principle from The National Statement[3] is:

> [A]lthough data or information may be publicly available, this does not automatically mean that the individuals with whom this data or information is associated have necessarily granted permission for its use in research. (The National Health and Medical Research Council (Australia) et al., 2007 (Updated 2018), p. 36)

In their systematic review of 17 studies[4] investigating the public's attitudes toward the ethical implications of research using social media data, Golder et al. (2017) found strong support for research that was perceived to be for the "good" (p. 11) of society. Participants also cautioned that certain groups of social media users including children, adolescents and members of vulnerable populations are at greater risk of harm, thus holding researchers to account for attending to ethical protocols. Similar findings emerged from Fiesler and Proferes (2018). At the same time, readers might question whether researchers can be certain about the age, competency or comprehension of individuals from which netnographers gather social media data. Notably, Walther (1992) explains that in traditionally accepted methods like telephone surveys researchers are faced with similar dilemmas.

---

[3] The National Statement also advises researchers to seek guidance on the use of data from the Office of the Australian Information Commissioner and state and territory equivalents, the Australian Bureau of Statistics and the Australian National Data Service.

[4] The smallest study involved 26 participants and the largest 2,260.

Contemporary research ethics are based on a consequentialist view where the morality of an action is judged through estimation of and arguments for the likely consequences of the action (Kozinets, 2019, p. 170). What is the probability of obtaining consent from individuals sharing content about #BlackLivesMatter versus #barnlife? Would studying the vast amount of users' #BlackLivesMatter posts outweigh the need for consent? Not only is there a practical difficulty to obtaining informed consent from members of some online communities, the fact that membership and the posts themselves may be in constant flux provides additional pitfalls for seeking informed consent (Sugiura et al., 2017). If a young person posts something that they are later embarrassed by or ashamed of, have they lost the right to make and/or rectify a mistake once their post has been captured without consent? Current ethical guidelines do not seem to reflect the realities of conducting research in the online space where advice for ethically obtaining informed consent is sparse and not clearly defined (Lehner-Mear, 2020; Sugiura et al., 2017).

Furthermore, consent may not end with the users. For example, I am a member of a private chemistry teachers' group on a popular social media site. As a researcher, I have an interest in better understanding the professional learning needs of chemistry educators in similar ways to Yildirim's (2019) netnographic study of a Turkish mathematics teachers' Facebook group hosting over 13,000 members. Through my membership and the posts that I have read, there appears to be a rich data seam to mine. The posts and participant membership, however, are moderated by select individuals who have assumed duties to ensure participation reflects the values of the wider community of chemistry educators. To conduct research through this site, I would also require consent from site moderators.

The situation becomes more complex when cyberspace breaks down international borders. Having a keen interest in teaching Earth science, I was drawn to a group of American teachers sharing and discussing resources. Gaining access to posts to analyse and disseminate findings not only means complying with the social media platform's policies but also with ethical governance and legislation applicable to my international friends and followers. As such, researchers must reflect on additional questions: Where are these online conversations being held? Are there additional laws or regulations stipulating whether and how social media data can be accessed and used? Notably, the European Union's General Data Protection Regulation is the strictest privacy and security law internationally.

Kozinets (2019) advocates for taking an anticipatory approach when planning ethical netnographic studies. At the outset, he cautions that no matter what the circumstance, netnographers should never engage in identity deception. Researchers are recommended to disclose their interests in their social media profiles and in regular status updates. Although a noble endeavour, this can be methodologically problematic. In her study of UK mothers' perspectives on primary school children's homework, Lehner-Mear (2020) highlights the methodological dilemma of obtaining informed consent and the potential to undermine the purpose of the research. She explained:

> The risk of only expressing socially acceptable viewpoints was already inflated in this study by existing moral imperatives for maternal compliance with school tasks. Gaining insider accounts was more likely through observation of mothers' natural interactions without disclosure of the research... (p. 127)

Notably, the ethical considerations on which this section reflects may make some data sites less attractive for some researchers (Kozinets, 2019). This has methodological implications because important aspects of the social landscape may inevitably remain unexplored. However, Nind et al. (2013) highlight that those responsible for research innovation are cautious and creative as they navigate the procedural ethical regulations where methodological development emerges from managing rather than taking risks.

Cloaking the data

When consent is obtained or the institutional review board is satisfied that the benefits of the research outweigh the need for consent and all ethics approvals have been granted, netnographers can collect and analyse social media data (Kozinets, 2019). In qualitative studies, exemplar data can be integrated into the writing in ways that contribute to the narrative and exemplify the approach to data analysis. At the same time, using quotations can enable readers to back-trace the data. Back-tracing occurs when verbatim quotations are inserted into a search engine to expose the original data source. When data are back-traced, the anonymity of participants is threatened as either their screen name and/or legal name may be revealed.

Perhaps back-tracing would not be problematic for the characters of a famous children's book made into a blockbuster movie. However back-tracing is particularly problematic when sensitive topics are discussed, such as eating disorders (Strand & Gustafsson, 2020), or when members of the online community belong to vulnerable populations, like the military (Schuman et al., 2021). The work of Schuman and colleagues charted evidence from seven online studies of the military and found that scholars did not implement sufficient ethical safeguards to protect this vulnerable population. As such, potential research participants may prefer to participate in studies that make use of their social media data through statistical reporting (Golder et al., 2017). For qualitative studies, however, researchers must take care to plan for greater risk mitigation during incarnation, i.e., representing the findings in concrete form.

To avoid back-tracing, Kozinets (2019) advises *cloaking* the data by changing some of the words and removing contextual elements such that users are untraceable. The extent to which cloaking the data occurs depends on a number of factors. Figure 9.3 presents a summary of three levels of cloaking. Notably, researchers must take account of the social media platform's terms and conditions, including whether these communities permit the removal of the author's screen name or changes to the wording of a post.

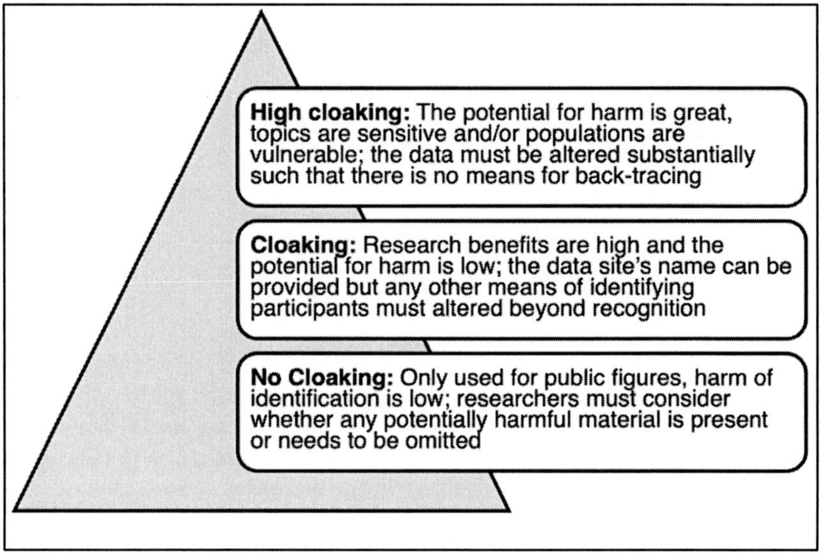

Figure 9.3: Levels of Cloaking for Social Media Data, adapted from Kozinets (2019)

High cloaking of the data would become necessary for studies investigating forums where users discuss illegal activity such as illicit drug use (e.g., Childs et al., 2021) or sensitive topics of a medical nature (e.g., Lee & Cooper, 2019) or where members of the community are part of a vulnerable population. Most netnographic studies, however, involve data that require mid-level cloaking where verbatim quotations are subtly rephrased or altered such that participants cannot be identified (Kozinets, 2019). To what extent would rephrasing or altering quotations affect the authenticity of the social situation unfolding? Kozinets (2019) alerts netnographers to this dilemma as "the more we remove of the context, the less the data have the authentic conversational and interactional elements that we are looking for in cultural research" (p. 398).

Returning to Charlotte and Wilbur's *Barker* conversation. Figure 9.4 presents what is perhaps the most important datum that speaks to the aim of our anticipated research. Here are the perceptions of an actual spider who may be speaking on behalf of many arachnids. In order to preserve Charlotte's anonymity yet retain the style and voice of the communicator, how would you cloak Charlotte's post? Which methodological considerations might you need to consider to meet this ethical expectation?

Figure 9.4: Charlotte's perspective on the beauty of spiders

Conclusion

Kozinets (2019) advises researchers to focus attention on the process and not the product, highlighting that ethical practice itself can also be considered a methodology (p. 186). This brief introduction to netnography echoes the sentiments of scholars who explain that when studying online communities a one-size-fits-all approach to research ethics is unlikely to emerge (Golder et al., 2017; Kantanen & Manninen, 2016; Sugiura et al., 2017; Tuikka et al., 2017). As such, the chapter serves to invite readers into the conversation and encourage intellectual debate reflecting on the ethical and methodological boundaries of online research case-by-case.

Notably the use of netnography is relatively new in education research (Lehner-Mear, 2020) yet it seems to be gaining traction. Recently netnography has been employed to study topics such as: mathematics teachers' use of Facebook to support professional development (Yildirim, 2019), the analysis of distance education contexts (de Oliveira, 2021), comparing underprepared teachers to fictional maverick educators portrayed in blockbuster movies (Henry, 2020) and examining issues of the digital divide in the wake of a global pandemic (Azionya & Nhedzi, 2021). This chapter has offered readers an introduction to a methodology that might be used to explore other fields of education research while highlighting some of the considerations for researchers seeking to design ethically and methodologically sound research designs.

Finally, readers may be asking: What was the outcome of the student researcher's ethics application that inspired this chapter? After a fruitful discussion by the members of the ethics committee, the application was returned to the researchers with a request for revision. In the meantime, I embarked on an interesting intellectual endeavour learning about the intricacies of online research, knowledge I was able to apply when the student researcher resubmitted the application with the necessary revisions to be approved. And thus, our story draws to a close with an online community being explored through an ethically and methodologically sound approach.

References

Azionya, C. M. & Nhedzi, A. (2021). The digital divide and higher education challenge with emergency online learning: Analysis of tweets in the wake of the COVID-19 lockdown. *Turkish Online Journal of Distance Education, 22*(4), 164-182.

Berdychevsky, L. & Nimrod, G. (2015). Let's talk about sex: Discussions in seniors' online communities. *Journal of Leisure Research, 47*(4), 467-484.

Charmaz, K. (2014). *Constructing grounded theory* (2nd ed.). Sage.

Childs, A., Bull, M., & Coomber, R. (2021). Beyond the dark web: Navigating the risks of cannabis supply over the surface web. *Drugs: Education, Prevention and Policy*, 1-12. https://doi.org/10.1080/09687637.2021.1916439

Costello, L., McDermott, M.-L., & Wallace, R. (2017). Netnography: Range of practices, misperceptions, and missed opportunities. *International Journal of Qualitative Methods, 16*(1), 1-12.

de Oliveira, É. T. (2021). Assessment, interaction and technology in distance education: A netnographic study at a Brazilian virtual university. *Open Praxis, 13*(3), [242]-252.

Fiesler, C. & Proferes, N. (2018). "Participant" perceptions of Twitter research ethics. *Social Media + Society, 4*(1), 1-14.

Golder, S., Ahmed, S., Norman, G., & Booth, A. (2017). Attitudes toward the ethics of research using social media: a systematic review. *Journal of Medical Internet Research, 19*(6), e7082. https://doi.org/10.2196/jmir.7082

Henry, J. (2020). The cinematic pedagogies of underprepared teachers. *Teaching and Teacher Education, 89*. https://doi.org/102990

Kantanen, H. & Manninen, J. (2016). Hazy boundaries: Virtual communities and research ethics. *Media and Communication, 4*(4), 86-96.

Kemp, S. (2021). *Digital 2021: Australia*. https://datareportal.com/reports/digital-2021-australia

Kozinets, R. V. (1998). On netnography: Initial reflections on consumer research investigations of cyberculture. *Advances in Comsumer Research, 25*(1), 366-371.

Kozinets, R. V. (2010). *Netnography: Ethnographic research in the age of the internet*. Sage Publications Limited.

Kozinets, R. V. (2015). *Netnography: Redefined*. Sage.

Kozinets, R. V. (2019). *Netnography: The essential guide to qualitative social media research*. Sage.

Lee, E. & Cooper, R. J. (2019). Codeine addiction and internet forum use and support: Qualitative netnographic study. *JMIR Mental Health, 6*(4). https://doi.org/10.2196/12354

Lehner-Mear, R. (2020). Negotiating the ethics of Netnography: Developing an ethical approach to an online study of mother perspectives. *International Journal of Social Research Methodology, 23*(2), 123-137.

Nind, M., Wiles, R., Bengry-Howell, A., & Crow, G. (2013). Methodological innovation and research ethics: Forces in tension or forces in harmony? *Qualitative Research, 13*(6), 650-667.

Rainie, H. & Wellman, B. (2012). *Networked: The new social operating system* (Vol. 10). Cambridge, MA: Mit Press.

Salmons, J. (2018). Getting to yes: Informed consent in qualitative social media research. In K. Woodfield (Ed.), *The ethics of online research* (Vol. 2, pp. 109-134). Emerald Publishing Limited.

Schuman, D. L., Highfill, C., Johnson, A., Henderson, S., & Ognyanova, P. (2021). Scoping review: Ethical considerations in online ethnographic

research with military populations. *Journal of Empirical Research on Human Research Ethics, 16*(4), 374-388.
Strand, M. & Gustafsson, S. A. (2020). Mukbang and disordered eating: A netnographic analysis of online eating broadcasts. *Culture, Medicine, and Psychiatry, 44*(4), 586-609.
Sugiura, L., Wiles, R., & Pope, C. (2017). Ethical challenges in online research: Public/private perceptions. *Research Ethics, 13*(3-4), 184-199.
Tavakoli, R. & Mura, P. (2015). Journeys in Second Life: Iranian Muslim women's behaviour in virtual tourist destinations. *Tourism Management, 46*, 398-407.
The European Parliament and of the Council of 27 April 2016 (2016). *General Data Protection Regulation*. Retrieved from https://eur-lex.europa.eu/legal-content/EN/TXT/PDF/?uri=CELEX:32016R0679
The National Health and Medical Research Council (Australia), the Australian Research Council., & Australia, U. (2007 (Updated 2018)). *National statement on ethical conduct in human research 2007* (Rev. ed.). Commonwealth of Australia. Retrieved from http://www.nhmrc.gov.au/publications/synopses/_files/e72.pdf http://www.nhmrc.gov.au/publications/synopses/_files/e72.pdf connect
Tuikka, A.-M., Nguyen, C., & Kimppa, K. K. (2017). Ethical questions related to using netnography as research method. *The ORBIT Journal, 1*(2), 1-11.
Walther, J. B. (1992). Interpersonal effects in computer-mediated interaction: A relational perspective. *Communication Research, 19*(1), 52-90.
White, E. B. (2014). *Charlotte's Web*. Penguin.
Yildirim, I. (2019). Using Facebook groups to support teachers' professional development. *Technology, Pedagogy and Education, 28*(5), 589-609.

Emily Rochette (BSc, DipEd, MTeach, PGCertEdRes, PhD) - The University of Melbourne, Melbourne Graduate School of Education, Kwong Lee Dow Building, 234 Queensberry Street, Parkville, VIC 3052, Australia. emily.rochette@unimelb.edu.au
Emily Rochette is a lecturer in science and chemistry education at the Melbourne Graduate School of Education (The University of Melbourne) Her research interests are situated within science and technology education, curriculum development and the out-of-field teaching experiences of science educators. Her methodological interests lie in the application of constructivist grounded theory and positioning theory. She is keen to explore new/complementary methodologies.
https://orcid.org/0000-0001-6769-1211

CHAPTER 10

THE PRAXEOLOGICAL SOCIOLOGY OF KNOWLEDGE–AN INTRODUCTION TO THE DOCUMENTARY METHOD AND A SKETCH OF AN EMPIRICAL IMPLEMENTATION

TERESA BECK

Abstract

In order to find a suitable approach to research areas relevant to schools and teaching, tailored research methods are required. Each research topic needs a tailored approach, because not every instrument is capable of finding the right answers, especially when it comes to deeper insights. The 'Documentary Method' worked out in the 1980s by Bohnsack, is to be understood as a reconstructive methodology as well as a methodical approach, rooted in the tradition of the Sociology of Knowledge. In the reconstruction of social reality, which is produced in social practice, the action-guided knowledge of the individual actors represents a reflective understanding of action practice. The method is used to search for orientation frames in the material, which provide information about typical patterns in the processing of topics and problems. The methodology is an instrument for practice and aims at a comparative analysis of different cases. It reflects what is common, how the practice of action of a group is constructed and what the common framework or the common orientation is, so that one can define the typical characteristics of the milieu in question and to find a meaningful answer, elaborated from the narratives of the people themselves and not from those who speak about them. In this chapter, I would like to examine the genesis of this special method established by Ralf Bohnsack, its individual methodological steps and possible applications in more detail, as it allows numerous possibilities of application, for example in teaching research and didactic research. Finally,

I will present methodological and basic theoretical considerations on my dissertation project.

Keywords: Qualitative research, documentary method, sociology of knowledge, action practice, professionalisation

The Praxeological Sociology of Knowledge

This chapter will focus on a basic methodological understanding that can provide an access to social and interactive contexts. The chapter has a theoretical core, which focuses on methodological and methodological approaches, explores them in depth, and then presents implementations. This approach is intended to show appreciation for methodology as well as method and to be an example of how tools for studying schools and classrooms as well as their actors, are also always an investigation of social societal phenomena of interaction and, in my opinion, reach into theoretical and deep-rooted foundations of the social sciences.

Therefore, I would like to provide some insight into how exactly this may work–specially in terms of a fundamental theory–and how exactly methodological implementations can be carried out. To open this topic, I will start by talking about the basic principles of the Documentary Method, or rather the guiding idea of the Sociology of Knowledge (Bohnsack, 2017, 2020; Bohnsack et al., 2007). In order to transfer these methodological considerations into a research context, I will briefly highlight relevant aspects of my dissertation in the context of the methodology. In doing so, this chapter focuses on basic theory and aims to give readers an insight into a promising qualitative research approach, which can be used for both educational science and subject didactic classroom research.

Basic interrelations

The Documentary Method is rooted in the tradition of the Sociology of Knowledge according to Mannheim and the Ethnomethodology (Bohnsack et al., 2007; Garfinkel, 2009; Garfinkel & Rawls, 2005; Mannheim, 1980), worked out by Bohnsack in the 1980s (Bohnsack, 1983, 1989). The methods of analysis open up access to reflexive but above all also to action-guiding knowledge of people. While Garfinkel focused on the analysis of Ethnomethods, in which people constructed reality. Bohnsack pursued how action competencies become visible in direct interaction. At the beginning of Bohnsack's research, the research approaches overlapped strongly.

Garfinkel and colleagues were investigating the way instances of social control interpreted deviant behaviour. Bohnsack wrote his thesis in 1973 on agency and juvenile delinquency. From the very beginning, Bohnsack was concerned with adjusting to the perspective of the other person and being able to perceive the interaction situation in its full meaning. According to Garfinkel, this is the Ethnomethod, i.e., the method of defining the situation as it is used in everyday life. Bohnsack moves close to the basic assumptions of Garfinkel as well as to Mannheim's idea of Sociology of Knowledge, but differentiates his basic approach from both. Bohnsack establishes very early his basic assumption of Expressive Sense, Object Sense and Document Sense[1] (Loos et al., 2013, p. 11). This will be taken up in more detail when it comes to the sub-steps of the method and how we distinguish between interpreting and understanding or between the Immanent Level of Sense and the Documentary Level of Meaning. In this chapter, basic assumptions according to Bohnsack, as a pioneer of the method, are explicated, while more recent approaches to the Documentary Method are also included (Bohnsack, 2017, 2020; Bohnsack et al., 2013; Loos et al., 2013; Nentwig-Gesemann, 2013; Nohl, 2007, 2017).

A qualitative-empirical research approach is important for identifying classroom practice and the involved actors. For didactical analyses especially, this might be an advantage in order to understand who is involved in classroom negotiations and how interaction takes place. In this context I consider it valuable to introduce the Documentary Method, which is one of the qualitative reconstructive approaches best suited to a variety of research interests (Bohnsack, 2020; Bohnsack et al., 2013; Mannheim, 1980; Nentwig-Gesemann, 2013; Nohl, 2017). The Documentary Method is both methodology and method at the same time. This practical research method can be used to interpret texts/transcripts, images and video data. Texts, images, interviews or group discussions enable a methodically controlled understanding of their protagonists, the situation and of what emerges from it. The reconstruction of habitus and milieu is analysed comparatively, while types are formed and generalisations developed (Asbrand & Martens, 2018; Bohnsack et al., 2013; Loos et al., 2013; Nentwig-Gesemann, 2013).

The epistemological premises that determine the Documentary Method following Mannheim (1980) and Garfinkel's Ethnomethodology (Garfinkel, 2009; Garfinkel & Rawls, 2005), seek access to the knowledge of actors

---

[1] Meant here are: “Ausdruckssinn”, “Objektsinn” und “Dokumentsinn” (Bohnsack, 2017; Loos et al., 2013).

guiding their actions. "The reconstruction of the practice of action aims at the underlying habitualized and partly incorporated orientational knowledge, which structures this action relatively independently of the subjectively intended meaning"[2] (Bohnsack et al., 2013, p. 9). It is always about expressions of everyday communication, which are seen as documents for underlying patterns. Intersubjectivity is considered the basis for the production of everyday communication, i.e., the existing (and not always renegotiated) agreement within a social group or milieu. This agreement does not have to be made explicit, but is based on a shared space of experiences in which people interact and, as it were, create and justify this 'space' (Berger & Luckmann, 1969; Bohnsack, 2017, 2020). The method aims less at hypothesis-testing and standardised procedures and turns away from fixed and close-meshed research sites. According to Bohnsack, no communication possibilities of a person should be curtailed, communication must not be understood narrowly. According to the principle of openness, the conduct of the conversation is left to the other person; only in this way can milieu-specific rules of communication emerge at all–which then lie in the research interest (Bohnsack, 2003).

In addition, the core of the Documentary Method is anchored in Mannheim's Praxeological sociology of knowledge, whereby it is concerned with the reconstruction of tacit knowledge elements. In doing so, the method also makes an attempt to distinguish subjective attitudes from actual implicit elements. "Mannheim's sociology of knowledge offers a perspective for observation and interpretation, in which–although there is a clear cut difference between the observer's perspective and the subjective meaning attributed by the actors–the actors' (those under research) knowledge is still the basis of analysis" (Bohnsack et al., 2010, p. 100). Furthermore–and this becomes clear in the further course of the dichotomy of the method–two central levels of knowledge are distinguished.

> Essential to this specific stance of observation is the distinction between two different sorts or levels of knowledge: the reflexive or theoretical knowledge on the one hand, and the practical or incorporated knowledge on the other. It is the latter kind of knowledge which gives orientation to action. This is implicit knowledge. Mannheim also called it 'atheoretical knowledge'. In English it is Michael Polanyi (1985) who has coined the term 'tacit knowledge'. This implicit or tacit knowledge forms a sort of structure, by which action is orientated mostly independent from the subjective meaning,

[2] Translated by the author

> and has insofar a certain objectivity opposed to it. (Bohnsack et al., 2010, p. 100)

How people interact

The Sociology of Practice means first and foremost to deal with a sociological model of society. It analyses how people act and interact with each other, an approach that can be very important for the practice-oriented actions of teachers and students in schools. Therefore, it is about reconstructing how people behave towards something. In doing so, the theory always understands action as something that relates to norms surrounding us and furthermore constituting society. That means one cannot act without underlying norms, regardless of whether it happens consciously or implicitly as a routinised action. Hence, a central feature in accordance with the Documentary Method is the genetic approach to analysis, which goes beyond the cultural sociology of Bourdieu (Bourdieu, 1974, 1984) and the explanation of habitual differences. It is oriented towards Mannheim's maxim implying "listening in on what always already exists"[3] (Mannheim, 1980, p. 9).

In this context, the aim is to gain access to the knowledge that guides the actions of the individual actors, cumulating a reflexive understanding of action practice. The reconstruction of action practice "aims at the habitualized and partly incorporated orientational knowledge underlying this practice, which structures this action relatively independently of the subjectively intended meaning"[4] (Bohnsack et al., 2013, p. 9).

Many individual people interact in their lived social practice, thereby shaping it. The reconstructed biographies or the actors of the action practice thus become representatives of their lived practice. The shared practice is then also called the collective space of experience. That means, the individual persons do not remain singularly arrested, but always interacts in lived practice, regardless of whether it is related to everyday life or to professional occupational spheres (Bohnsack, 2017, 2020). In this process, the document-meaning[5] of the subjects or their practice of action, determined on a habitual level and the genesis of atheoretical knowledge[6] is made visible or is reconstructed. These patterns can be located in group-

[3] Translated by the author
[4] Translated by the author
[5] Translated by the author. In German: "Dokumentensinn".
[6] Translated by the author. In German: "atheoretisches Wissen"

specific, gender-specific and generation-specific spaces of experience (Mannheim, 1964). The concept of habitus, incorporated by Bohnsack, whose knowledge-theoretical location also contributes to the analysis of my research, stems from Bourdieu's habitus-theoretical social structure analysis and is a knowledge-sociological category "with which the significance of symbolic representations constitutive for the reproduction of social structures is opened up"[7] (Meuser et al., 2013, p. 223).

Based on this understanding, Bourdieu brings the agent back into social structure analysis and in doing so, no longer considers action on the metaphorical level of the incorporation of social structures only (Bourdieu, 1974). "It is in the inventive, yet habitually bound interpretations of the actors–and only in them–that social structures 'live'" (Meuser et al., 2013, p. 224). This means that elements are only incorporated into the respective practice by their way of execution, i.e., in the performative performance, and thereby also condition it. Bohnsack writes at this point that "stuff as stuff only comes into existence through this integration into the practice of action" (Bohnsack, 2017, p. 71).

According to Heidegger, this is the relation of the actors to the things themselves in their entanglement with the practice of action detaching social structures (Heidegger, 1967). "In the structure of the worldliness of the world is that the others are not initially present as free-floating subjects alongside other things, but show themselves in their special environmental being in the world from what is present in it" (Heidegger, 1967, p. 123).

Conjunctive space of experiences

Likewise, Mannheim understands the social affiliation to a social group as a conditionality of world experience. The fact that individuals belong to the same 'social camp' (groups), that they live in or have grown up under similar conditions, creates a commonality of experience which Mannheim calls a 'conjunctive space of experience' (Mannheim, 1980). Conjunctive experiential spaces[8] endow a basic commonality that can also exist between people who have never been in interactive co-presence, in other words, who have never lived or interacted with each other. Mannheim explicated this in the generational context and in a similar way Bourdieu talks about the 'homogenisation of habitus forms'[9] of class (Bourdieu, 1993, p. 109;

---

[7] Translated by the author

[8] Translated by the author. In German: "kollektive bzw. konjunktive Erfahrungsräume"

[9] Translated by the author

Mannheim, 1964). For Bourdieu, homogeneity of a group already results in its actual existence, which in turn means that social practices can be adapted to each other without direct agreements and explicit interaction, as these 'habitus forms' embody exactly the same history (Bourdieu, 1993, p. 108). Therefore, actors who have incorporated the same habitus schemata can understand each other reciprocally and interact with each other (without constant negotiation). This understanding can be interpreted as a conjunctively grounded knowledge according to Mannheim and therefore differs from a communicatively produced understanding (Mannheim, 1980; Meuser et al., 2013).

This theoretical foundation is therefore important, because the research focus lies within reconstructing these conjunctive experiential spaces and the associated action-guiding knowledge of the person, thereby revealing the field of tension between performative logic and propositional logic. At this point, it is explicitly noted that the explored conjunctive experiential space considers both performative performativity and proponed performativity[10] without assigning weight to any of the two. Bohnsack states in this regard:

> The conjunctive experiential space of teachers as well as that of teaching is constituted–according to our general definition–by the tension between the performative experiential knowledge about a practice or in the practice on the one hand, and the knowledge about the normative expectations of the institutional or organizational programmatic associated with this practice, about role requirements as well as the identity norms of students and teachers, on the other hand.[11] (Bohnsack, 2017, p. 100)

Consequently, the tension and the research interest, as it were, lies within the consideration of precisely determining these collective experiences of habitualised practice and the institutional and societal norms that open up the tension alluded to them. These describe the collective experience space or the orientation framework in a broader sense and refer to the discrepancy to be explored between propositional and performative logic, consequently between norm and habitus. At this point, Bohnsack points to the conditionality of the two logics, as "the external experiences are constitutive for the conjunctive experiential space, since the external, the communicative dimension is only experienced as such at all against the

---

[10] Translated by the author. In German: "proponierte Performanz"
[11] Translated by the author

(implicit) background of the incorporated and habitualized practice–and vice versa" (Bohnsack, 2017, p. 103).

A distinction between conjunctive and communicative knowledge

For the interpretation of these incorporated practices, actions and utterances, a distinction is made between conjunctive and communicative knowledge according to Mannheim (1964, 1980), who differentiates between understanding and interpreting. In addition to experiential elements, there is also the later abstraction or construction, so there is always an artificial abstraction in pure thinking. "Every act of cognition is only a dependent part of an existential relationship between subject and object, an existential relationship which in each case establishes a different kind of commonality and an always specific unity between these two" (Mannheim, 1980, p. 206).

Of course, the inclusion of "theorizing-reflecting propositions (related to one's own actions), which, in addition to common-sense theories, include in particular the examination of societal identity expectations and identity norms as well as institutionalized norms, should nevertheless not be considered a risk to the validity of the reconstruction of action practice" (Bohnsack, 2017, p. 99). The focus on validity in the reconstruction of interpreted data, is the balance of proponed performance and theorising proposition. The "conjunctive experiential space of the actors to be investigated includes not only their action practice (directly observable in the form of performative performance or reconstructable on the basis of the proponed performance), but also its relation to the theorizing-reflecting propositions (directed at it)" (Bohnsack, 2017, p. 99).

The distinction between understanding and interpreting explains that actors with a shared space of experience understand each other without conflict and do not have to interpret each other or their actions first. There are two structures behind this, that of sociality and that of experience. On one hand, conjunctive experiences based on direct understanding and, on the other hand, communicative relations that have to be interpreted mutually (Bohnsack, 2003, p. 60). As can be seen in the following figure, the difference is between two levels, which is summarised by the *Methodological Distinction*. Once the level of the *Immanent Sense* (in the left column), here the *WHAT* is identified thematically. So the general meaning of an action. Often the *Expressive Sense* is also included here, i.e., subjectively meant intentions or motives. Then there is the *Document Sense* (in the right column). This asks about the HOW and the orientation in the described experience, i.e., how it is structured and documented. This can be reconstructed on the basis of the production process.

These two distinctions establish the double structure of everyday understanding and show that norms and habitus mutually condition it. The final interested lies in the tension between these two poles in the professional practice of actors.

Table 10.1: Methodological Distinction

Methodological Distinction	
Double Structure of Everyday Understanding	
Immanent Level of Sense	Documentary Level of Meaning
Understanding (communicative)	Interpreting (subjunctive)
WHAT	**HOW**
reflexive, describable, explicit	action-determining, not readily reflexive, implicit
Norm	**Habitus**
propositional logic	*performative logic*

In this table, the duality of the guiding difference becomes along the two columns clearer. It is important to notice that there is a reciprocal relationship between conjunctive and communicative knowledge and thus, a tension between habitus (performative logic) and norm (propositional logic). In order to make the distinction methodologically clearer, we will now approach the methodological implementation. The focus is on the distinction between the first and the second level.

From the methodology to the method

To do justice to this distinction also in the method, there are three major steps on the way to type formation. First, the Formulating Interpretation[12], which asks for WHAT and the immanent meaning[13] and second, the

[12] Translated by the author. In German: "Formulierende Interpretation"
[13] Translated by the author. In German: "immanenter Sinngehalt"

Reflective Interpretation[14], which asks for HOW and elaborates or makes visible the implicit and subjunctive knowledge, as you can see in Table 10.1. These two main steps can be seen in the next figure and are visualised in the first two boxes: Formulating Interpretation and the Expressive Sense, and Reflective Interpretation and the Object Sense (see Figure 10.1) This transition from the first to the second step is oriented towards the transition of first and second order observations according to Luhmann (Luhmann, 1990, p. 86).

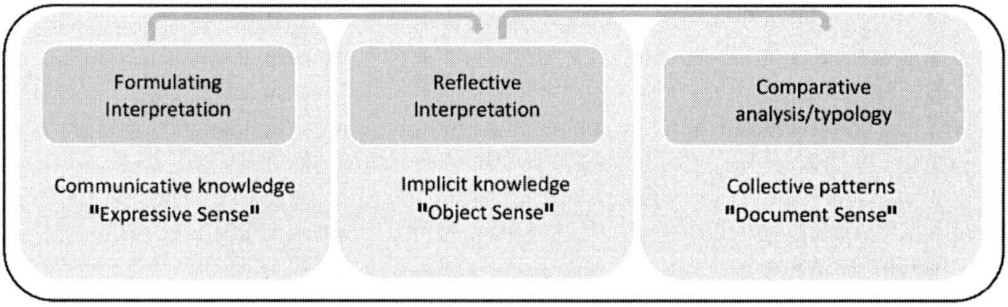

Figure 10.1: Methodical sequences

The analytical approach maintains a corresponding distance to the question and the criteria of validity of the statements, meaning it is not of interest what corresponds to the truth or what normative correctness ('what is said') a statement has. It is rather a matter of reconstructing which patterns of orientation can be made visible along a theme (Bohnsack, 2014b, p. 65). This goes hand in hand with the fact that documentary interpretation does not make an assessment of the immanent meaning, but illuminates how things are talked about and "how the researched interact with each other and how they associate themselves with the things"[15] (Asbrand & Martens, 2018, p. 24). Bohnsack grasps the methodological guiding difference along the bracketing of the validity character. This analyses the experiential production of reality, it is about the question of HOW and not WHAT, i.e., not what has produced this milieu-specific reality, but HOW it is produced. "This is at the same time the epistemological difference to everyday practice with its pragmatic constraints as well as to the claims of an objectivist approach to reality and the associated aporias of the juxtaposition of 'objective reality' and 'subjective experience'" (Matthes, 1992, p. 99). One could therefore also say that the research focus is first on the second step, which then leads into the third step. Thereby, the first

[14] Translated by the author. In German: "Reflektierende Interpretation"
[15] Translated by the author

step serves as a pre-orientation, and also as a demarcation to motives and subjective attitudes within the reconstruction.

The third step, Document Sense, as seen in Figure 10.1 in the third box, is the comparative analysis of the reconstructed cases. Once the implicit meaning of a single case is identified, cross-case comparisons are made. By analysing the cases step by step, commonalities can be identified and reconstructed as they progress. These are then combined into a typology by establishing horizons for what is 'typical' and delineating them by distinguishing them from other typical cases. The Tertium Comparationis is formed as a kind of analytical third case to highlight the similarities and differences of a subject (Bohnsack, 2005; Bohnsack et al., 2007; Nohl, 2007). By delimiting the empirical cases and the reconstructed orientation framework, a maximum significance can gradually be reached, so that in the comparative analysis, a suitable degree of conjunctive abstraction can be achieved (Nohl, 2007, p. 272). Here, Nohl follows Luhmann, who states when referring to Luhmann's theory on the observation of the observer, that a certain similarity can be seen, because "the respective Tertium Comparationis observes, so to speak, the previously applied Tertia Comparationis, by absorbing them into itself. As soon as the circle of cases and typologies is closed and the typology is developed, there is–according to Luhmann (1990, p. 83)–a system in the form of a 'reconstructive network'"[16] (Nohl, 2007, p. 273).

The methodical implementation

The methodical implementation, consisting of the three mentioned steps, and its application will now be explained in more detail. Applying the Formulating Interpretation seen in Table 10.2 in the left column, it will be analysed what is immanently expressed in everyday communication (e.g., videotaped lessons) or in the survey situation (interview, group discussion, etc.) by the research subjects. This means that the Formulating Interpretation deals with the communicative knowledge of the subjects, for which also Table 10.1 is a good example. We can also say that the interpreter still remains within the orientation framework of the group or the interviewee and does not yet make this the object of conceptual-theoretical explication. Instead, they first try to get an overview of the text and the

[16] Translated by the author

topics dealt with (with upper and lower topics[17)][18] (Bohnsack, 2014b, p. 136; Bohnsack & Nohl, 2007, pp. 303-304). Here, the passage is first labelled with an overarching theme (upper topic), then divided into individual specific-sub themes (lower topics). "It is necessary to distinguish what becomes thematic and as such is the object of formulating interpretation from how a topic, i.e. in which (orientation-) framework or according to which modus operandi it is treated, which is documented in what is said about the group or the individual." (Bohnsack & Nohl, 2007, p. 303) At this level, the interpreter always remains on the immanent level of sense as you can see in the left column in the table.

Table 10.2: The methodical steps: From WHAT to HOW

WHAT	**HOW**
Immanent Level of Sense	Documentary Level of Meaning
formulating interpretation	*reflective interpretation.*
Norm	**Habitus**

In order to be able to move on to the right column, the perspective changes. The reflective interpretation, which can be seen in the right column of Table 10.2, aims to reconstruct additionally how the topics are dealt with. This step of interpretation approaches the implicit, atheoretical knowledge that emerges and is produced conjunctively in social practice with the question of 'how'. This means that it is not a singular event, but a sequence of process events that enter into a connection with themselves, and many other things, that generate themselves or are generated in social practice. As Bohnsack points out,

> The objects of reflective interpretation are thus the reflexive, circular, or recursive relation between an action, utterance, or incorporated practice, on the one hand, and an overarching existential social context, the context produced by the actors themselves, on the other. Likewise, the modus operandi of the production of this relation is the object of reflective interpretation. However, access to the context produced by the actors themselves is only possible if we can distinguish the knowledge generated

---

[17] Translated by the author. In German: "Ober- und Unterthemen"

[18] Note of the author: see also formulating practices from Garfinkel and Sacks (Garfinkel & Sacks, 1970).

> on the basis of the reconstruction of this system or context from our own prior knowledge or contextual knowledge as observers, i.e., if we can step out of our own circle and thus suspend or exclude areas of our own prior knowledge in a controlled manner.[19] (Bohnsack, 2017, p. 80)

Consequently, it aims at explicating the framework in which a topic is dealt with and patterns of orientation can be reconstructed (Bohnsack, 2014b, p. 137). The regularity of the habitus, which is explicated in the reflective interpretation, is a social and epistemological one prior to the subjectively intended meaning of the actors. Thus, when utterances become meaningful in the context of the other actors' reactions, a regularity is constituted in this relation, which has to be explicated. Bohnsack adds here that these utterances and reactions are empirically observable, hence part of the reflective interpretation (Bohnsack & Nohl, 2007, pp. 303-304). With the pursuit of these regularities, a class of homologous or functionally equivalent reactions is sought. Thus, the methodological procedure becomes a comparative analysis. In addition, not only are homologies sought after, but also contrasting cases.The reconstruction of orientation frameworks is therefore about finding certain patterns that are unfolded in the discourse as well. This may happen through narratives and descriptions or through argumentations and evaluations in the interviews. In the best case, comparisons can be found within a frame through positive and negative counter horizons. So-called enactment potentials are essential components of the experiential space of a group. Orientation figures are stretched between these components, and the experiences are carried within them. The orientation figures are embedded in a representation of experience; in interaction processes. Since the different experiential spaces are superimposed, there are also differently nested orientation figures (Bohnsack, 2014b, p. 138).

In further course, orientation patterns and frameworks are thus worked out. In the processual representation of the course of discourse and the reconstruction of the framing, a condensed case description can then be written. A discourse description becomes more rounded and denser the more comprehensively orientation patterns and framing components can be summarised in an overall characteristic. According to Bohnsack, it remains central that the language levels of the persons and those of the interpreters remain clear; the strangeness relation must be visible (Bohnsack, 2014b, p. 143).

---

[19] Translated by the author

Within an orientation framework, dimensions are then reconstructed and structured by the Tertium Comparationis. In this way, commonalities can be specified in further course, which then form a typology through the Tertium Comparisons. In comparative analyses, the Tertium Comparationis is a kind of thinking space in which methodological reflections are abstracted. Thus, in the best case, two cases can be compared with each other meaningful (Bohnsack et al., 2013, pp. 268-274). Along these comparisons, further efforts towards type formation and a typification become possible.

With the help of the special and unique features of the individual case, the 'typical' is extracted and transferred into a larger overall context. Finally, the results are generalised and contextualised. In the course of type formation, the generation of a type is highlighted. In this way, Bohnsack tries to work out, for example, what the milieu type is in relation to specific orientation problems. In the process, individual experience backgrounds are interwoven with common socio-spatial confrontations. For a deeper and differentiated understanding, it should also be mentioned that at this point, the methodological procedures for group discussions (Bohnsack, 2014b, 2017) and narrative interviews (Nohl, 2013, 2017) are quite different.

A collective orientation in my material lies in the topic of the examination of 'one's own teaching', i.e. the experience of the teaching situation as a career entrant. Whereby the orientation is attached to a normative idea of proper teaching, which is oriented by drawing on normative school ideas of good teaching and is documented in the empirical material. For this I would like to share a short example from my material, which clarifies the methodical two-step. My Tertium comparisonis is the action practice 'teaching' or the experiences in teaching from the lateral entrants. This is the thinking space in which I search for orientation figures in the material.[20] In doing so, I try to recognise figures or patterns of how in this case Ms. Gau constructs herself as a teacher in order to be able to compare them across cases.[21]

---

[20] Examples can be found in the chapter by Bohnsack in The SAGE Handbook of Qualitative Data Analysis (Bohnsack, 2014a, pp. 217-233)

[21] This is a very brief excerpt chosen to illustrate the method. Of course, longer passages are used otherwise, whereby the context can be included more.

349 – 351

Mrs Gau: *If I succeed in managing the subject from all sides, that is what I like best of all. (1.5) Otherwise. (.) I actually want to work close to reality. (1.5) and quite often. (.) and reach the heart. (Interview Dec. 2020)*

Formulating Interpretation

First, an interpretation takes place on a formal level, the Formulating Interpretation. Thereby the excerpt is described. The upper topic is 'own teaching', the lower topic is 'managing the subjects'.

Managing the subjects: to succeed

349 Ms. Gau continues her explanation of teaching and begins this statement with 'when I succeed'. She says that she likes it best when she can manage the subject from all sides.

Managing the subjects: the heart

350 In addition, she wants to work close to reality. Very often she wants to reach the heart.

Reflecting Interpretation

The second step–the Reflective Interpretation–follows. Here, statements are interpreted in more detail so that common orientations can be worked out in further views of this case and other cases. In this case, tensions regarding one's own teaching can be documented. For further orientations it will be exciting to examine which norms become apparent and how they are in tension with habitualised action practice.

349 – 351 Narration in the mode of description and evaluation

In the mode of describing her idea of teaching, an evaluation is also integrated. This is also referred to as a background construction in the mode of evaluation. Here, Ms. Gau introduces her explanation of her teaching with the premise "if I succeed", marking the vagueness of not succeeding at times. What 'that' is exactly is not made explicit. What becomes clear is that she is supposed to work on the topic from all sides.

Here, an expectation becomes clear that she has of herself as a teacher. With the adverb 'otherwise' she marks another addition, which gives the passage an enumerative character. She wants to work close to reality; this pedagogical norm for teaching is not further explicated or classified in terms of subject didactics. She then adds that she often wants to reach the heart with her lessons. In this context, a tension becomes clear between, on the one hand, doing justice to a school norm: that is, to cover a topic well and to design it realistically and on the other hand, to do justice to her attitude of reaching the children in their hearts.

However, this is not explained further, in which form emotions play a role for her–it only becomes clear that she wants to include them and marks it as part of her teaching concept.

Research example

In order to show a more concrete implementation following the methodological representation, I will present thoughts and applications from my dissertation project in the following. First and foremost, some basic information is outlaid followed by the application of the Documentary Method in excerpts.

Desideratum and Actors

In order to specify the term, I would like to first briefly explain the background of the target group, namely lateral entrants (Seiteneinsteiger*innen in Germany) who are career changers with a side entry into school and no formal qualification or teacher training but a higher educational degree. First, some information concerning the German struggle with a massive shortage of teachers for several years is provided.

According to Klemm and Zorn's 2017 forecast of pupil numbers a total of 105,000 new primary school teachers will need to be recruited across Germany by 2025 due to the primarily age-related retirement of 60,000 teachers, the additional need for around 26,000 teachers to provide instructions as student numbers rise, and the need to hire 19,000 more teachers to provide additional coverage for pedagogical tasks at all-day schools (e.g., through inclusion, all-day supervision, remedial measures) (Klemm & Zorn, 2018). According to the Statistical Publication of the Standing Conference of the Ministers of Education and Cultural Affairs [KMK] (2019), there has been a steady increase in recruitment for the

Eastern Federal States. While only about 600 teachers were hired in 2008, the number initially levelled off between 1,000 and 1,800 hires by 2012 and continued to grow in the following years to peak at 5,600 new hires in 2018 (KMK, 2018, p. 11). The Free State of Saxony topped all other states by hiring 1,174 lateral entrants (Klemm, 2019). Therefore, state's lateral entrant program was conceptualised on the basis of the KMK law on the 'Design of Special Measures to Attract Teachers for the Supply of Teaching Staff' (Resolution of the Standing Conference of the Ministers of Education and Cultural Affairs of December 5, 2013) (KMK, 2013). Three locations were linked to universities for this purpose.

The lateral entrants who participated in this program are qualified academics who usually have a higher university degree (Master, Magister or Diploma), but did not study teacher training. Therefore, they do not have a teaching certification. According to their very different studied subjects and missing teaching background, they are employed in schools without preparatory service (for 12 months or longer) and attend an additional pedagogical and didactical qualification program while teaching regularly in schools. This program takes place 2-3 days per week (Beck, 2022). Officially, they are immediately hired to maintain the provision of instruction "in specific subjects, types of schools and regions with special lack" (Klemm, 2019; KMK, 2018, p. 32), which usually means in the countryside. The KMK (Conference of Ministers of Culture) released updated public-school hiring figures in March 2019, showing a slight increase from 2018 to 13.3% (4,798 of 36,084 total hires). The number of new recruitments via lateral entrant is continuously rising every school year. The federal state of Saxony, as the front runner of all other federal states, covers 50.6% of the vacancies in schools with lateral entrants. This puts Saxony far ahead in a nationwide comparison of the employment rates (KMK, 2018). Therefore, the concept of lateral entry and of teaching in-, between- and out-of-field in Saxony should become more prominent in school research.

Looking at recent years and knowing that the special measure works as an intermediate solution with long-lasting results, the goal of the project was to recognise the changing school culture in primary schools. Furthermore, the increasing number of recruited lateral entrants shows that the idea of recruiting teachers formally and regularly is not contemporary anymore. In summary, teaching in-, between- and out-of-field is unavoidable, since we have schools, particularly in rural areas (e.g., Saxony), where more than 90% of the vacant positions are filled with out-of-field-teaching teachers as lateral entrants. For the project, it is therefore important to include this growing target group in the controversial discussion about teaching out-of-field as a lateral entrant and to carry out research efforts from a reconstructive perspective towards the concept of in- and out-of-field-teaching (Beck, 2022).

The specific choice of the research method used, lies in the fact that it was not primarily a matter of a declarative sketch of the people (this was already done in advance with questionnaires), but rather of the very specific experience of the actors themselves in a role that is still very new for our school culture, namely that of lateral entrants. The German teacher training system generally regulates access to school or becoming a teacher, but here these regularities were suspended, people shifted into the teaching profession without formal qualification. For me, it was clear at this point, in addition to numerous critical debates, that a further perspective was needed, namely that of the actors themselves. The Documentary Method succeeds in bundling collective experiences while still doing justice to individual biographies and condensing them into common experiences. Documentary reconstruction aims at (the second step), i.e., at the conjunctive, action-determining and implicit knowledge about the practice of action, less at that which is directly communicated. It goes beyond common sense theories, norms and institutional role models, but tries to understand this tension in which the actors move. This tension is evident in Table 10.2 between the columns; the tension between norm and habitus in a brand new role as an career entrant.

Methodological implementation

The methodological-theoretical embedding of my research work and dissertation project 'SeLe an Grundschulen' (lateral entrance into the teacher profession in primary school) is based on the Praxeological Sociology of Knowledge and uses the methodological understanding of social practice as well as the method to understand the narratives of the interviewed person. It aims at gaining valid access to the actors' experiential knowledge guiding their actions. In doing so, it is necessary to look at the proponed performance, i.e., the narratives or the disposal of communicative knowledge. Including this propositional logic of knowledge, it is furthermore about the action practice and the inclusion of the concept of habitus or the incorporated habitus schema[22] of the actors and how they externalise their modus operandi of action practice (Bohnsack, 2017).

The research interest lies on the reconstruction of how teachers who teach out-of-field incorporate these performative routines of habitualised practice and reconstruct them in tension with their educational biographical breaks at the level of propositional logic and individual narratives. Consequently, and briefly stated, of the central question is how teachers who teach-out-of-field or across-specialisation, have experienced their professional biographical breaks and transition into the new teaching profession (as a primary school teacher). It analyses how they reconstruct their teaching and their teaching practice as well as what conclusions can be drawn from this for further (qualifying) measures. The aim is also to better understand teaching and the conditions for its success. The point is to look at their professional biographies, their transition, career entry, experiences in teaching practice and their didactic convictions of teaching which all appears in a collective orientation and is determined by their constructions of being in or out of the field. To understand their collective orientation of 'the field' in which they are teaching as lateral entrants, I use 15 interviews and additional data from a standardised questionnaire of four groups (n=114) from lateral entrants in a special qualification measure at one of the locations in Saxony. The narrative-biographical interviews give an insight into the teaching profession with low to no prior teaching experiences as primary teachers (Beck, 2022).

Furthermore, I aim to contribute with the empirical studies to further progress concerning the professionalisation of teachers. In doing so, I draw from Bohnsack's understanding of professionalisation, in which the constituent framing of school as a shared experiential space has an effect on the habitualised practice of action of those entering the profession. The school as an institution is a place that is conditioned by multidimensional conjunctive spaces of experience and brings them to unfoldment. It is about interaction between people arising within the practice of action itself and marks existentially shared spaces of experience of the actors (Bohnsack, 2017, 2020; Mannheim, 1980). By distinguishing between understanding and interpreting, it becomes clear that people in the same milieu who have a shared space of experience understand each other without conflict and do not need to continuously interpret themselves or their actions. According to Bohnsack, this is a milieu-specific interaction, which is a practice of

[22] Translated by the author. In German: "inkooperiertes Habitusschemata"

routinised decisions and is always subject to a management of notorious discrepancy. This discrepancy exists between normative demands on the role and the constitution of a shared practice of action with the clients–in the sense of a shared space of experience (Bohnsack, 2020, pp. 31-32). This can also be related to the process of professionalisation. In this process, experts internalise a routinised sequence of actions in constant interaction with their counterparts. This happens without having to reinterpret and negotiate every sequence, action or utterance. Professionalisation thus evolves along the constituent framing of the institution, which represents a kind of process structure in which tensions emerge and are sustained.

> I see a central task or function of professionalized action in managing the tension between the system of interaction unfolding in discourse between the clientele and the professional actors* on the one hand and the general normative factual programmatics and role relations of the organization on the other hand in an action-practical and ultimately habitualized way, thus integrating and modifying the requirements of normative organizational programs and designs into a conjunctive experiential space or establishing it in the course of implementing this requirement. (Bohnsack, 2020, p. 30)

Consequently, the tension between school-based and action-based norms conditions the very discrepancy in which actors operate. Thus, the programmatic, role relationship, and practice of coping is notoriously discrepant with the practice of action. The constituent framing is thereby the structural precondition for all decisions. It is a process structure that unfolds interactively within organisational conjunctive experiential spaces and at the same time co-constitutes them. Through the generation of this framing, professionalisation takes place (Bohnsack, 2020).

The conjunctive experiential space in the school milieu is also an association of interactions in which the milieu professionalises itself in interaction. Following the social worlds of the Chicago School (Bohnsack, 2017; Goffman, 1990), interaction grounds the organisation equally with normative, institutionalised programs (Strauss, 1978; Unruh, 1980). Consequently, through this structural framing, the institution of school can succeed in "making decisions in a routinized way and in a way that is expectable for the clientele and connectable for further decisions and thus able to ensure continuity and create a system trust" (Bohnsack, 2020, p. 33). As a rule, the decisions remain unspoken and are not discussed; this is the only way to maintain the routines of action practice in the milieu. According to Bohnsack, professionalised action is consequently "appropriating the relevant environment i.e., the actions of the clientele and their biographies, in a reformulating way that makes them decidable with reference to organizational norms and role relations" (Bohnsack, 2020, p. 33). This creates consequences for the construction of biography and identity of the clientele. According to the constituting external framing, through decisions made in the past by an institution, further decisions and consequences follow for the actors* of the participating milieu. Garfinkel speaks–when Bohnsack means the constituent framing–of a normalisation of the relationship between the professionals and the clientele (Garfinkel, 2009; Garfinkel & Rawls, 2005). As a result, different norms affect professionalised action. One can also speak of a doubling of norming here. "For the practice of action in organizations has to relate not only–as is also true for those outside of organizations–to the social and the institutionalized norms, role and identity expectations, but also to the codified norms, programs of the organization itself" (Bohnsack, 2020, p. 39). Thus, the school milieu is a space for role relations and different existing norms. With this sociological understanding, the school and its milieu conditions people's identity and forms the professional biographically influenced identity of being a teacher (Bohnsack, 2020).

This tension between institutionalised norms, role and identity expectations in the institution of school spans the professional biographical narratives of my interviewees and makes this approach perfectly suited for me to identify this tension and to be able to make statements about experienced action practice along the shared spaces of experience. In doing so, I reconstruct typical structures or orientation figures through the formulating interpretation and the reflective interpretation, which allow a typology to appear in a comparative analysis. Typical structures of career starters–the lateral entrants–are revealed, which are not so typical at all (Beck, 2022). Thus, a new picture becomes clearer, to which more appropriate measures can be linked. At this point, I am not yet able to give a complete picture of what the typical structure in an overall typology looks like. I can only refer to a first sketch of the typology (Beck, 2022; Hobbs & Porsch, 2022), and hope that I can make a contribution to contemporary approaches to research.

In addition, some initial conclusions are ready to be drawn. My initial overarching goal was, and still is, to shed light on an area that has been little researched in order to generate concepts applicable to teachers in Germany who do not teach a particular subject and who meander between in-, between- and out-of-field-teaching. Research has been conducted in the field of subject didactics and educational science for a number of years, but there is still little specific focus on lateral entrants.

Moreover, the concept of lateral entry is upsetting canonised teacher education. This is also reflected in the heated debate in the German educational landscape (Gehrmann, 2020; Miller & Martschinke, 2017; Puderbach & Gehrmann, 2020). The previously phased education, which extends over five years until the state examination, receives strong competition in the direct employment of lateral entrants. The discourse of de-professionalisation is steadily increasing. At this point, research also attempts to contribute to this complex discourse. After all, teacher shortages need to be met and quality losses do not necessarily have to be factored in. In this context it seems to be very important that:

- Education policy promotes the integration of teachers into the school via alternative approaches and supports them accordingly

- Education policy is also connected to seeing and understanding the coherence between support systems and school development.

- The high relevance of the school entrance phase of the teachers is considered, therefore special support is needed. Career starters need mentoring and targeted support structures.

- Subject-specific training is provided; this becomes apparent, for example, when new teachers recognise a need but are not yet able to work out their own options for action, but can expand on them with, for example, a mentor.

- Collegial structures are provided, supported, expanded and established as a basic premise for teachers.

And while these points present first conclusions, on the other hand it is essential to always include the actors in the discourse and to open the research field further. My contribution aims to present the reconstruction of shared school experiences from career entrants to examining about individual experiences, challenges and teaching practice, so that support systems can be tailored in the best possible way. In doing so, the possibilities are far from exhausted.

References

Asbrand, B. & Martens, M. (2018). Dokumentarische Unterrichtsforschung. Springer VS.

Beck, T. (2022). 'Teaching was not in my head' – Narratives from lateral entrants of their experienced biographical transition into the teaching profession in Germany. In R. Porsch & L. Hobbs (Eds.), *Out-of-field teaching across teaching disciplines and contexts* (pp. 221-240). Springer.

Berger, P. L. & Luckmann, T. (1969). Die gesellschaftliche Konstruktion der Wirklichkeit: Eine Theorie der Wissenssoziologie (M. Plessner, Trans.; 22. Auflage). Fischer Taschenbuch.

Bohnsack, R. (1983). Alltagsinterpretation und soziologische Rekonstruktion. Westdeutscher Verlag.

Bohnsack, R. (1989). Generation, Milieu und Geschlecht: Ergebnisse aus Gruppendiskussionen mit Jugendlichen. Leske + Budrich.

Bohnsack, R. (2003). Dokumentarische Methode und sozialwissenschaftliche Hermeneutik. Zeitschrift für Erziehungswissenschaft, 6(4), 550-570. https://doi.org/10.1007/s11618-003-0057-7

Bohnsack, R. (2005). Standards nicht-standardisierter Forschung in den Erziehungs- und Sozialwissenschaften. Standards und Standardisierungen in der Erziehungswissenschaft, 63-81. https://doi.org/10.1007/978-3-322-80769-4_5

Bohnsack, R. (2014a). Documentary Method. In U. Flick, The SAGE Handbook of Qualitative Data Analysis (pp. 217–233). SAGE Publications Ltd. https://doi.org/10.4135/9781446282243.n15

Bohnsack, R. (2014b). Rekonstruktive Sozialforschung: Einführung in qualitative Methoden (9., überarbeitete und erweiterte Auflage). Verlag Barbara Budrich.

Bohnsack, R. (2017). Praxeologische Wissenssoziologie. Verlag Barbara Budrich.

Bohnsack, R. (2020). Professionalisierung in praxeologischer Perspektive: Zur Eigenlogik der Praxis in Lehramt, Sozialer Arbeit und Frühpädagogik. Verlag Barbara Budrich.

Bohnsack, R., Nentwig-Gesemann, I., & Nohl, A.-M. (Eds.). (2007). Die dokumentarische Methode und ihre Forschungspraxis: Grundlagen qualitativer Sozialforschung (2.Aufl.). VS Verlag für Sozialwissenschaften. https://doi.org/10.1007/978-3-531-19895-8

Bohnsack, R., Nentwig-Gesemann, I., & Nohl, A.-M. (Eds.). (2013). Die dokumentarische Methode und ihre Forschungspraxis: Grundlagen qualitativer Sozialforschung. VS Verlag für Sozialwissenschaften. https://doi.org/10.1007/978-3-531-19895-8

Bohnsack, R. & Nohl, A.-M. (2007). Exemplarische Textinterpretation: Die Sequenzanalyse der dokumentarischen Methode. In R. Bohnsack, A.-M. Nohl, & Nentwig-Gesemann (Eds.), Die dokumentarische Methode und ihre Forschungspraxis: Grundlagen qualitativer Sozialforschung (2., erw. und aktualisierte Aufl). VS Verlag für Sozialwissenschaften.

Bohnsack, R., Pfaff, N., & Weller, W. (Eds.). (2010). Qualitative analysis and documentary method in international educational research. Barbara Budrich.

Bourdieu, P. (1974). Zur Soziologie der symbolischen Formen (W. Fietkau, Trans.). Suhrkamp.

Bourdieu, P. (1984). Distinction: A social critique of the judgement of taste (1. print). Harvard Univ. Press.

Bourdieu, P. (1993). Sozialer Sinn: Kritik der theoretischen Vernunft (G. Seib, Trans.; 11. Auflage). Suhrkamp.

Garfinkel, H. (2009). “Good” Organizational Reasons for “Bad” Clinical Records. In K. Krippendorff & M. A. Bock (Eds.), *The content analysis reader* (pp. 45-54). Sage Publications.

Garfinkel, H. & Rawls, A. W. (2005). Seeing sociologically: The routine grounds of social action. Paradigm Publishers.

Garfinkel, H. & Sacks, H. (1970). On formal structures of practical actions. In J. D. McKinney & E. A. Tiryakian (Eds.), Theoretical sociology (pp. 337-366). Appleton-Century-Crofts, Educational Division.

Gehrmann, A. (2020). Hat die Erziehungswissenschaft das Thema “Seiteneinstieg in den Lehrerberuf” verschlafen? Zur Tendenz der (Selbst-)Marginalisierung in Zeiten hohen Ersatzbedarfs. Erziehungswissenschaft, Heft 60(31/2020), 59-66.

Goffman, E. (1990). The presentation of self in everyday life (1. Anchor Books ed., rev. ed). Anchor Books.

Heidegger, M. (1967). Sein und Zeit (14., durchges. Aufl. mit d. Randbemerkungen aus d. Handex. d. Autors im Anh). Niemeyer.

Hobbs, L. & Porsch, R. (Eds.). (2022). Out-of-field teaching across teaching disciplines and contexts. SPRINGER.

Klemm, K. (2019). Seiten- und Quereinsteiger_innen an Schulen in den 16 Bundesländern. Versuch einer Übersicht, März 2019. Netzwerk Bildung, 14.

Klemm, K. & Zorn, D. (2018). Lehrkräfte dringend gesucht: Bedarf und Angebot für die Primarstufe. https://doi.org/10.11586/2017048

KMK. (2013). Gestaltung von Sondermaßnahmen zur Gewinnung von Lehrkräften zur Unterrichtsversorgung (Beschluss der Kultusministerkonferenz vom 05.12.2013).

KMK. (2018). Statistische Veröffentlichung der Kultusministerkonferenz Dok. Nr 218, März 2019, Einstellung von Lehrkräften 2018 (pp. 1-66).

Loos, P., Nohl, A.-M., Przyborski, A., & Schäffer, B. (Eds.). (2013). Dokumentarische Methode: Grundlagen - Entwicklungen - Anwendungen. Budrich.

Luhmann, N. (1990). Die Wissenschaft der Gesellschaft (8. Auflage). Suhrkamp.

Mannheim, K. (1964). Beiträge zur Theorie der Weltanschauungsinterpretation. In Wissenssoziologie. Auswahl aus dem Werk. (Soziologische Texte 28) (1.Aufl., pp. 91-154). Luchterhand. https://www.amazon.de/Wissenssoziologie-Auswahl-Werk-Soziologische-Texte/dp/8474757878#detailBullets_feature_div

Mannheim, K. (1980). Strukturen des Denkens (D. Kettler, V. Meja, & N. Stehr, Eds.; Nachdr.). Suhrkamp.

Matthes, J. (Ed.). (1992). Zwischen den Kulturen? Die Sozialwissenschaften vor dem Problem des Kulturvergleichs. Schwartz.

Meuser, M., Bohnsack, R., Nohl, A.-M., & Nentwig-Gesemann, I. (2013). Repräsentation sozialer Strukturen im Wissen. Dokumentarische Methode und Habitusrekonstruktion. In Die dikumentarische Methode und ihre Forschungspraxis. Grundlagen qualittiaver Sozialforschung (3. Aufl.). Springer.

Miller, S. & Martschinke, S. (2017). Stellungnahme zur Einstellung von Personen ohne erforderliche Qualifikation als Lehrkräfte in Grundschulen (Seiten- und Quereinsteiger) [Stellungnahme der DGfE Sektion Schulpädagogik - Kommission für Grundschulforschung und Pädagogik der Primarstufe].

Nentwig-Gesemann, I. (2013). Die Typenbildung der dokumentarischen Methode. In Die dokumentarische Methode und ihre Forschungspraxis. In Grundlagen qualitativer Sozialforschung (pp. 295-323). Springer.

Nohl, A.-M. (2007). Komparative Analyse: Forschungspraxis und Methodologie dokumentarischer Interpretation. In R. Bohnsack, A.-M. Nohl, & I. Nentwig-Gesemann (Eds.), Die dokumentarische Methode und ihre Forschungspraxis: Grundlagen qualitativer Sozialforschung (2., erw. und aktualisierte Aufl, pp. 255-307). VS Verlag für Sozialwissenschaften.

Nohl, A.-M. (2013). Relationale Typenbildung und Mehrebenenvergleich: Neue Wege der dokumentarischen Methode. Springer VS. https://doi.org/10.1007/978-3-658-01292-2

Nohl, A.-M. (2017). Interview und Dokumentarische Methode: Anleitungen für die Forschungspraxis. Springer Fachmedien Wiesbaden. https://doi.org/10.1007/978-3-658-16080-7

Puderbach, R. & Gehrmann, A. (2020). Quer- und Seiteneinstieg in den Lehrerinnen- und Lehrerberuf. In C. Cramer, J. König, M. Rothland, & S. Blömeke, Handbuch Lehrerinnen- und Lehrerbildung. Verlag Julius Klinkhardt. https://doi.org/10.35468/hblb2020-041

Strauss, A. (1978). A social world perspective. *Studies in Symbolic Interaction*, *1*, 119-128.

Unruh, D. R. (1980). The nature of social worlds. *The Pacific Sociological Review*, *23*(3), 271-296. https://doi.org/10.2307/1388823

Transcription guidelines (TiQ: Talk in Qualitative Research)[23]

(.)	short break, about 1 second
(3)	seconds that a break lasts
no	emphasised
no	loud (in relation to the speaker's usual volume)
°no°	very quit (in relation to the speaker's usual volume)
.	strongly dropping intonation
;	Weakly dropping intonation
?	strongly rising intonation
,	weakly rising intonation
mayb-	interruption of a word
wou::ld	Extension of a word, the frequency of: corresponds to the length of the extension
( )	word(s) not understood, according to length
(well)	uncertain in transcription
((moans))	Vents beyond language
@(.)@	laughing
//mhm//	for biographical interviews: listener signal of interviewer if "mhm" is not overlapping
└	Overlapping of speech acts

[23] For a more detailed description see Bohnsack, Ralf (2014). Rekonstruktive Sozialforschung. Einführung in qualitative Methoden. Opladen: Verlag Barbara Budrich.

Teresa Beck (M.A., PhD) - Technical University Chemnitz, Teacher Training Center, Carolastraße 4-6, 09111 Chemnitz, Germany.
teresa.beck@zlb.tu-chemnitz.de
Teresa Beck is a research assistant and PhD student at the Chair of Primary Didactics in Mathematics at the Centre for Teacher Education–with an educational and sociological focus. Her dissertation project deals with professional biographical breaks and transitions of career starters (with alternative approaches) and the phenomenon of teaching out-of-field and teaching-across-specialisation in primary school. Her research focuses on qualitative-reconstructive social research, especially the Documentary Method resp. the Praxeological Sociology of Knowledge.
https://orcid.org/0000-0003-3781-3243

CHAPTER 11

RESEARCHING FOR SYSTEM REFORM: GIVING VOICE TO THE AUTHORITATIVE OUTSIDER

RUSSELL TYTLER

Abstract

There are many research methods tracts discussing the role of informants in phenomenological or ethnographic studies probing insider experiences of and views about educational processes at classroom, school and system level. For researchers who advocate reform processes, however, there is a problem with such inside informants who may have limited perspectives on possible alternative educational purposes and practices. Advocating for reform of science education can involve acknowledging voices from the wider ecological system surrounding the purposes and practices of school science. In this chapter I trace the methodological innovations of a reform-oriented researcher, David Symington, over a number of projects involving external commentators on school science education. I trace David's own history that brought an outsider's perspective, and the consistency with which he strove to identify and give voice to critically informed and wider perspectives on the appropriate purposes and practices in science education. The chapter examines the nature of research practices that aim to provide critical, reform-oriented perspectives through the strategic selection of research informants. I trace a consistent history of research design of a reform-oriented researcher to tease out the logic of research designs advocating change in science education at a system level, including a set of principles that might guide critical research methods in science education.

Keywords: Science Education, critical methodologies, Science Education policy reform, research design, stakeholder voice

Introduction

A dominant concern, if not the dominant concern for science education researchers over recent decades is to develop insights into and ways of better developing student knowledge of key concepts around which the science curriculum is organised. These concepts are many, but attempts have been made to organise these into fundamental concepts such as energy transfer and transformation, adaptation, the genetic basis of inheritance, material properties and chemical reactions, interactions of earth systems, the structure of the solar system, and stellar evolution. These analyses guide curriculum reform within well understood parameters that underpin the curriculum focus on content knowledge, and knowledge of inquiry skills such as questioning, planning and conducting experiments, analysing, and communicating scientific findings. Mainstream research in science education has focused on themes consistent with this curriculum focus such as: students' conceptions in these topics and ways to bring them closer to scientific conceptions; the multimodal nature of learning in science and how best to support this; whether inquiry approaches or direct teaching are more effective for supporting learning; and students' attitudes and identities in relation to science. Another set of themes relate to teacher professional learning of effective pedagogies to support current perspectives on student learning focused on these themes. These foci can be seen in the contents pages of major science education research handbooks (e.g., Lederman & Abell, 2014).

Research around these themes is extremely valuable and necessary for informing effective ways to conceptualise and support student conceptual learning and affect in relation to science and science education. This research, however, operates within established curriculum paradigms built around the primacy of scientific knowledge for future generations of science professionals or a science public knowledgeable about scientific concepts and traditions. The research operates within the culture of school science and schooling systems, and while it can lead to radical reappraisal of the nature of learning and of teaching, does so within existing views of the shape and purpose of the curriculum. Some of this research, however, opens wider questions about the porosity of boundaries between schooling and wider systems, such as research investigating learning science more effectively through interactions with local industry, or including topics exploring the relevance of science to students' lives as a way of engaging interest.

The methodologies employed in science education research are many and varied, including ethnographic studies of classrooms to investigate framing of and responses to teaching and learning, interview or mixed method studies of teacher response to innovation, interview or focus group studies of student or teacher perceptions and attitudes about teaching and learning science, experimental studies comparing different teaching approaches, studies involving assessing of progression in student learning, or design based research focused on the development and refinement of curriculum innovations. These studies constitute a rich tradition in science education research that have opened up our perspectives on the nature of learning in science, on student responses to science, on effective pedagogies to establish conceptual knowledge and positive affect in relation to science. However, the bulk of these studies focus on improving teaching and learning within a system that is accepted, including, in broad terms, the focus of assessment regimes and of curriculum structures. The questions that are asked amount to: how can we better support student learning and teacher effectiveness, within the established curriculum practices of science education?

The nature and purpose of the science curriculum has from the beginning been contested. In Victoria, contestation of the place of a 'general science' curriculum began during the war years and continued into the 1960s (Fawns, 1985, 1996, 1998; Fensham, 1993). The debate contrasted an academic, discipline-focused view of the science that justified its place within a traditional classics curriculum, with a view that school science should be framed to serve purposes of democracy and social reconstruction, meeting the needs of all students and not just an elite minority (Fawns, 1998; Jenkins, 1979). These debates have continued with continuing advocacy of 'Science for All' (Fensham, 1985) and with the championing of scientific literacy as a reframed purpose of science education (Linder et al., 2007). Beyond this, advocacy for a school science serving socialist conceptions of the purposes of science have a long history, from Haldane in the UK (Haldane, 1939) to advocates of 'liberation science' (Riley & Claris, 2009) aligned with the ideas of Paulo Freire (2021). Alongside this research focused on the social and political purposes of school science there have been pressures on the school science curriculum from reconceptualisations of the nature and purposes of science itself. Writers such as John Ziman (1980, 2000) or David Layton (1991) draw on social studies of science to argue for a reconceptualisation of its nature and the need to teach science alongside its technologies and social uses. Understandings about how science is practised and produced continue to change, with implications for the school science curriculum. Ravetz (1999) coined the term 'post-normal science' (see also

Funtowitz & Ravetz, 2018) to emphasise the uncertainty of science in conditions of turbulence, implying a need to further consider the nature of science presented in school.

These pressures on the nature and purpose of the school science curriculum can be thought of in terms of Bronfenbrenner's (1989) ecological systems theory in which local or micro effects are described as influenced within a nested ecological system through meso-, exo- and macro-systems increasingly distal from those immediate effects. Figure 11.1 interprets this ecological layering in the case of school science practices.

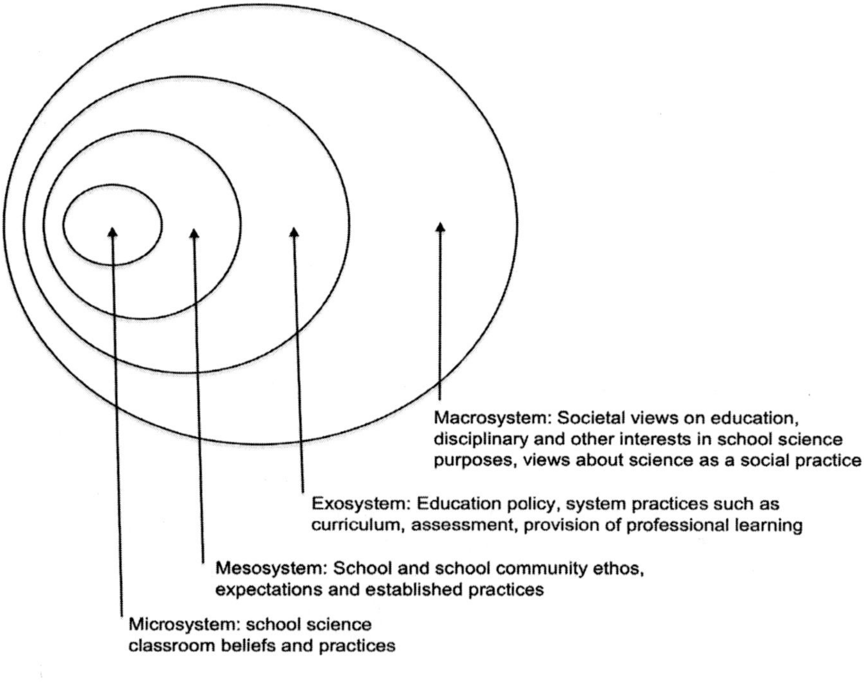

Figure 11.1: The socio-ecological system impacting school science classroom beliefs and practices

Bronfenbrenner's system provides insights into the widening systems that impact on teachers and students but has been criticised for masking more complex network relations within and across the different systems (Neal & Neal, 2013). We can see from the foregoing discussion that the macro system consists of a range of elements, each of which may have differential effects on system policy, school and parent expectations, and teachers and students themselves. In framing critical research into the wider influences that exist, and which should properly inform the improvement or reform of science education at the wider policy (exosystem) level, I explore in this chapter a research program driven by one researcher, David Symington, over two decades, to give voice to elements of the macrosystem that have a legitimate interest in the framing and practice of school science. In the chapter I will focus on the question:

By what means do we bring in convincing, fresh perspectives that can form the basis of critical judgments of the prevailing school science traditions?

I look particularly at the methodological and design elements of David's approach that effectively do this. I do this by reviewing and analysing the various studies I have been involved in, initiated by David, over 18 years to 2021.

David Symington

David had a long and productive career first as a secondary science teacher and principal in rural schools in Victoria, Australia, then as a research academic at a Teachers College focusing on primary science education. He was Dean of Primary Education at Deakin University for several years, until 1997. He then worked in a communication unit at the Commonwealth Scientific and Industrial Research Organisation (CSIRO), an influential, Australia wide government funded body, until 2001. Since that time, he was part of the 'STEME' (Science, Technology, Environmental and Mathematics Education) research group at Deakin University as Adjunct Professor, working on a range of influential projects. He passed away in 2021.

Figure 11.2: David Symington

David's early research in primary science education focused on student learning and alternative conceptions from constructivist perspectives (Biddulph et al., 1986; Hayes et al., 1994; Symington & Spurling, 1990). He also wrote about strategies for learning from museum visits (Griffin & Symington, 1997). From his time with CSIRO, he gained a solid understanding of the practices of contemporary science (Upstill & Symington, 2002). His championing of the inclusion in the science curriculum of authentic practices and school-science community partnerships can arguably be linked to the insights and connections he gained in that final phase of his formal career. At Deakin, he was involved in some significant projects in rural science and mathematics education such as the professional learning needs of rural science and mathematics teachers, and school-community partnerships which proved particularly productive for rural schools. He was involved in researching successful projects in the national 'Australian School Innovation in Science, Technology and Mathematics' (ASISTM) project focused on school and community partnerships. He was part of an evaluation in 2015 of the CSIRO STEM Practitioners in Schools initiative. From these projects we have gained a greater understanding of the conditions under which such partnerships can prosper.

Research methods: Analysis

In the chapter I present an analysis of the projects David was involved in during his time with the STEME group (Table 11.1), focused on those features of the research design that bear on our interest in bringing in critical perspectives on traditional school science, based on voices from outside the immediate schooling system. While David was not first author on most of the publications from this work, for the most part he initiated the idea underpinning the study, and in many cases used his wider contacts to identify suitable informants who would provide legitimate perspectives from the macrosystem; representing wider interests and perspectives. This methodological approach of seeking outsider perspectives on school science was a hallmark of David's interest, and commitment to a critical perspective. I will summarise each of these projects in turn to emphasise the nature of the informants, their 'authority' as stakeholders in school science, and the nature of the challenge to orthodox school science practices.

1. Community leaders' views: David had the idea that, if we were to gain an authoritative community perspective on the purpose and outcomes of a science education, as a potential challenge to the perspectives of governments intent on producing a future STEM workforce, or scientists themselves intent on reproducing themselves, then interviews with community leaders represented in 'Who's who in Australia', who were not associated with a science-related profession, would be a novel way to provide this. A random selection process, modified to ensure a range of occupations, led to a series of interviews. Two things emerged: most informants were interested in and engaged with science and scientists in an informal way, such as leading an environmental group or interacting with engineers in producing toys, and the strong message they gave was that school science should prepare students for 'science in life' with strong motivation to maintain involvement in using science, rather than being 'about' science. The findings thus challenge the predominant focus on knowledge accumulation in orthodox science education.

2. From his experience in CSIRO David was familiar with groups of science and technology practitioners working at the cutting edge of Australia's national science priorities, and saw this as a challenge to the depiction of science in schools as a settled, formal knowledge system. He organised a series of focus groups representing Australia's research priorities at the time; such topics as infectious diseases, contemporary materials, climate change–and put to them the questions: 'What does your field need from the public in order to support your work and your

findings?'; and 'What are the implications for school science education?'. Our conclusion was: "The participants in these groups had a great deal to say about how important the understandings and attitudes of members of the community were to their field of activity, as well as the way science is practised in contemporary settings. On the basis of the evidence, the authors of this article argue that school science should take as its focus the development of understandings of, and attitudes to, science for citizens generally. The authors suggest that this means, for both future citizens and scientists, that practice in school science needs to change to better represent contemporary science practice" (Tytler & Symington, 2006, p. 15).

3. In framing research for the design of a new science degree structure at Deakin University, David suggested that the content of a science degree should serve the needs of students destined to work outside the discipline (about 50% of those graduating) as well as future science professionals. We therefore interviewed graduates working outside their specialisation, to ask what made them choose a science course and what competencies they use in their jobs that were developed in their degree. From analysis of the data we argued that "attention be paid to widely used skills, such as communication and problem solving, and to developing an understanding of science within its social and ethical context".

4. In 2007 I was commissioned to write a monograph based on the Australian Council of Educational Research (ACER) 2006 annual conference which was about 'boosting science learning'. David, alongside me, was involved in monitoring the presentations and organising a final panel discussion on the insights produced. We decided to present a proposition to the floor that science education needed to be 're-imagined' based on the quite fundamental challenges that were being presented. In the vote that David put, 80% strongly agreed with the proposition. On the panel was the then Australian Chief Scientist, Jim Peacock, who agreed to write a foreword to the resulting monograph. He argued in that foreword that the nature of science had changed sufficiently that school science no longer adequately represented contemporary perspectives on science and should be 're-imagined'. David was thus instrumental in bringing an authoritative scientist's voice to the call for reform.

Table 11.1: The research projects and the nature of their critical standpoint

Research project	Interest group, and method	Focus of critical standpoint
2003/4 Interviews with community leaders from who's who (2003/4: Symington & Tytler, 2003, 2004)	Interviews with community leaders	Purposes of school science
2006 Focus groups of scientists and science communicators about what their areas needed from citizens (Tytler & Symington, 2006)	Focus groups of scientists and science communicators representing key Australian research areas	Purposes of a scientific literacy, school science
2007 'Lost voices'– interviews with science graduates who didn't take up science careers (Rodrigues et al., 2007)	Interviews with science graduates who did not take up science careers	Purposes and use of a science degree
2006-7 Re-imagining science education– Boosting Science Ed ACER conference (Tytler, 2007)	Critique from within the field of science education and the science community	Practices and purposes of school science
2008 Case studies of school-community projects, with an emphasis on rural schools (Tytler et al., 2008)	Case studies of best partnership practice	Nature and purpose of orthodox school science
2008-11 ASISTM best practice case studies (Tytler, Symington & Smith, 2011)	Perspectives of science and education community	Closed nature of orthodox school science

2009-11 Rural teachers of mathematics/science professional learning needs (Tytler et al., 2009; Tytler, Symington, Darby et al., 2011)	Interviews with rural teachers of mathematics and science	Identifying challenges of localised PD for science and mathematics teachers
2015 Rural schools and teachers (Tytler & Symington, 2015)	Interviews with rural teachers, parents, students	Challenging deficit perspectives on rural education
2015 CSIRO STEM practitioners in schools evaluation (Tytler et al, 2015; Tytler, Symington, Williams et al., 2017)	Interviews with STEM practitioners, teachers and CSIRO managers	Critically evaluating partnership model
2017 Boundary issues in school-science community projects (Tytler, Symington & Cripps Clark, 2017; Cripps Clark et al., 2014)	Case studies of partnerships	Challenging uninformed partnership perspectives
2020 Popular writing as a window into science compared to school science (Symington et al., 2021)	Analysis of 'best Australian science writing'	Contrasting school science with the richness of science stories

5. From his experience of scientific research practices within CSIRO David was very much a champion of bringing school science practices more in line with contemporary and local community scientific practices. We had funding at the time from the national SiMERR (Science, ICT and Mathematics Education in Rural and Regional Australia) project and undertook case studies of school partnerships with the scientific community in rural areas. Here, the intention was to study the conditions under which these partnerships were successful, and the ways in which interaction with community scientists and their research opened up new pedagogies and fresh engagement with learning. Partly on the basis of these insights, David orchestrated a major symposium

inviting teachers, science education researchers, and speakers from science, from future studies, and science communication, to bring fresh perspectives into school science

6. Following this work on school-science community partnerships we were funded to produce a set of exemplar case studies of the national ASISTM (Australian School Innovation in Science, Technology and Mathematics) project which involved substantive funding for school-STEM industry partnerships. We took the opportunity to widen our approach and interview schools, students and the scientific/engineering partners about the nature of the partnership, the pedagogies that arose, the conditions of success. From this we developed an Innovation Framework, drawing on Actor Network Theory, that emphasised how successful projects involved an alignment of needs and of agreed principles, and the enlistment of local conditions around which the projects would be structured. Here we argued for the enlistment within schools of the expertise and involvement of STEM practitioners to open the school science curriculum to contemporary practices.

7. Again, as part of SiMERR, we undertook a study of rural teachers of science and mathematics to investigate their experience of professional learning. Particularly with teachers of senior school science and mathematics, we found professional learning opportunities were inadequate, and argued (Tytler, Symington, Darby et al., 2011) that these teachers needed to have access to senior science and mathematics 'discourse communities' as important support for discipline based teaching and teachers. While these perspectives were from within the system at the micro level they had implications for exosystem level policy changes concerning rural teachers' professional learning.

8. Furthering this focus on rural schools, we also interviewed students and parents in rural schools to throw a critical perspective on the proposition that disadvantage is the dominant condition of rural schools and their teachers. We found many stories relating to lifestyle, safety and community engagement from parents, students and teachers which challenged this perspective.

9. 2015 saw a return to our interest in school-science community partnerships. Based on our previous work, we were engaged to evaluate the CSIRO Scientists and Mathematicians in Schools (now STEM Professionals in Schools) program as a model of school-STEM community partnerships. Again, the perspectives of a range of outsiders

to the system were involved, and the report emphasised the value for students and teachers in being involved with scientists and their practices.

10. Further to this, we undertook other case studies of such partnerships, gathering the perspectives of teachers and the science community on the determinants of successful partnerships. Our findings were analysed using a 'boundary crossing' theoretical lens, highlighting the need for teachers of science to better understand scientists and their practice, and vice versa.

11. In the last project David undertook, he was struck, in reading 'Best Australian Science Writing 2019' (Nogrady, 2019), how different the view was of science presented in the book, compared to school science concerns. The book included stories of major projects bringing scientists together with geographers, economists, mathematicians, and diverse themes such as the role of tradition and bias in science, involvement of wider professions in space exploration, the role of governments in providing evidence-based advice, and physical and cultural challenges in pursuing field studies. For David, these high-status writings represented the kind of alternative voice we had privileged in our empirical studies, and he proposed an analysis of their content as a challenge to normal school science conceptions of the discipline. We used Ravetz's (1999) construct of post-normal science to make sense of the nature of science represented in these stories, and showed how the school science curriculum as interpreted in Australian schools fails to acknowledge these contemporary settings. We argued from there a need to better represent these views of science in the school science curriculum.

Discussion

In considering David's approach to arguing for a widening or 're-imagination' of the school science curriculum we can see a consistent search for different and alternative voices representing legitimate interests in the purposes and practices of school science. In these cases we are working with beliefs and practices in the macrosystem of education that should provide legitimate perspectives for policy action. The 'authoritative outsiders' sit in different relations to school science, but all have views that we consider legitimate in the wider framing of the nature and purposes of a science curriculum. To list these, David has brought into play:

Members of the community:

- High status community leaders outside science
- Science graduates working in a variety of occupations
- Local community members such as rural families

Authoritative members of the science community:

- Science and technology practitioners working in key areas
- High status science researchers with a perspective on the nature of contemporary practices
- Local STEM professionals engaged in partnerships with schools

Voices who speak with some authority about scientific practice in social settings:

- Writers interpreting scientific practice through case stories
- Researchers on the nature of science and scientific practices
- Teachers with experience of the implications of new perspectives on classroom practice

Figure 11.3 represents these three different types of outside voice represented in David's research.

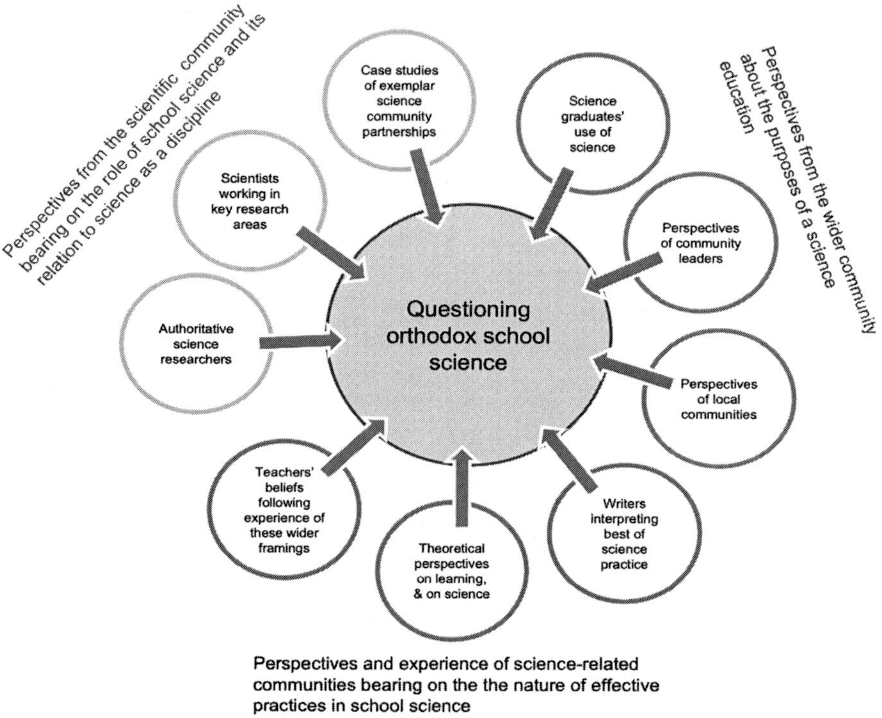

Figure 11.3: Questioning policy settings from outside–different types of voice

From a methodological perspective there are a number of characteristics of the research reporting in these projects that I see as important in framing these outsider voices in a convincing way. These are:

- The informants need to be chosen strategically so that their interests are clear and legitimate
- The relevance and status of the informants' perspectives needs to be established, to lend authority to the account
- The argument needs to be theoretically informed so that the perspectives are seen in relation to legitimate curriculum principles

A further device used by David in advancing these perspectives was the use of metaphor to sharply identify the nature of the challenge to orthodox classroom practice. Examples of titles for the papers included:

- Science learning in rural Australia–Not necessarily the poor cousin
- School-community collaborations: Bringing authentic science into schools
- What would they know about school science? (The community leaders' paper)
- The usefulness of a science degree: The 'lost voices' of science trained professionals
- Science in School and Society (The focus group study of national science priorities)

We can see from these titles the implied fresh perspectives and challenge represented by the publications.

In terms of the usefulness of Bronfenbrenner's nested socio-ecological systems model (Figure 11.1) we can also see that these voices, particularly in relation to the work on school-science community partnerships, are not necessarily filtered down through the socio-ecological layers but also operate directly on teachers, students and classroom practices. These 'voices' can also be actors in innovations involving real classrooms, as somewhat of a 'proof of concept' needed for establishing the nature and viability of system policy changes relating to 'bringing authentic science to school'. The pathway to reform, through such studies as these, involves a network of interactions between players in these different systems. Figure 11.4 illustrates a possible model of influence to change the culture of school science through school-science community partnerships.

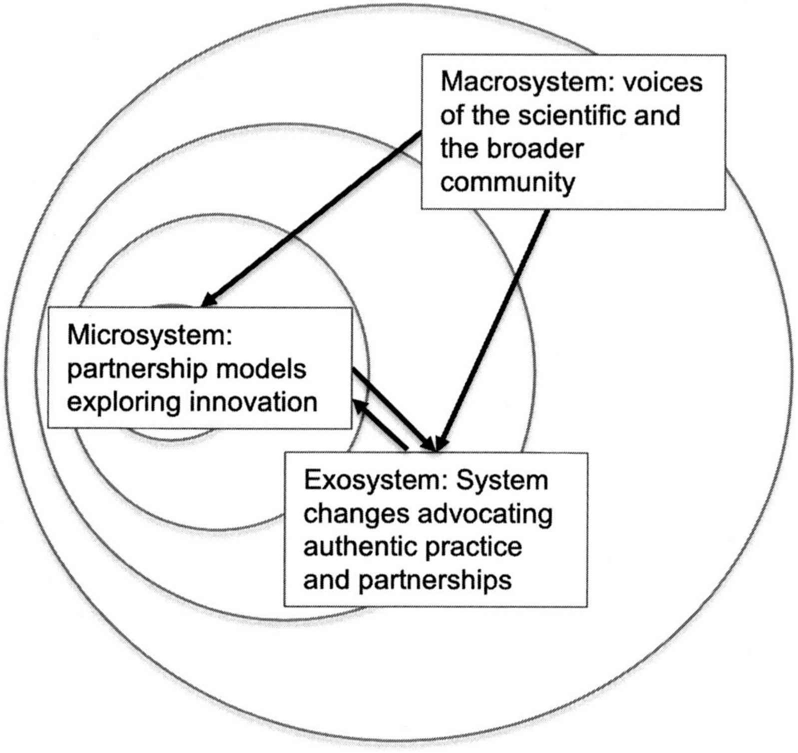

Figure 11.4: Network of influence bringing authentic science into classroom practice

In this networked model we see the voices from the scientific community that were established in the earlier studies are studied and supported in tangible ways through partnership practices. Policy framings in the exosystem are then potentially influenced both by authoritative voices from the scientific and broader communities and by the evidence from classroom practices. These policy changes, which indeed have been in evidence over the last two decades, relate to advocacy of the representation of authentic science (and STEM) in school curricula, and advocacy of and support for school-science community partnerships. The policy settings thus in turn can influence science classroom practice. Figure 11.4 does not include a further important node in these networked relationships; that of the science education research community. As part of that community, we can only hope that our research and associated communications serve to open up the pathways represented in Figure 11.4. David's vision and work provide evidence of the important role of the science education research community in offering a critical perspective on the wider relationships relevant to framing the purposes of school science, and ways of strategically bringing these voices into a meaningful dialogue.

Conclusion

The school science curriculum has always been a contested entity, with advocacy of its nature and purposes framed by wider socio-ecological influences. The vast bulk of research into science education understandably focuses on developing sharper understandings of current settings and on the nature and effectiveness of improvements within these settings. For a reform-oriented researcher such as David Symington, the framing of research designs aimed at critically questioning these settings follows a logic that is traceable through two decades of research–that of identifying and giving voice to authoritative outsiders to the school science system who have a legitimate case for influencing policy and practice. Such judgements require the development of criteria to determine the value of particular perspectives and reflect critical judgments concerning the purposes of science education. David's design practice included identifying those legitimate voices and framing them theoretically to justify their validity.

Ecological systems theory enables us to see more clearly the different systems involved that determine reform processes. We see, through David's research practice, the advantages of focusing on the voices themselves and their legitimacy, and also cases of practice that give expression to these same voices of reform to bring science classroom practices closer to practices in the field. This dual approach provides two pathways to influence the exosystem of policy reform.

From a research methodology perspective this case of a researcher pursuing a vision over two decades provides lessons for 1) the importance of strategically framing research questions and informants in order to give legitimacy to voices for reform, and 2) the integrity of a longitudinal research agenda focused on a critical perspective on science education orthodoxy.

References

Biddulph, F., Symington, D., & Osborne, R. (1986). The place of children's questions in primary science education. *Research in Science & Technological Education*, *4*(1), 77-88.

Bronfenbrenner, U. (1989). Ecological systems theory. *Annals of Child Development, 6*, 187-249.

Cripps Clark, J., Tytler, R., & Symington, D. (2014). School-community collaborations: Bringing authentic science into schools. *Teaching Science*, 60(3), 28-34.

Fawns, R. (1985). Negotiating an Australian general science: The professional dilemma 1939-45. *Research in Science Education.* Retrieved from https://search.informit.org/doi/10.3316/aeipt.176406

Fawns, R. (1996). The struggle for general science in Australia: The final campaign in the technical schools of the state of Victoria. *Research in Science Education*, *26*(1), 1-22

Fawns, R. (1998). The democratic argument for science curriculum reform in Britain and Australia: 1935–1945. *Research in Science Education*, *28*(3), 281-299.

Fensham, P. (1985). Science for all: A reflective essay. *Journal of Curriculum Studies, 17*(4), 415-435.

Fensham, P. (1993). Academic influence on school science curricula. *Journal of Curriculum Studies, 25*(1), 53-64.

Freire, P. (2021). Pedagogy of hope: Reliving pedagogy of the oppressed. Bloomsbury Publishing.

Funtowicz, S. & Ravetz, J. K. (2018). Post-normal science. In N. Castree, M. Hulme & J. D. Proctor (Eds.), *Companion to Environmental Studies* (pp. 443-447). London, UK: Routledge

Griffin, J. & Symington, D. (1997). Moving from task-oriented to learning-oriented strategies on school excursions to museums. *Science Education*, *81*(6), 763-779.

Haldane, J. B. S. (1939). *Science and you* (Key Books No. 1). London: Fore Publications.

Hayes, D., Symington, D., & Martin, M. (1994). Drawing during science activity in the primary school. *International Journal of Science Education*, *16*(3), 265-277.

Jenkins, E. (1979). The general science movement. In E. Jenkins (Ed.), *From Armstrong to Nuffield* (pp. 70-107). London, UK: John Murray/ASE.

Layton, D. (1991) Science education and praxis: The relationship of school science to practical action, *Studies in Science Education, 19*(1), 43-79, DOI: 10.1080/03057269108559992

Lederman, N.G. & Abell, S.K (Eds), *Handbook of research on science education*, Volume II. New York, NY: Routledge

Linder, C., Ostman, L., & Wickman, P-O (Eds.) (2007). *Promoting scientific literacy: Science education research in transaction*. University of Uppsala. Retrieved from https://eprints.qut.edu.au/12883/1/Linnaeus_Tercentenary_Symposium.pdf

Neal, J. W. & Neal, Z. P. (2013). Nested or networked? Future directions for ecological systems theory. *Social Development*, *22*(4), 722-737.

Nogrady, B. (Ed.) (2019). *The best Australian science writing 2019*. Sydney, NSW: New South Publishing.

Ravetz, I. R. (1999). What is post-normal science. *Futures-the Journal of Forecasting Planning and Policy*, *31*(7), 647-654.

Riley, D. & Claris, L. (2009). From persistence to resistance: Pedagogies of liberation for inclusive science and engineering. *International Journal of Gender, Science and Technology*, *1*(1).

Rodrigues, S., Tytler, R., Darby, L., Hubber, P., Symington, D., & Edwards, J. (2007). The usefulness of a science degree: The 'lost voices' of science trained professionals. *International Journal of Science Education*, *29*(11), 1411-1433.

Symington, D. & Tytler, R. (2003). What would they know about school science? *Australian Science Teachers Journal, 49(*2), 28-34.

Symington, D. & Tytler, R. (2004). Community leaders' views of the purposes of science in the compulsory years of schooling. *International Journal of Science Education*, *26* (11), 1403-1418.

Symington, D., Tytler, R., & White, P. (2021). Methodological complexity in analysis of popular writing to critique the nature of school science. In P. White, R. Tytler, J. Ferguson, & J. Cripps Clark (Eds.), *Methodological Approaches to STEM Education Research Volume 2*. Cambridge Scholars Publishing.

Symington, D. & Spurling, H. (1990). The 'Draw a Scientist Test': Interpreting the data. *Research in Science & Technological Education, 8*(1), 75-77.

Tytler, R. (2007). *Re-imagining science education: Engaging students in science for Australia's future. Australian Education Review No. 51.* Camberwell, VIC: Australian Council for Education Research, ACER press. Retrieved from www.acer.edu.au/research_reports/AER.html

Tytler, R. & Symington, D. (2006). Science in school and society. *Teaching Science, The Journal of the Australian Science Teachers Association, 52*(3), 10-15.

Tytler, R. & Symington, D. (2015). Science learning in rural Australia- Not necessarily the poor cousin. *Teaching Science, 61*(3), 19-25.

Tytler, R., Symington, D., & Cripps Clark, J (2017). Community-school collaborations in science: Towards improved outcomes through better understanding of boundary issues. *International Journal of Science and Mathematics Education, 15*(4), 643-661. DOI: 10.1007/s10763-015-9711-9

Tytler, R., Symington, D., Darby, L., Malcolm, C., & Kirkwood, V. (2011). Discourse communities: A model for considering professional development for teachers of science and mathematics. *Teaching and Teacher Education, 27*(5), 871-879.

Tytler, R., Symington, D., Kirkwood, V., & Malcolm, C. (2008). Engaging students in authentic science through school – community links: Learning from the rural experience. *Teaching Science, The Journal of the Australian Science Teachers Association, 54*(3), 13-18.

Tytler, R., Symington, D., Malcolm, C., & Kirkwood, V. (2009). Assuming responsibility: Teachers taking charge of their professional development. *Teaching Science, 55*(2), 9-15.

Tytler, R., Symington, D., & Smith, C. (2011). A curriculum innovation framework for science, technology and mathematics education. *Research in Science Education, 41,* 19-38.

Tytler, R., Symington, D., Williams, G., & White, P. (2017). Enlivening STEM education through school-community partnerships. In R. Jorgensen, & K. Larkin (Eds.), *STEM Education in the Junior Secondary School* (pp. 249-271). Dordrecht, The Netherlands: Springer.

Tytler, R., Symington, D., Williams, G., White, P., Campbell, C., Chittleborough, G., Upstill, G., Roper, E., & Dziadkiewicz (2015). Building productive partnerships for STEM Education: Evaluating the model and outcomes of the Scientists and Mathematicians in Schools program. Melbourne, VIC: Deakin University. Retrieved from https://www.csiro.au/en/Education/Programs/STEM-Professionals-in-Schools/How-the-program-works/Program-evaluation

Upstill, G. & Symington, D. (2002). Technology transfer and the creation of companies: The CSIRO experience. *R&D Management*, *32*(3), 233-239

Ziman, J. M. (1980). *Teaching and learning about science and society.* Cambridge University Press.

Ziman, J. (2000). *Real science: What it is, and what it means.* Cambridge University Press.

Russell Tytler (MSc, MEd, PhD, FASSA) - Deakin University, School of Education, 221 Burwood Highway, Burwood, 3125, Victoria, Australia. russell.tytler@deakin.edu.au
Russell Tytler is Alfred Deakin Professor and Chair in Science Education at Deakin University, Melbourne. He has researched and written extensively on student learning and reasoning in science. His interest in the role of representation as a multimodal language for reasoning in science extends to pedagogy and teacher learning. He researches and writes on student engagement with science and mathematics, aesthetics and identity in learning, school-community partnerships, and STEM curriculum policy and practice. He is widely published and has been chief investigator on a range of Australian Research Council and other research projects.
https://orcid.org/0000-0003-0161-7240

CHAPTER 12

MAKING EXPLICIT THE POLITICAL AGENDA WITHIN RESEARCH METHODOLOGIES

R. SCOTT WEBSTER

Abstract

Genuine democracy is on the decline throughout Western societies and unfortunately we are witnessing that some educational research is playing a role in this–although usually inadvertently. Educational research is political because its findings, if acted upon, directly affect educational practices, policies and specifically what sort of persons students become. When educational concerns in research are reduced to being myopically focused on narrow technical 'challenges' or 'problems', then only narrowly defined 'solutions' are pursued without reference to their political implications. Such a reduction of ignoring the political context is symptomatic of the tendency to overlook how the military-industrial-academia nexus is aggressively serving an authoritarian agenda of monopoly capitalism. This chapter argues that educational researchers ought to be cognisant of how this military-industrial-academia nexus threatens democracy by arguing that we ought to make the effort to 'reimagine democracy' in our research. For research methodology, this requires an articulated political position within the rationale which places the potential political implications front and centre. It also involves a careful choice of language where 'education' and *being* democratic are valued over narrower concepts such as teaching and learning.

Keywords: Education, democracy, political, methodology

The social nature of knowledge

In order to highlight why the political should be made explicit within research methodology, this section will demonstrate how the knowledge obtained from education research is inescapably political. On the quest for 'truths' (perhaps better understood as 'knowledge'), a central concern for all research methodology is the issue of validity. For example, scientific research into various physical phenomena has been greatly assisted by Popper's hypothetical-deductive approach and his clarification that scientific knowledge is to be falsifiable rather than verifiable. Popper clarified that the weaker approach of verification is very much related to inductive thinking, that is, with reference to raw data an idea or theory is attempted to be supported. Instead, Popper (1992, p. 139) stressed that "there are no uninterpreted... data… whatever is 'given' to us is already interpreted, decoded." His insight enables us to appreciate that it is difficult to make a claim of validity by simply engaging with 'raw' data, as per inductive reasoning, but instead we actively test the robustness of theories by seeking to find data which falsifies them. Pulling a particular fossil from the ground and conducting some preliminary analysis on it, and then claiming that it constitutes evidence which supports a particular creationist/evolutionary theory, is an example of the inductive approach, where one's worldview forms the backdrop or framework within which all of ones' *preferred* ideas and theories are founded.

In contrast, the deductive approach of scientific research begins by predicting what sort of data/findings would constitute evidence that the hypothesis can be falsified, and then proceeding to seek/discover such evidence. Popper's contributions also allow us to clarify the difference between validating knowledge-claims which have been produced scientifically (by seeking data which might falsify theories) in contrast to knowledge-claims which are only produced empirically, which are inductive in nature (Webster, 2009). Unfortunately, the predominant approaches to evidence-based research in education are inductive and empirical rather than deductive and scientific (see Bridges et al., 2009; Thomas & Pring, 2004; What Works Clearinghouse, 2018, 2020). For example, the meta-analyses of Hattie (2009) and Hattie and Yates (2014) are empirical and inductive, *not* scientific, because they never demonstrate their theories as being able to *cause* particular effects nor do they 'test' their theories via falsification, even although Hattie (see Toscano & Hattie, 2018) himself claims to be familiar with the work of Popper. This contrast between empiricism and science was identified by Dewey (1989, p. 275) many years ago, who argued:

> The empirical method inevitably magnifies the influences of the past; the experimental method throws into relief the possibilities of the future. The

> empirical method says, '*Wait* till there is a sufficient number of cases'; the experimental method says, '*Produce* the cases.' [original emphasis]

While Feyerabend (1993) claims there is no 'method' to do physical science research, and therefore there is no 'method' for the social sciences such as education to try and emulate, we nevertheless appreciate that the common challenge for researching into both physical science and social science is to pursue validity. What we do gain from the likes of Feyerabend (1993) and Dewey (1989) is that method, more broadly understood as methodology, should not be seen as being mutually exclusive of the actual understanding of the phenomenon being investigated nor of the actual knowledge which is produced about such a phenomenon. For research methodology in general, this results in serious attempts to pursue validity as much as possible within one's context, without ever being able to declare that one has accessed the knowledge, truth or essence of the matter 'out there' in the world of pure objectivity which is somehow beyond social constructs. This has been famously explored by Gadamer (1989) in his book *Truth and Method* where he juxtaposed the terms 'truth' and 'method' to indicate that there is not a method available to us which is able to give us access to absolute truth. Similarly, Richard Rorty (1979) in his *Philosophy and the Mirror of Nature* explains how problematic it is for us to assume that our mind or our knowledge can simply 'mirror' the essence of nature. Therefore, for researchers, we must settle for methodologies and findings which are considered valid rather than true.

Some researchers in education may be inclined to object to the claim that there is an inescapable link between their published findings and the political dimensions by making their research specifically apolitical (see Apple, 2000, 2004; Giroux, 2014). This objection depends upon the assumption that there is a clear distinction between knowledge which is about the physical 'real' and 'objective' world compared with knowledge about the social world which is more 'subjective' in nature. A consequence of this distinction is to then presume that knowledge claims of physical phenomena can be regarded as having nothing to do with the social world in which we expect politics to be limited within. However, a separation between knowledge of the physical world and knowledge of the social world is not tenable because, as Pring (2015, p. 139) points out:

> One cannot avoid the common-sense framework through which experience must be organized (*objects* existing in space and time, *causal relations* between objects, *persons* explained in terms of intentions and motives, *social settings* of rule-governed behaviour, *shared meanings* of

communicating communities, *values* in terms of which choices are made). [original emphases]

All knowledge has been constructed through social communities of practice and they interact with each other in a social *and* physical environment.

One past attempt to gain access to objective reality or the essence of phenomena was made by Edmund Husserl using his theory of phenomenology (1931). Here he asserted that one could let the actual phenomenon reveal itself to us, if we could lay aside our own biases, expectations and paradigm by which we view the world. This theory has parallels with the sort of inductive thinking that Popper was against. Husserl termed his method of laying aside our predetermined socially constructed views as 'bracketing out', or epoché, and therefore his theory became known as transcendental phenomenology (Heidegger, 1982, 1996) because the researcher was considered able to transcend his/her own socially constructed prejudice and bias. However, both Heidegger and Gadamer challenged Husserl's theory with their own theories of hermeneutical phenomenology which acknowledge that transcendence from one's personal bias is not possible. Consequently, one must critically cycle between interpreting the meaning (rather than the essence) of phenomena as well as one's own socially constructed paradigm of viewing the world in general. In other words, one must engage in deductive thinking in a critical manner, being willing to question the assumptions that one tends to take for granted, perhaps assuming them to be a matter of 'common-sense'.

This is why researchers, and PhD researchers in particular, are encouraged to identify their own bias, expectations or paradigm before venturing into formulating a methodology. A valid methodology ought to be designed to include opportunity to have one's bias challenged. Our 'normal' disposition in the world is to give stable and predictable meanings to the various stimuli which we encounter. William James (1918, pp. 272, 459) referred to this as "sameness" and Dewey (1989, p. 240) referred to it as an "attitude of anticipation." Such a disposition influences us towards conformational bias, that is, being prepared to find or see what we already expect to see. Clarifying and articulating our bias is not easy to do because typically our framework, paradigm or 'enframing' as Heidegger (1977) would describe, is invisible to us in our everyday lives. As Gadamer (1989, p. 299) acknowledges, "[i]t is impossible to make ourselves aware of a prejudice while it is constantly operating unnoticed, but only when it is, so to speak, provoked." Often our bias is embedded in what we assume is a matter of common-sense, but which, on closer inspection, draws upon particular ideological understandings which, as Apple (2004) explains, have become unnoticed and therefore hegemonic in our everyday view of the world and ourselves. Therefore, this is why we can all benefit from engaging with our colleagues and/or supervisors to help us draw some of what is often hidden to ourselves. When the nature of knowledge is understood to be social (and subject to prejudice and bias) rather than a reflection of physical reality, it can be easier to appreciate the political dimension that is intrinsic to it.

Education and educational research are inescapably political

From the section above we can appreciate that the knowledge which emerges from education research can never be absolutely objective or mirror nature because it is inescapably social in its construction while also being reflective of the physical environment. In this present section it shall be considered how this socially constructed knowledge has political aspects, especially when we consider the phenomenon of 'education' rather than the narrower aspects of teaching and learning.

Many philosophers of education recognise that education is an intrinsic aspect of political life. However, contemporary education research is heavily invested into studying the dynamics of teaching and learning while ignoring 'education'. This has become so hegemonic that Biesta (2017, p. 22) has even suggested that these narrower terms can be blurred into the one word "*teachingandlearning*" because of the dominant assumption that together these two key concepts are assumed to be inseparable and form the important core of education practice. To assist with the critique of this narrow focus he has invented the term 'learnification' to accentuate how the concept of 'learning' has come to dominate and even replace 'education'. Once this domination has been achieved then both teaching and learning become reduced to technical activities which are presumed to be apolitical. However, as Biesta (2013, p. 3) reminds us, "[t]he education concern rather lies… what can justifiably be desired… [and] always requires engagement with what or who is other (which makes the educational question also a question about democracy...)." One of the important outcomes of this current dominating approach of focusing on teaching and learning is that it has created a dichotomy between means and ends, that is, between the 'best' teaching and learning techniques–as means–but having no end purposes available for re-evaluation. Consequently, we can investigate various approaches and strategies that are 'effective' but we are unable to question or determine what purposes they are 'good' for. Efficiency provides its own criterion–i.e., minimising inputs and maximising outputs, and is therefore restricted to processes or means. However, when considering end purposes, the educational concern, as outlined by Biesta, is upon what is desirable and 'good' for individuals and especially society, and hence engages with what is politically desirable such as the promotion of democracy. Democracy pertains to an aspirational ideal (see Fairfield, 2008; Schostak & Goodson, 2020) rather than a simple means of governing.

This dichotomy between means and ends has been identified earlier by Neil Postman (1995, p. 26) in his book *The End of Education*, in which he laments that "there was a time when educators became famous for providing reasons for learning; now they become famous for inventing a method." Since the time of the publication of Postman's book we have observed an ever-narrowing focus upon methods, techniques and strategies for teaching and learning, often under the banner of business terms such as 'quality'. However, terms such as 'quality' should cause us to be aware of focusing too much upon processes while overlooking the potential promotion of anti-democratic dispositions amongst students (Webster, 2017). Dewey (2008a) referred to this limiting sort of focus upon the process of learning as 'miseducation' because it fails to give serious consideration to the long-term implications. For example, particular strategies such as methods of direct instruction, may be identified in a particular context as promoting desired learning outcomes more quickly and effectively than other approaches. However, in the process students may develop the habit of uncritically accepting materials from their teachers as 'givens' which are approved by authorities (such as teachers and government departments of education) and which should therefore not be questioned. This is a habit that is more suited to living in an authoritarian community rather than in a democratic one. So even though teachers might be teaching and students might be learning through an efficient process, the capacity of students for participating in life democratically may become compromised because of the habits they develop 'collaterally' as Dewey has described.

One of the important aspects for education researchers is to be clear as to what makes the field of our research 'education' (Webster & Pring, 2018). Very broadly understood, education involves an intervention in the lives of others, usually regarded as students who often are comprised of young minors. This intervention is considered by teachers, parents, researchers and the public in general, to offer value for both the individual students (private goods) as well as for society as a whole (public good). Irrespective of which ideological perspective might be dominant (e.g., progressivism, social efficiency, social reconstruction, neoliberalism or traditionalism), education has an impact upon *both* the private and the public, that is, both the lives of individual students *and* society at large are benefitted in some manner through education. Even for science education there is an inescapable link between the growth in the capacity for critically knowing and ethical capacity in a political context (Webster, 2008). Therefore, when scholars such as Shor (1992, p. 11) announce that "education is politics", they are not offering a marginal or radical perspective but are simply acknowledging the inescapable nature of how society is impacted by educational interventions. This is where we understand 'politics' as not pertaining to the views of any particular party, but rather, as Mouffe (2005) explains, it is simply the nature of how society is structured and functions. Specifically, she differentiates this understanding of 'politics' from 'the political' which refers to how individuals are disposed to being engaged in social life.

Similar to education practices, the activity of researching into education is also inescapably involved in these very same ethical and political dimensions. That is, if one's research is read and adopted by those involved in the decision making processes of policies and/or the enactment of practices, then it will have a direct impact on the lives of students and of society. For example, if one researches how to teach algebra more efficiently through a particular strategy, then the experiences which are constructed to incorporate such a strategy will impact on students, specifically how they engage with the world and what sorts of understandings, habits and attitudes they will develop, as well as their appreciation of the nature of knowledge. These long-term influences will then contribute to how students participate in society more broadly and what sort of dispositions they will adopt, ranging perhaps from conforming credulity at one end of the spectrum to critical, investigative scepticism at the other.

Dewey (1985, p. 103) reminds us how dependent knowledge and understandings of education are upon political views by clarifying that "the conception of education as a social process and function has no definite meaning until we define the kind of society we have in mind." In other words, he is identifying that we are unable to provide a definition for education, including the aspect of what is 'education' in our own educational research, until we first clarify our own political aspiration. That is why many education researchers, such as Peters and Besley (2021) and Schostak and Goodson (2020), are calling for the need to refocus and reimagine democracy in the work that we do. They clearly understand the links between education research and the political and are warning us all that currently, education and educational research are failing to address the anti-democratic disposition being developed within students in schools and universities and also amongst educators.

As has been stated above, all education practices are inescapably political. Actually, this has remained the case since Ancient Greece where they conducted philosophy in the attempt to pursue an understanding of what constitutes a 'good life' (*eudaimonia*) for individuals and for society *together* while appreciating that education is the pursuit of such a communal life. This is partly why Dewey (1985, p. 338) suggests that "philosophy may even be defined as the general theory of education" because of this very close connection which relates to human flourishing which is always communal and hence political. For the Ancient Greeks, philosophy, education, politics, ethics and religion were all clearly interconnected into a unified whole, imbued with purpose and often understood as religious. As Gadamer (1999, p. 39) explains:

> The religious vocabulary that they [i.e., the Ancient Greeks] took over… is not intended to make a statement about god or the gods, but rather to designate the order of being about which they are inquiring; the whole, the all, being.

Therefore, in our time we might refer to this holistic or totalising view as a 'spiritual' view rather than a 'religious' one. On this very issue and in his own time Dewey (2008b) made the effort to distinguish between being 'religious' in attitude (i.e., emphasis upon the adjective) from having 'a religion' (where the emphasis is upon the noun) which designates a particular creed and/or denomination. However, in our contemporary secular society even the terms 'religious' and 'spirituality' tend to be viewed with suspicion, and so perhaps we can consider one's holistic view which gives totality and unity to how a society might be oriented towards the 'good' (i.e., *eudaimonia*) as one which is political.

How education research is being misused politically

In this section we shall consider how some of the impact of educational research knowledge can be understood as undesirable from a political perspective. There is an ever-increasing amount of government intervention into schools and universities, including the recent doubling of fees to study the humanities thereby discouraging the population from pursuing an Arts degree. In addition to controlling fees and policies, Pring (2015, p. 187) also observes governments having "greater control over the direction, content and dissemination of [educational] research". This trend has been growing steadily over recent decades. As reported by Simons (1995) regarding the situation in England:

> The policy context which we have to conduct educational research sponsored by government in England at the current time has made the politics of research more explicit and contestable than ever before… so much so that the quality of educational research itself is under threat… (cited by Pring, 2015, p. 187)

Schostak and Goodson (2020, p. 5) add that recent education research "has become an instrument of social manipulation, exploitation and control." Critiquing the propaganda which has been pouring through our media for years and by drawing upon Walter Lippman, Edward Bernays, Edward Herman and Noam Chomsky, these education researchers make the claim that both education and education research are harnessed to manipulate public opinion rather than educating and enabling the public to be critically engaged in public life and to carefully interrogate claims made by various authorities with intellectual rigor.

Rather than enabling individuals to be worthy of democracy, the sorts of teaching, learning and curriculum work being encouraged through contemporary education research (see Hattie & Yates, 2014; Heward & Twyman, 2021; Oliver, McConney & Woods-McConney, 2021) and policies tend to encourage anti-democratic outcomes and practices. One influential organisation in the U.S. which researches and promotes 'best practices' for teachers is the 'What Works Clearinghouse' Institute of Education Sciences. The recommended 'best' strategies might demonstrate some short-term benefits, such as students learning an aspect of algebra more efficiently in a manner that raises test scores or demonstrating better scientific 'literacy' skills in a test, but through habitual recipe-following behaviours they may also have their potential for critical thinking and experimentation stifled. The political implication would then be that if this disposition was to be manifested then they would make suitable citizens for an authoritarian society but not for a democratic one.

Due to the massive propaganda continuously being promoted through our media outlets, governments and especially their agencies, Schostak and Goodson (2020, p. 2) provokingly ask "[a]re research and education blind to what is going on… [regarding the] war that is effectively waged against 'truth'"? Key to their concerns is an insight from those who investigate the nature of public relations which specifically recognises that the "management of opinion is key to the production of pseudo-democracies for the government of people in the interests of the elites" (Schostak &

Goodson, 2020, p. 6). These education researchers are challenging the rest of us to give serious consideration to the potential political impact of our own work.

Too often influential elites, primarily through grant-giving, dictate a discourse of research methodology which reduces issues to closed-system 'problems' which need 'solutions'. Perhaps some of these 'problems' include a country's reported failure to be on the top of a particular league table or some other declared high-stakes normative distribution, or to increase the representation of perceived disadvantaged or minority groups into particular programs and settings. When presented in such a closed-systems manner it is far easier to identify when 'solutions' have been discovered and/or implemented because there tends to be tangible and easy-to-measure outcomes that unequivocally demonstrate that the problem has been 'solved' because we have discovered 'what works' in addressing only the 'problem' at hand. This approach, which is now considered to be dominating much of the research in education, is identified by Schostak and Goodson (2020, p. 63) who report that:

> Research methodology is drawn into the ambit of the agendas constructed by power elites as soon as it is limited to the puzzle-solving techniques deployed to bring about answers to elite defined problems.

As an example of these elite defined problems, recently at least one Australian university advertised and encouraged its academics to consider applying for National Intelligence and Security Discovery Research Grants (NISDRG), and it is reasonable to assume that many other universities have similarly advertised this program to their staff. Some of the specific 'intelligence challenges' that the NISDRG (Australian Research Council 2022) is funding for 2021 and 2022 include:

Covert collection challenges
Identity management challenges
Emerging biological and material science exploitation challenges
Cyber security, protective security and offensive cyber challenges
Human behaviour and influence challenges
Disinformation Detection and Counter Narratives
Human Identification and Vulnerability Assessments

These should not be accepted as simply closed-system 'challenges' and 'problems' to be 'solved' because they are an extension of a political agenda which disturbingly appears to promote anti-democratic interests. Therefore, to engage with any of these projects is to participate in promoting a particular political climate which is difficult, if not impossible, to reconcile with a democratic way of life. It demonstrates the link that academia has in the anti-democratic and global corporate interests of the military-industrial-academia complex (Chomsky, 2016; Peters & Besley, 2021; Schostak & Goodson, 2020) which, according to Sargent (2007), has been apparent for at least the last two centuries.

If instead we consider these activities through a holistic perspective which incorporates a political concern as to what sort of society and people we and our students are becoming, then we may quite rightly become concerned. This is because the sorts of activities promoted here, such as *covertly* collecting information, managing information on people's identity (without their knowledge), *exploiting* biological materials (such as viruses perhaps as per Wuhan and Ukraine), designing *offensive* cyber capabilities, influencing (and perhaps *controlling*) *people's behaviours* and undertaking counter narratives (i.e., *propaganda*) then we may appropriately be alarmed that our government is actively funding research for activities that are more useful to an authoritarian society. This is because in contrast to democracies where the people ought to know all the details about the government and its activities, totalitarian governments are characterised as knowing all the details about individual people and their activities.

What is emerging are discourses of research methodology which are strongly promoting the notions of 'what works' where recommended solutions are 'evidence-based' (see What Works Clearinghouse, 2018, 2020). As a consequence, it has become accepted too often that both educational research and education practice are merely technical or clinical, and therefore apolitical. But this is not true. If one assumes that one can limit oneself to a closed-system of understanding, often with clearly defined (and measured) inputs and outputs, then the case might be made to look convincing until it is recognised that the world does not consist in an array of closed-systems existing alongside one another. So, despite Voltaire's (2016) call to simply tend to our own garden, Dewey (1988a, pp. 122-123) perceptively points out; "there is no fence about this garden: it is no sharply marked-off enclosure. Our garden is the world." Hence each of us, including our research, are inescapably part of the greater political environment.

Everything is interrelated including education research and politics. This is why Peters and Besley (2021, p. 1) warn that education researchers (particularly those working in philosophical aspects):

> …must start by recognizing the state of peril faced by world democracy. [They] must take on board the empirical reality of a world no longer built for democracy, democracy promotion or even largely comprised of democratic regimes.

They warn of the 'military-industrial-academia' complex which, as a network, primarily serves the interests of elites rather than of the public. The agenda of the elites is to monopolise markets primarily through influencing political policies which serve the interests of the public and environment (Wolin, 2017). However, they also have a very long history of influencing and taming what universities and researchers are able to do (e.g., Sargent, 2007; Vidal, 2014). Recently, some branches of the established elites are even announcing that they will label any activities to which research is to be included, which contradicts their narrative and consequently "undermine[s] public trust in U.S. government institutions" as acts of terrorism (Department of Homeland Security, 2022). Hence educational researchers are having to choose between trying to depoliticise their work to 'safe' projects, which do not threaten the status quo of the government's agenda, or to undertake work which may challenge official narratives in the pursuit of truths by embracing a clearly articulated political aspiration which values the 'good' offered through truths which transcends the comfort of conformity.

Understanding that education research does impact the moral and political dimensions ought to alert us to the seriousness that we ought to invest into the political dimension of our own research to ensure that our work offers value to the private goods of students' lives and the public good of political life overall. While the discussion above referencing the NISDRG serves as a very concerning example which potentially has profound implications for the political life of our society, we need to be cognisant of the political impact that our own research might be having. To assist with this reflection, it is considered of value to clarify how we might understand 'democracy' in such a way that we can see more clearly as to how education might be more supportive of this aspirational ideal.

Democracy

In light of the call to make our political aspirations clear as we embark upon educational research and methodology in particular, this section seeks to explain how democracy might be understood, and how this might orient one's research trajectory. While democracy is not a panacea to cure all the ills in the world, it is nevertheless a most valuable aspiration because it enables genuine freedom and flourishing for both individuals and society, where everyone can have the opportunity to freely participate and have a say in how things ought to function rather than simply being led in an authoritarian manner (Fairfield, 2008).

John Dewey is well-recognised for linking both education and democracy together, particularly through his book *Democracy and Education* which was originally published in 1916. In this he explained:

> A democracy is more than a form of government; it is primarily a mode of associated living, of conjoint communicated experience. The extension in space of the number of individuals who participate in an interest so that each has to refer his own action to that of others, and to consider the action of others to give point and direction to his own, is equivalent to the breaking down of those barriers of class, race, and national territory which kept men from perceiving the full import of their activity. (Dewey 1985, p. 93)

In addition to being a form of government with its institutions and procedures, Dewey also argued that democracy is a way of living. Understanding democracy as a way of living broadens our vision of democracy beyond the official structure and function of society's governing institutions. It is 'the political' rather than 'politics' as explained by Mouffe (2005), who draws attention to the disposition of individuals. Interestingly, she argues that there is "too much emphasis on consensus, together with aversion towards confrontations" which she claims, "leads to apathy and to a disaffection with political participation" (Mouffe, 2013, p. 7). This emphasis on consensus, in addition to being characteristic of our mainstream media, is also often prevalent within classrooms where students become conditioned to the importance of pursuing consensus in the form of the right answers. Consequently, while there is some merit to engaging in critically developed views as instances of consensus, educators should remain mindful as to what sort of persons they and their students are becoming and whether they value the potential of dissenting perspectives for validity, especially for attempting to falsify ideas. This was how Dewey understood the meaning of democracy which he considered as offering a particular holistic view which gives unity to *all* our understandings and manners of participating in life generally.

By attending a Congregationalist church as a youth, Dewey came to appreciate how a community might conduct itself in a very horizontal fashion, where a hierarchy of authority is absent. He later likened this to the ideal scientific community where hierarchical authority does not determine knowledge, but rather all members can actively experiment and test ideas.

Dewey (1985) warned that authoritative or uncritical acceptance of knowledge “swamps thinking” (p. 165) and therefore an educated democratic community is characterised by thinking and experimentation because for him “thinking is the method of an educative experience” (p. 170).

Here we can appreciate that *being* democratic as a way of life is something that a person or persons might be able to embrace, irrespective of the particular form of government that might happen to be governing at the time. For education this is profound; attempting to perhaps change the way society’s governing institutions function and/or are structured is a very daunting and often overwhelming task at the best of times. However, what is far more achievable for teachers and educational researchers is to invest far more thought and energy into enabling students–and indeed society at large–to *be* more democratic in their way of living. Drawing on Dewey, we can appreciate that there is a great deal in common between *being* democratic and *being* educated.

One of the first aspects of *being* educated is that it is distinctively different from simply *having* an education. Through the mode of *having* we can understand that it is possible, through formal education, for one to acquire things such as knowledge and skills, which have value for undertaking many activities, particularly those suited for participating in the economy. A focus on this mode of *having* would influence the role of educational research, to be primarily concerned with whether students are receiving or developing the content or ‘things’ which teachers teach and seek to impart. As such, the ‘success’ of learning is often limited to a demonstrable acquisition of skills and knowledge. However, the limitation of this focus on the mode of *having* is that it ignores the personhood of the students. Consequently, it can be argued that while students are gaining the knowledge and skills for being able to participate in the economy, as an example, they simultaneously might be becoming suited to uncritically accept living in an authoritarian culture. As was mentioned earlier, this is what Dewey referred to as miseducation.

From Dewey we can recognise that *being* democratic is necessarily also a moral way of living. This is because one must respectfully engage with others who will be affected by one’s activities and/or when one will be affected by the activities of others. As such, being democratic means that one must desire (Biesta, 2013) what is most appropriate to be desired for those who constitute a democracy. It is not just enough to be knowledgeable, skilful and thoughtful, but one must also be *motivated* by the community 'spirit' of democracy, and as such remain mindful of the public/global good. It is contended therefore, that there are clear overlaps between *being* democratic and what it means to *be* educated and even what it means to *be* scientific, and these overlaps shall now be considered in the following section.

Being educated, *being* democratic and *being* scientific

In this final section, I wish to draw the three ideas together of *being* educated, *being* democratic and *being* scientific, as a possible way for making the political explicit in our own educational research. To accentuate the difference between the modes of *being* and *having* (Fromm, 1976; Marcel, 1949; Webster, 2021) with regards to science, we can identify that the acquisition of formal scientific knowledge and skills for conducting scientific investigations are examples of the *having* mode. In contrast, when the *being* mode of students becomes educated through science, they develop a scientific disposition or 'temper' as described by Bertrand Russell (1922).

Russell (1922, p. 41) purported that in his time "everything in newspapers is more or less untrue". In order to avoid becoming manipulated and led astray by mainstream media and political spin, he argued that people ought to be educated to develop a 'scientific temper' or attitude. Importantly, he distinguished this sort of temper from "the knowledge of scientific results" which he clarified "is an altogether different thing" (Russell, 1922, p. 44). Similar to Feyerabend (2011), Russell (1932) recognised the tyranny that can be exercised through "the governmental use of science" and so he argued that people ought to be educated with a "scientific spirit" characteristic of what he understood to be an "anarchic state of mind" (p. 11). This mindset is essentially free from authoritarian manipulation and therefore enables people to see through how governments manipulate the masses by *using* science rather than *being* scientific.

For Dewey (1988b, p. 7) he likened a scientific attitude to providing "a light to the eyes and a lamp to the feet… [it is] an organ of personal illumination and liberation." He did not regard science to be primarily a body of knowledge but rather as an approach to inquiry. It doesn't offer a method as a recipe might do, but instead offers an experimental 'trial and error' habit of investigation. Putting this into practice in the social realm, he was opposed to cultural norms and moral traditions simply because they had been of value historically. Instead, he sought to have such norms and traditions demonstrate their value in the present environment to then appreciate their potential relevance.

This aspect of fostering a scientific attitude or temper through the *being* mode, is offered as a suggestion as to how, as educational researchers, we might make the political explicit in our work, and it connects well with what it means to *be* educated and to *be* democratic. These connections might provide the end purpose or warrant for particular phenomena that we might be researching. If we are to avoid falling into the apolitical narrow focus of 'learnification' as Biesta (2010) has warned, then we might consider how being scientific is a crucial aspect for being educated in such a way as to be worthy of democracy. To appreciate this interrelation, it might be helpful to reflect on what it might mean to *be* scientific.

Being scientific involves appreciating that knowledge, even scientific laws, are not absolute or fully settled. Consequently, rather than uncritically accepting and memorising content, one needs to appreciate that many claims are contentious and therefore one must come to understand the warrants that support various assertions. Consequently, *being* scientific involves the characteristics of critical *thinking* (not just problem solving), *understanding* the warrants behind assertions, appreciating what constitutes validity of evidence and how this needs to be treated carefully and critically. It is superior to philosophical reasoning, which can remain in the abstract, because it is willing to embody such reasoning to test ideas via trial and error in an experimental temperament in the material environment. One also greatly values and actively seeks dissenting views, knowing that these provide opportunities to re-evaluate claims to knowing, being forever curious rather than dismissive towards points of view which differ to the norm or to one's own beliefs. In addition to being curious, one also cares for and desires what one ought to desire as a responsible member of a community and its ecosystem.

Throughout our research methodology, we can make our political aspirations explicit by identifying what the implications might be for the 'stakeholders' (usually students) in terms of how they are better enabled to *be* democratic. This could incorporate the links that the implications offer for *being* educated and *being* scientific rather than just *having* or attaining particular levels of competencies. Importantly, by making such aspirations explicit this will also help prevent embarking on research projects that are presented as closed-system 'problems' or 'challenges' but which promote anti-democratic ways of *being*. Through this approach of making the political explicit as argued here, we are better able to ensure that our research is contributing towards 'reimagining democracy'.

References

Apple, M. W. (2000). *Official knowledge*. New York, NY: Routledge.

Apple, M. W. (2004). *Ideology and curriculum*. New York, NY: RoutledgeFalmer.

Australian Research Council. (2022). ARC Research Grants Services: National Intelligence and Security Discovery Research Grants Program. Retrieved from https://www.arc.gov.au/research-grants-services/arc-research-grants-services-rms-nisdrg

Biesta, G. J. J. (2010). *Good education in an age of measurement*. Boulder, CO: Paradigm.

Biesta, G. J. J. (2013). *The beautiful risk of education*. Boulder, CO: Paradigm.

Biesta, G. J. J. (2017). *The rediscovery of teaching*. New York, NY: Routledge.

Bridges, D., Smeyers, P., & Smith, R. (Eds.) (2009). *Evidence-based education policy: What evidence? What basis? Whose policy?* Oxford, UK: Wiley-Blackwell.

Chomsky, N. (2016). *Who rules the world?* New York, NY: Henry Holt.

Department of Homeland Security. (2022). *DHS Issues National Terrorism Advisory System (NTAS) Bulletin*, released February 7, 2022. Retrieved from https://www.dhs.gov/news/2022/02/07/dhs-issues-national-terrorism-advisory-system-ntas-bulletin

Dewey, J. (1985). Democracy and education. In J. Boydston (Ed.), *John Dewey the middle works Vol. 9, 1916* (pp. 1-370). Carbondale and Edwardsville, IL: Southern Illinois University Press.

Dewey, J. (1988a). Individualism old and new. In J. Boydston (Ed.), *John Dewey the later works Vol. 5, 1921-1930* (pp. 41-123). Carbondale and Edwardsville, IL: Southern Illinois University Press.

Dewey, J. (1988b). The sources of a science of education. In J. Boydston (Ed.), *John Dewey the later works Vol. 5, 1921-1930* (pp. 1-40). Carbondale and Edwardsville, IL: Southern Illinois University Press.

Dewey, J. (1989). How we think. In J. Boydston (Ed.), *John Dewey the later works Vol. 8, 1933* (pp. 105-352). Carbondale and Edwardsville, IL: Southern Illinois University Press.

Dewey, J. (2008a). Experience and education. In J. Boydston (Ed.), *John Dewey the later works Vol. 13, 1938-1939* (pp. 1-62). Carbondale, IL: Southern Illinois University Press.

Dewey, J. (2008b). A common faith. In J. Boydston (Ed.), *John Dewey the later works Vol. 9, 1933-1934* (pp. 1-58). Carbondale, IL: Southern Illinois University Press.

Fairfield, P. (2008). *Why democracy?* Albany, NY: State University of New York Press.

Feyerabend, P. (1993). *Against method*. London, UK: Verso.

Feyerabend, P. (2011). *The tyranny of sci*ence. Cambridge, UK: Polity.

Fromm, E. (1976). *To have or to be?* London, UK: Continuum.

Gadamer, H-G. (1989). *Truth and method 2nd Ed.* New York, NY: Continuum.

Gadamer, H-G. (1999). *Hermeneutics, religion and ethics*. Tr. J. Weinsheimer. New Haven, CT: Yale University Press.

Giroux, H. A. (2014). *Neoliberalism's war on higher education*. Chicago, IL: Haymarket Books.

Hattie, J. (2009). *Visible learning*. London, UK: Routledge.

Hattie, J. & Yates, G. (2014). *Visible learning and the science of how we learn*. London, UK: Routledge.

Heidegger, M. (1977). *The question concerning technology*. Tr. W. Lovitt. New York, NY: Harper & Row.

Heidegger, M. (1982). *The basic problems of phenomenology*. Tr. A. Hofstadter. Bloomington & Indianapolis, IN: Indiana University Press.

Heidegger, M. (1996). *Being and time*. Tr. J. Stanbaugh. Albany, NY: State University of New York Press.

Heward, W. L., & Twyman, J. S. (2021). Teach more in less time: Introduction to the special section on direct instruction. *Behavior Analysis in Practice, 14*, 763-765.

Husserl, E. (1931). *Ideas: General introduction to pure phenomenology*. Tr. W. R. B. Gibson. Abingdon, VA: Routledge.

James, W. (1918). *The principles of psychology, Vol. 1*. New York, NY: Dover Publications Inc.

Marcel, G. (1949). *Being and having*. Glasgow, UK: Robert MacLehose & Co.

Mouffe, C. (2005). *On the political*. London, UK: Routledge.

Mouffe, C. (2013). *Agonistics*. London: Verso.

Oliver, M., McConney, A., & Woods-McConney, A. (2021). The efficacy of inquiry-based instruction in science: A comparative analysis of six countries using PISA 2015. *Research in Science Education, 51* (Suppl 2): S595-S616. DOI: 10.1007/s11165-019-09901-0

Peters, M. A. & Besley, T. (2021). Making democracy safe for the world? Philosophy of war, peace and democracy. *Educational Philosophy and Theory*. DOI: 10.1080/00131857.2021.1960503

Popper, K. (1992). *Unended Quest.* London: Routledge.

Postman, N. (1995). *The end of education*. New York, NY: Vintage Books.

Pring, R. (2015). *Philosophy of educational research 3rd Ed*. London, NY: Bloomsbury.

Rorty, R. (1979). *Philosophy and the mirror of nature*. Princeton, NJ: Princeton University Press.

Russell, B. (1922). *Free through and official propaganda*. London, UK: Watts and Co.

Russell, B. (1932). *Education and the social order*. London, UK: Routledge.

Sargent, P. (2007). *War and education*. Boston, MA: Brewster Press.

Schostak, J. & Goodson, I. F. (2020). *Democracy, Education and Research*. London, UK: Routledge.

Shor, I. (1992). *Empowering education*. Chicago, IL: The University of Chicago Press.

Simons, H. (1995). The politics and ethics of educational research in England: contemporary issues. *British Educational Research Journal, 21*(4), 435-449.

Thomas, G. & Pring, R. (2004). *Evidence-based practice in education*. Maidenhead, UK: Open University Press.

Toscano, M. & Hattie, J. (2018). John Hattie on interpretation, the story of research and the necessity of falsifiability: In dialogue with Maurice Toscano. In J. Quay, J. Bleazby, S. A. Stolz, M. Toscano & R. S. Webster (Eds.), *Theory and philosophy in educational research* (pp. 85-100). London, UK: Routledge.

Vidal, G. (2014). *History of the national security state*. Scotts Valley, CA: CreateSpace Independent Publishing Platform.

Voltaire (2016). *Candide*. New York, NY: Dover Publication Inc.

Webster, R. S. (2008). How a Deweyan science education further enables ethics education. *Science & Education, 17*(8), 903-919.

Webster, R. S. (2009). How evidence-based teaching practices are challenged by a Deweyan approach to education. *Asia-Pacific Journal of Teacher Education, 37*(2), 215-227.

Webster, R. S. (2017). Education or quality of teaching? Implications for Australian democracy. *Australian Journal of Teacher Education, 42*(9), 59-76.

Webster, R. S. & Pring, R. (2018). Richard Pring on making research *educational* research: In dialogue with R. Scott Webster. In J. Quay, J. Bleasby, S. A. Stolz, M. Toscano & R. S. Webster (Eds.), *Theory and*

philosophy in education research (pp. 129-143). London, UK: Routledge.
Webster, R. S. (2021). *Caring confrontation for education and democracy*. London, UK: Routledge.
What Works Clearinghouse. (2018). *WWC Procedures and Standards Handbook* (Version 2.0). Washington, DC: Institute of Education Sciences. Retrieved from http://ies.ed.gov/ncee/wwc/
What Works Clearinghouse. (2020). *WWC Procedures and Standards Handbook* (Version 4.1). Washington, DC: Institute of Education Sciences. Retrieved from http://ies.ed.gov/ncee/wwc/DocumentSum.aspx? sid¼19
Wolin, S. (2017). *Democracy incorporated*. Princeton, NJ: Princeton University Press.

R. Scott Webster (Dip.T, BEd, BA, MSc, PhD) – Federation University.
sswebster@federation.edu.au
Scott has worked for 20 years in universities, primarily in Education Studies, Philosophy of Education, Curriculum theory, Pedagogy and Spirituality, and is currently working for Federation University. Before this he was a secondary high school teacher working in science, maths and HPE. His research publications focus mostly on the philosophy of education and curriculum studies. Recent book projects include: *Caring Confrontations for Education and Democracy* (2021 Routledge), *Rethinking Reflection and Ethics for Teachers* (2019, Springer) and *Understanding Curriculum: The Australian Context, 2nd Ed* (2019, Cambridge University Press).
https://orchid.org/0000-0002-5253-4894

CHAPTER 13

TENSIONS BETWEEN THEORY AND PRACTICE: A MATHEMATICS EDUCATION PERSPECTIVE

JILL P. BROWN, CARLY SAWATZKI
AND WANTY WIDJAJA

Abstract

Since education research is complex, methodological decisions often produce tensions for researchers and teacher participants that are not easily overcome. The focus of this Chapter is the design and implementation of a bespoke *Graduate Certificate of Secondary Mathematics* course specifically designed for out-of-field teachers of mathematics in Years 7-10 in Victoria, Australia. Out-of-field teaching is increasingly becoming the norm and in mathematics this raises challenges and tensions specific to the discipline. In the field of mathematics education research, it is taken as a given that deep mathematical understanding by the teacher is a necessary but not sufficient prerequisite for quality teaching and learning. Fractions and decimal fractions are an important yet challenging area of mathematics. We detail how using a research-based interview raised tensions related to mathematical understanding, assessment and ways of teaching. Additionally, engagement with the interview provided opportunities for productive learning about the teaching of mathematics to emerge.

Keywords: Assessment, mathematics education, productive learning, out-of-field teaching of mathematics, tensions

Introduction

Globally, school leaders are solving the problem of limited supply of mathematics teachers by assigning teachers qualified in other disciplines to the mathematics classroom (Price et al., 2019). The trend of "out-of-field teaching of mathematics" is an emerging field of research with the potential to inform policy and system-level responses (Bosse & Törner, 2015, p. 4). There are a growing number of studies whose findings suggest implications for the recruitment of teachers to the profession, as well as their ongoing education and retention (e.g., du Plessis, 2016; Weldon, 2016). It is due to this trend that the authors' own work at Deakin University's School of Education has recently come to include the design and delivery of a bespoke *Graduate Certificate of Secondary Mathematics* (GCSM) in partnership with the Victorian Department of Education and Training (DET). This is part of DET's *Secondary Mathematics and Science Initiative* (SMSI) for out-of-field teachers of mathematics and science. More than 200 out-of-field mathematics teachers of Years 7-10 will have completed this 12-month course by the end of 2022.

In this chapter, we reflect on the methodological underpinnings and guiding principles that shaped the design and delivery of the course, which aims to address the problem of disengagement and declining mathematics achievement in secondary schools by giving out-of-field mathematics teachers time to study and develop the knowledge and skills required to teach mathematics. We describe the distinguishing characteristics of educational design research methodology (Brown, 1992; Tytler et al., 2021; Walker, 2006), and examine tensions we experienced as we were called upon to be 'agile' in response to the various discomforts associated with learning research-informed approaches to teaching mathematics while balancing the demands of work and family life during the COVID-19 pandemic. Through one example–an assessment task requiring teachers to complete the *Fractions and Decimals Online Interview* with two of their students–we highlight the challenges for us as academics and mathematics teacher educators working to disrupt the status quo and improve the conditions for secondary school students to learn mathematics.

A philosophy of mathematics teacher education

Scholarship is at the heart of academic work. Academics read, research and write over many years to establish a track record and reputation within their field. Such scholarship is slow and deliberate. Yet the reality of academic life at the modern university is the requirement to engage in strategic, collaborative and impactful research partnerships with government, industry and end-users (Australian Research Council, 2015). Increasingly, this expectation sees academics applying for competitive tenders for projects that might meet market-driven demands and specifications within tight timelines. This neoliberal context has been further complicated by the COVID-19 pandemic, which has restricted access to schools, teachers and students for the purpose of research and professional learning activities.

Within this context, the focus of our work, be it supporting preservice or inservice teachers, is the improvement of opportunities for students to learn and value the discipline of mathematics. We believe that this is best achieved by improving teacher knowledge and practice by exposing long-held misconceptions and challenging problematic teaching practises that are commonplace and may not have been questioned before. Lovric (2017) argues that such work inevitably produces tensions "that require modifications to our cognitive models" (p. 1) and a rearrangement of core beliefs about mathematics, mathematics teaching and learning.

We argue that outsiders to mathematics teaching (including out-of-field teachers) often perceive the discipline to be precise and well-defined. This belief about mathematics leads many to view learning mathematics as "fluency in recalling facts and performing skills" (Swan, 2006, p. 42). This view is perpetuated in the media and similarly to reporting about climate change, both 'sides' are given equal space as if they are equally well supported by evidence. For example, the online article by Lu (2022) regarding mathematics teaching suggests that the draft revised Australian Curriculum for mathematics proposed problem solving as a pedagogy rather than problem solving being an essential mathematical activity; and this is a concern. Once again, explicit teaching and inquiry learning are presented–both in the media and by those outside mathematics education–as a dichotomy and as if they mean the same thing to all readers and generally through the psychological lens that explicit teaching should be the (sole) pedagogy of choice.

In fact, the Australian Curriculum describes what students should learn (including problem solving), not how teachers should teach. Quality mathematics teaching involves a range of student-centred pedagogies, that is, learning of the student is central. Explicit teaching is in fact one of the ten recommended high impact teaching strategies (DET, 2020) for Victorian teachers to consider in their "repertoire of effective strategies" (p. 4). As a result of both media reports and personal school experiences, classroom mathematics is too often characterised by teaching that tends to be transmissive and procedural (e.g., Stigler & Hiebert, 1999) and this is challenging to change.

Tensions

Tensions are a natural part of boundary crossing and transformative practice (Berry, 2008; Galleguillos & Borba, 2018). According to the Macquarie Dictionary (Butler, 2017), tension is the act or state of being stretched or strained. This includes mental or emotional strain and both anxiety and excitement. The etymology is from the Latin *tensio*, the act of stretching. "The ability to support a tension that can occasionally become almost unbearable is one of the prime requisites for the best sort of scientific research" (Kuhn, 1977, p. 226). Moreover, such tensions are essential and in fact implicit in changing one's thinking and or practice. Tensions are an increasing focus in research and teaching (Anthony et al., 2014; Berry, 2008; Carrejo & Marshall, 2007; Galleguillos & Borba, 2018). The Organisation for Economic Co-operation and Development (OECD) in presenting a vision and principles for learning proposes three categories of what is described as, "'transformative competencies', that together address the growing need for young people to be innovative, responsible and aware" (OECD, 2018, p. 5). These competencies are creating new value, reconciling tensions and dilemmas, and taking responsibility, with the third being "a prerequisite of the other two" (p. 6).

It is interesting to note that some tensions may be discipline related. Park Rogers et al. (2021) in their recent special issue on the *Professional Growth of Science and Mathematics Teacher Educators* note that whilst both mathematics and science teacher educators value collaboration–a well-established "educational strategy for supporting teaching and learning" (p. 5168)–nonetheless there are "disciplinary specific aspects of collaboration" (p. 5167), thus foreshadowing a tension for science teachers, for example, in the course for teachers of mathematics. It is hence important to consider the in-field subjects when thinking about various tensions, noting these include science, physical education and primary school teachers.

Tensions related to mathematics, and mathematical teaching and learning are critical to teacher change. In our work, we aim to transform teachers' mathematical knowledge for teaching along the continuum from out-of-field towards in-field. The extent of transformation along the continuum is impacted by myriad factors. These include motivation (e.g., Lünne et al. 2021), beliefs (e.g., Beswick, 2017) and the local context of the teacher (for example, the nature of their experience teaching out-of-field, and school-level support available). As teachers shift along the continuum from out-of-field to in-field, and hence more to being 'insiders' in understanding what it means to be a capable and confident teacher of mathematics, tensions arise as the realisation that mathematics is not fixed evolves. A challenge for us as mathematics teacher educators is to ensure that tensions are productive (Christensen et al., 2020; Risen & Goldstein, 2021; Valsiner, 1997), respectfully acknowledged, and viewed as opportunities, not barriers (Lovric, 2017). Ingram et al. (2021, p. 279) describe this approach as more likely to result in "self-initiated change and serve, in turn, to transform the 'change environment' itself" (in this case, teachers of mathematics classrooms).

The extent and impact of out-of-field mathematics teaching in Australia

Varied definitions of out-of-field teachers and teaching are used in different studies and reports, making it challenging to interpret and compare results and claims. Care is also needed when comparing the proportion of out-of-field teachers across countries. Not only do definitions of out-of-field vary, so do the requirements to practice as a teacher.

In Australia, initial secondary teacher education courses accept and prepare candidates to be qualified to teach between one and three disciplines (for example, mathematics and accounting, or English and history). Whether they undertake a Bachelor of Education or Master of Teaching, preservice teachers have completed disciplinary units of study to at least second year university level, and units focused on pedagogy specific to those disciplines, by the time they graduate. This model assumes a healthy supply of candidates to initial teacher education programs, and smooth entry to a career that will involve teaching in the discipline of one's qualification. Families, caregivers and students expect that secondary classroom teachers (teachers of Years 7-12, students approximately 12-17 years of age) are qualified specialists who have been prepared to plan, teach and assess disciplinary learning.

Weldon (2016) drew on the 2013 Staff in Australia's Schools (SiAS) survey to compile a report on out-of-field teaching across Australian secondary schools. In this survey, in-field teaching was defined as "either studied the subject at second year tertiary level or above, *or* [emphasis added] trained in teaching methodology for that subject at tertiary level" (p. 2). We note that the use of 'or' rather than 'and' in this definition decreases the number of all teachers classified as out-of-field. Weldon (2016) reported that around one in four (26%) of all teachers of Year 7-10 in Australia are teaching out-of-field, impacting approximately 16% of classes. The problem is more felt in remote schools (26% versus 14% in metropolitan schools), and schools serving low socioeconomic communities (19% versus 13% in high SES locations). Graduate and early career teachers are more likely to be assigned to teach out-of-field (37%) compared with their more experienced counterparts (>5 years, 25%) (Weldon, 2016).

Out-of-field teaching of mathematics is particularly pronounced and problematic. Being a compulsory subject in Years 7-10, mathematics has a high number of teachers, and around 38% of these are out-of-field (Weldon, 2016, p. 3). Weldon reports that around one in five (21%) mathematics teachers are teaching the subject out-of-field, with one in six (17%) Years 7-10 mathematics classes being taught by an out-of-field teacher. The Australian Council of Educational Research's (ACER) most recent analysis of the 2018 Trends in International Mathematics and Science Study (TIMSS, an international standardised assessment of mathematical knowledge and skills), reported that around one in four (23%) Year 8 students are being taught by out-of-field mathematics teachers, a similar proportion to the 2015 TIMSS (Thomson et al., 2021, p. 47). This proportion is substantially higher than the international figure (10%) (Thomson et al., 2021). An interesting paradox is that the data revealed that over 15% of qualified or in-field teachers of mathematics were not teaching the subject, despite working in schools and presumably being available to do so.

While a lack of formal data collection regarding teacher qualifications and what teachers are teaching in Australia makes it difficult to know the extent and impact of out-of-field mathematics teaching, it is clear the phenomenon can be detrimental to both student learning and teacher well-being, and therefore needs addressing (e.g., Hobbs & Porsche, 2022; Van Overschelde, 2022). We note nuances in our reading of available reports. For example, TIMSS report four categories of teacher qualifications: having a major in both mathematics **and** mathematics education ('expert' teachers); having a major in only one **or** both of these; and having a major in **neither**. In Australia, a qualified secondary mathematics teacher (as described earlier) includes the expert teacher category but could also be a teacher with a major in mathematics education and a minor in mathematics (i.e., a second method area with only a minor in the discipline of mathematics). This means that some of the teachers that TIMSS categorises as having a major in mathematics education but not in mathematics could in fact be qualified mathematics teachers. Weldon (2016) reported early career teachers leaving the profession in record numbers. It is reasonable to suggest some correlation between teacher attrition and the one-in-three chance (Weldon,

2016) of being expected to teach out-of-field in their first two years of teaching.

The TIMSS data shows notable achievement differences between Year 8 students taught by in-field compared with out-of-field mathematics teachers (Thomson et al., 2021). The mean TIMSS score of the 46% of students taught by in-field teachers (expert teachers) was 531. This dropped to 523 when taught by teachers with a major in mathematics but not mathematics education (strong content knowledge and pedagogical knowledge for disciplines other than mathematics) (14% of students), 511 when taught by teachers with a major in education but not mathematics (17 % of students) and 501 when taught by teachers with neither a major in mathematics nor in mathematics education. These data additionally confirm the benefits for teachers and students when both mathematical content knowledge (MCK) and mathematical pedagogical content knowledge (MPCK) can be drawn upon to inform planning, teaching and assessment decisions. This is consistent with Shulman's (1987) argument that teacher knowledge, specifically MCK and MPCK, and curricula knowledge, enables teachers to take risks, try new strategies, adjust and develop approaches, hence, improving the quality of teaching and learning of mathematics.

In the TIMSS student survey, "40% of Year 8 students reported *high clarity of instruction* in their mathematics lessons. Noticeably, these students had higher achievement in mathematics than those who reported *low clarity of instruction*" (Thomson et al., 2021, p. 59). Clarity of instruction is related to teacher quality and in TIMSS, this information is ascertained by items related to students 'knowing what is expected', and having teachers that are 'good explainers' including when posed questions by students and when asked to re-explain something that has not been understood, providing a variety of ways to help students learn, and linking ideas to prior learning (p. 62). Furthermore, and of notable relevance to our research and course design, for Year 8 students, regular access to digital technology to support learning "during mathematics lessons was associated with significantly higher achievement in mathematics than not having access" (p. 59). Alongside TIMMS reports, sit consecutive OECD Programme for International Student Assessment (PISA, an international standardised assessment of the ability to apply mathematical knowledge and skills to real world contexts) reports showing a decade of decline in 15-year-old students' mathematical literacy and a learning gap between high and low socioeconomic students of up to 3 years (e.g., Thomson et al., 2017).

It is clear out-of-field teaching is detrimental to both the profession and students. Fast forward to 2022, and teacher supply issues have only been exacerbated by the COVID-19 pandemic and the pressures it has produced for schools and teachers. There is a stronger impetus to support out-of-field mathematics teachers in developing their MCK and MPCK so they can provide quality mathematics learning for their students.

Supporting the learning of out-of-field mathematics teachers

Teacher supply issues mean the need for out-of-field teachers seems set to continue (Hobbs & Quinn, 2020). In fact, du Plessis (2016) suggests the phenomenon will always exist. At the same time, teaching mathematics is complex and demanding (e.g., Kazemi et al., 2018). As Hobbs and Quinn (2020) argue, the challenges are greater for out-of-field teachers, who need support to expand their teaching practice. The need for research-informed interventions identifying, recognising and addressing needs of out-of-field teachers can only improve students' educational opportunities.

Du Plessis (2016), used semi-structured interviews to consider "the lifeworld of out-of-field teachers, their real-life experiences in classrooms, and what they need to make a success of their out-of-field" (p. 43). Stress, lack of confidence and concerns regarding their professional identity were identified as issues for out-of-field teachers. Furthermore, a lack of both MCK and MPCK were highly related to these issues (du Plessis, 2015, p. 98). A lack of knowledge and confidence often led to omitting "challenging sections of the curriculum…[reluctance to include] group work…and teaching to the 'middle' group of students" (p. 98), hence neglecting the needs of low and high attaining students.

According to du Plessis (2016) in some environments, out-of-field teaching has positive outcomes, while in others it produces negative concerns. Regardless, studies are showing that it is possible to manage the out-of-field phenomenon in such a way that negative influences on quality education are reduced (du Plessis, 2016). We are of the view that the needs of out-of-field mathematics teachers, whilst bearing some commonalities with the needs of other out-of-field teachers, are in fact distinct. These professional learning needs include access to opportunities to develop MCK and MPCK, access to opportunities to develop vertical knowledge of the mathematics curriculum, and opportunities to apply newly learned approaches in mathematics classrooms (Shulman, 1987). Strategies to reduce the number of out-of-field teachers of mathematics must be paired with strategies to provide genuine and on-going support for those in existing out-of-field teaching arrangements. This dual approach can only improve teacher preparation and quality (du Plessis, 2016), reduce the attrition of teachers from the profession (NEitA & ACE, 2021; Perryman & Calvert, 2020), and enhance student learning opportunities.

Internationally, systems are adopting the approach of funding and supporting the participation of out-of-field teachers in professional learning, as well as formal university courses (e.g., Lünne et al., 2021). A study of 61 out-of-field mathematics teachers in Germany by Lünne et al. (2021) explored the motivation to enrol in a year-long course with certification (i.e., qualification). The course was in the format of one-day professional learning each week for 40 weeks, with a focus "on school-related mathematical content knowledge, and pedagogical content knowledge, including the planning and reflection of lessons in mathematics" (p. 689). We note with interest that Lünne et al. (2021) explain that in Germany, mathematics is viewed as "easy to teach out-of-field" (p. 689). They studied the motivations and experiences of participants from three iterations of the course and identified three categories of applicants for their course. There were: (a) *intrinsically motivated* teachers who reflect on their practice, considered professional learning as an ongoing part of being a teacher and wanted to learn and improve through participation in the course; (b) the *inexperienced, extrinsically motivated*, who are somewhat interested in learning but see this as following along with the professional learning rather than learning through and beyond it (typically struggling to describe what they want or need to learn or improve); and (c) the *extrinsically motivated* teachers, who place value on the qualification or certification, but see little need to improve their practice (p. 703).

Lünne et al. (2021) found there appeared to be a correlation between teachers with only extrinsic motivation and the length of time teaching out-of-field. With regard to experience in out-of-field teaching, all participants in the second category had fewer than three years' experience (7 of 7 participants) and 20 out of 32 in the third category. The lack of stated interest in learning from the professional development apparent in the second and third categories, despite wanting to teach mathematics, can be explained in part by out-of-field teachers not knowing "what they do not know and what they are not able to do" (Bosse, 2017, p. 332, quoted in and translated by Lünne et al., 2021, p. 703). With the caveat that the numbers in the study are small, it appears that for these teachers, recognition that interest in teaching out-of-field was not sufficient and that specific 'knowledge' was required to successfully teach in a subject outside their 'expertise' was yet to be appreciated. However, 25% (9 of 36) of the participants had fewer than 3 years' experience in out-of-field teaching of mathematics and were intrinsically motivated to learn and develop their knowledge and expertise in the out-of-field subject. Hence, time teaching out-of-field is not the only factor in motivating a desire for learning. This might also relate to teacher over-confidence in teaching out-of-field, something that is under-studied and worthy of exploration.

Our methodological bearings

Stemming from the work of Brown (1992) and Collins (1992), educational design research methodology (also termed design-based research methodology, e.g., Anderson & Shattuck, 2012) is recognised as a robust and practical approach to understanding and influencing improvements in teaching and learning. In educational design research, principles from theory and research, together with thoughtful analysis of a problem, shape quite specific and informed design ideas for interventions (Walker, 2006) which are subject to iterative cycles of testing and refinement (Gravemeijer & Cobb, 2006). The aim is to produce "new theories, artefacts, and practices that account for and potentially impact learning and teaching in naturalistic settings" (Barab & Squire, 2004, p. 2). Studies guided by this methodology provide rich descriptions of the contexts in which the work occurred, the challenges of implementation, the development processes involved in creating and administering the interventions, and the design principles that emerged (Anderson & Shattuck, 2012, p. 22). Iterations are not a result of flaws in the original design, rather the researchers are always looking for ways to improve the design (e.g., Anderson & Shattuck, 2012).

The assumptions and principles that guided the course design

The Graduate Certificate of Secondary Mathematics was the intervention being designed and studied. Together, the design team brought decades of experience teaching mathematics in Years 7-12 (the first author), as well as educating preservice and inservice teachers of primary and secondary mathematics (all authors). Substantive knowledge of mathematics education theory and research, together with the wisdom of practice, guided and informed course design.

The Department of Education and Training specified the course be designed for Victorian government teachers teaching Years 7-10 mathematics out-of-field (and be completed within one calendar year). It was therefore assumed that the teachers who enrolled in the course would bring general knowledge of the structure of the Victorian Curriculum (that is, the Mathematics *learning area's interrelated strands and proficiencies,* as well as the requirement that all teachers develop general capabilities including *critical and creative thinking*), together with a working knowledge of at least one year level. We soon discovered that while most teachers enrolled in the course had been teaching mathematics out-of-field for several years, not everyone was, in fact, an out-of-field mathematics teacher. This meant that familiarity with the Victorian Curriculum: Mathematics and access to a mathematics classroom to apply their learning varied.

Constraints beyond our control influenced course design and teachers' learning experiences in ways that were less than ideal. For example, in 2021, wholly online delivery was necessitated by the COVID-19 pandemic. Regardless, the course aimed to develop teachers' knowledge for teaching, including their MCK, their MPCK, and their vertical curricular knowledge (Shulman, 1987). The 2021 design format consisted of 20 intensive days of class. Interactive learning via keynote addresses exploring mathematics curriculum and concepts, and workshops where students worked together on learning activities involving "low floor and high ceiling" mathematical tasks (Andersen et al., 2018, p. 3). Such tasks have many entry and exit points and different ways of solving are valued. Thus they are more likely to result in teachers experiencing success and engagement. Collaborative group work allowed for differentiated learning to occur, depending on prior mathematical knowledge and skills and the direction the group wanted to take the learning opportunity. A single task could be used to develop or deepen existing knowledge, regardless of teachers' access points.

Teachers were introduced to conceptual frameworks for teaching and learning mathematics by Smith and Stein (2018) and Wood (2002) in support of the development of knowledge. Smith and Stein's (2018) practises for orchestrating productive mathematical discussions include a series of questions to consider at various stages of planning and implementation of learning activities. Wood's ideas relate to fostering the development of mathematical thinking by changing the expectations of teachers and students in class discussions. Consider the task shown in Figure 13.1.

Comparing a pair of fractions Which is greater $\frac{7}{8}$ or $\frac{4}{5}$? Convince me.

Figure 13.1: Task to promote and elicit fraction thinking

Teachers thinking about this could focus on the mathematical knowledge required to solve the problem. They would solve the task themselves and discuss alternative approaches and thinking involved. They might consider flexibility of thinking between representations of fractions and decimal fractions. Common misunderstandings (i.e., the numbers are equal as both are 'one' away from 1) might be a focus or the location of the task in the curriculum. Teachers might consider related content in the curriculum or how this task might be used in a particular learning sequence (e.g., expected prior knowledge, future development and use of concepts). Teaching and learning approaches to support development of student understanding (e.g., use of linear models and partitioning) could be discussed. Ways to challenge incorrect thinking (e.g., *Compare* $\frac{1}{2}$ *and* $\frac{9}{10}$. *Are these two numbers equivalent? Convince me*) could also be a focus of teachers' intended discussion of the task. Note for this second pair, both might still be perceived as 'one' away from one but familiarity with one half may challenge this notion.

Beyond the virtual university classroom, additional time was to be dedicated to preliminary readings exploring research insights, classroom observation and experimentation, and assessment processes. Each of the four units included two assessment tasks, which were to be completed to a pass standard in order to satisfy the course requirements. This chapter explores teaching and learning associated with one of the first assessment tasks, which required teachers to complete the *Fractions and Decimals Online Interview* (FDOI) with two of their students. As expected, this generated discomfort for teachers and philosophical and practical challenges for us as academics and mathematics teacher educators working to disrupt the status quo and improve the conditions for secondary school students to learn mathematics.

Rationale for the inclusion of the Fractions and Decimals Online Interview

Fractions, including decimal fractions are well researched and known to be 'hard to learn and hard to teach'. We were mindful of the potential for a focus on rational number to contribute positively to the Victorian government's key target in STEM education, that by 2025 "for Year 9, 25 per cent more students will be reaching the highest levels of achievement in mathematics" (DET, 2016, p. 5). Rational number underpins higher mathematical understandings and important aspects of everyday life such as health-related and financial decisions.

The FDOI, developed by Clarke and colleagues (e.g., Clarke & Roche, 2009; Clarke et al., 2008, 2011) is a powerful tool for informing teachers on planning, teaching and assessing mathematics learning related to the study of rational number. The FDOI aligns with Years 5-8 of the Victorian Curriculum: Mathematics and can be used with students in Years 4-10 (ages 8-15). Initially implemented as a paper-based interview, in 2008 the interview was developed and made available to Victorian government schools via the DET website as part of the assessment information. DET also provides a mapping of each question in the FDOI to the overarching important mathematical ideas. A second, and older, clinical interview developed in the *Early Years Numeracy Research Project* (e.g., Clarke & Cheeseman, 2000) and now known as the Mathematics Online Interview (MOI) has been used by F-4 teachers in Victorian schools (and beyond) since 2001. While use of both the MOI and the FDOI are encouraged by DET to support mathematics teaching and learning, the uptake of the FDOI remains low. According to the Victorian Curriculum and Assessment Authority (VCAA, personal communication, 29 September 2021), between February and September 2021, approximately 140,000 students in government schools were assessed using the MOI but fewer than 1,000 using the FDOI. This speaks to the differences in assessment practices over the course of students' schooling. In the early years of primary school, annual assessment via the MOI is typical, whereas the FDOI is less likely to be used in the middle years of schooling.

Yet, research by Clarke et al. (e.g., 2005, 2008, 2011) shows that implementation and analysis of clinical interviews broadens teachers' knowledge as to differences in mathematics learning, developing, and improving both content knowledge and pedagogical content knowledge (Clarke et al., 2011). Studies have also demonstrated that interacting with students during the interview–where the teacher probes student thinking, but no teaching occurs–provides teachers with more sophisticated insights into individual students' thinking within diverse classroom environments, insights that can be used to encourage greater communication of reasoning within communities of mathematical inquiry (Clarke et al., 2011). Clarke and Roche (2009), "strongly support the provision of opportunities for teachers to interview at least a small number of their students" (p. 137). This evidence-based argument informed the decision to include teaching, learning and assessment related to the use of the FDOI within the course. Exposing teachers to quality educational research, and preparing them to engage critically with available evidence, are essential aspects of university courses. Teaching and learning that supports teachers to consider the implications of this knowledge base for the improvement of instructional practice within their settings has the potential to generate ongoing change.

Further, as the use of the FDOI was already licensed to and encouraged by DET, we saw its inclusion as having the potential to provide rich opportunities to build teacher knowledge of rational number and how to teach it, thus building their confidence to drive real improvements in their classrooms for this hard to teach and hard to learn mathematical content. Through its use, teachers gain not only deep knowledge of rational number, but also opportunities to learn about common misunderstandings or partial understandings and difficulties. Understanding approaches taken by students, especially benchmarking and residual thinking, provides a basis for teaching approaches "which may yield the kind of connected understanding promoted in various curriculum documents and required for the development of proportional reasoning in later years" (Clarke & Roche, 2009, p. 127).

We also expected that using the FDOI would generate an awareness for teachers of important pedagogical approaches in teaching this content strand of the mathematics curriculum including an enhanced awareness of the need to use mathematics terminologies and appropriate language, and opportunities to use multiple representations to develop procedural and conceptual understandings. We were confident that teachers' MCK and MPCK could be developed through use of the FDOI, and this would open conversations as to how to more effectively address learning of this content. In turn, more nuanced understandings of the Victorian Curriculum: Mathematics could be generated.

Emerging tensions for teachers and teacher educators

Almost immediately, discomfort and tensions related to the FDOI emerged for teachers and us. This resonates with Berry's (2008) notion of "tensions of practice" (p. 20). She described the fine balance between 'safety and challenge' when "engaging prospective teachers in forms of pedagogy that are intended to challenge ideas about teaching and learning, and pushing prospective teachers so far beyond their comfort zone that productive learning cannot occur" (p. 28). Tensions identified by Berry (2008) were: telling and growth, confidence and uncertainty, safety and challenge, planning and being responsive, action and intent, acknowledging and building on experience, and valuing and reconstructing experience. Being aware of discomforts and tensions, and having the maturity to hold in apposition is important to learning. Below, we describe the complex interplay between teachers and teacher educators that resulted from the use of the FDOI within the GCSM course.

Some teachers, accustomed to a culture of 'quick testing' in secondary mathematics, questioned the requirement to undertake the FDOI with two of their students. They identified that the task would be challenging and time-consuming and felt we as teacher educators lacked an understanding of the pressures associated with modern teaching. Having been indoctrinated in mathematics assessment that places emphasis on the correctness of answers, some teachers needed to be convinced that alternative approaches to assessment could be more informative as to the reasons why students make errors. Limited access to mathematics classrooms for some teachers to implement FDOI presented an unexpected practical challenge for the course team. The course team assumed that all teachers in the course were teaching mathematics out-of-field and might have underestimated the hurdles for teachers to implement the assessment task. We need to navigate our own expectations to match the reality and the practical constraints faced by teachers.

Related to this, another issue was that teachers requested a copy of the FDOI solutions. To us, this revealed a lack of confidence and perhaps anxiety as to the mathematical knowledge and skills associated with the FDOI. We felt that providing the solutions would alleviate the need for teachers to grapple with the cognitive demands of the task. The requirement to not only report their students' responses but also analyse what these revealed about students' mathematical knowledge and skills was another source of discomfort and tension for some teachers, as they came to terms with gaps in their own knowledge of fractions, decimal fractions, and percentages. So, while teachers' complaints and requests inevitably generated discomfort and tensions for us as course designers and teacher educators, we wanted to foster deep engagement with students' assessment data to inform teaching. We were adamant that the investment in time required and persisting through any struggles associated with undertaking the items in preparation, and analysing student responses, would be productive for teachers' learning.

Despite the practical challenges and the need to be resilient to pressures to adapt our expectations, we observed many teachers embrace the FDOI and report that the assessment experience added to their learning. These teachers tended to belong to more supportive workplaces, where school leaders had provided an appropriate workload allocation (e.g., 0.8 teaching plus 0.2 time release for study) and access to a mathematics class where teachers could experiment and apply their learning. They also tended to produce assessment submissions that revealed a growing appreciation of the need to devote more time to develop conceptual understanding of fractions, and to emphasise the use of multiple representations and appropriate mathematical language in support of this learning.

Concluding remarks

This chapter has considered the ways and means by which our research-informed course design produced discomfort and tensions related to mathematics, and mathematical teaching and learning. We have argued that discomfort and tensions are critical to reform initiatives and teacher change. The inclusion of *Fractions and Decimals Online Interview* was effective in challenging a common approach in mathematics assessment where the focus is placed more on evaluating the correctness of the answers. This accentuates the gap between the design of the course and the complexity of shifting mathematics teaching culture in secondary schools. In addition, the FDOI challenged and supported deep mathematical understanding by teachers and in turn their future students. Use of the FDOI, in the GCSM provided insights into both the teachers and their own students' struggles, misunderstandings, and innovative and flexible approaches

Returning to Lovric's (2017) ideas related to tensions, as teachers learn more mathematics, they in fact realise that mathematics is not as precise as initially thought. We have observed a shift in teachers' growing awareness of the need to focus on the process of learning mathematics and understanding students' strategies. Yet, a question remains on the impact of this course design intervention to improve the conditions and quality of mathematics teaching and learning in secondary schools.

Ensuring tensions experienced by teachers is a productive learning struggle puts the course team under pressure to be agile in responding and minimising the complaints about the administrative and cognitive loads required in the GCSM. Finding a fine balance between allowing time for teachers to experience discomfort in learning challenging topics and being responsive to feedback from teachers creates significant philosophical and practical challenges for us. The course design team grappled with our own tensions and discomfort while trying to be responsive and agile to the needs of the teachers, ensuring quality delivery of the course, and meeting expectations of external funding bodies. Reforming a mathematics teaching culture requires a long-term investment and supportive workplaces, which is in conflict with the pressure of tight timelines in competitive tenders. This raises a methodological question about the timescale of iterative design research as to what constitutes a cycle in a design research context.

References

Anderson, R. K., Boaler, J., & Dieckmann, J. A. (2018). Achieving elusive teacher change through challenging myths about learning: A blended approach. *Education Sciences, 8*(98), 1-33. https://doi.org/10.3390/educsci8030098

Anderson, T. & Shattuck, J. (2012). Design-based research: A decade of progress in education research? *Educational Researcher, 41*(1), 16-25.

Anthony, G., Hunter, R., & Thompson, Z. (2014). Expansive learning: Lessons from one teacher's learning journey. *ZDM Mathematics Education, 46*(2), 279-291. https://doi.org/10.1007/s11858-013-0553-z.

Australian Research Council. (2015). *About us*. Retrieved from http://www.arc.gov.au/welcome-australian-research-council-website

Barab, S. & Squire, K. (2004). Design-based research: Putting a stake in the ground. *Journal of the Learning Sciences, 13*(1), 1-14.

Berry A. (2008). *Tensions in teaching about teaching: Understanding practice as a teacher educator*. Springer.

Beswick, K. (2017). Raising attainment: What might we learn from teachers' beliefs about their best and worst mathematics students? In C. Andrà, D. Brunetto, E. Levenson, & P. Liljedahl (Eds.), *Teaching and learning in math classrooms. Emerging themes in affect-related research: Teachers' beliefs, students' engagement and social interaction* (pp. 95-106). Springer.

Bosse, M. & Törner G. (2015). Teacher Identity as a theoretical framework for researching out-of-field teaching mathematics teachers. In C. Bernack-Schüler, R. Erens, T. Leuders, & A. Eichler (Eds.), *Views and beliefs in mathematics education.* Wiesbaden, Germany: Springer Spektrum. https://doi.10.1007/978-3-658-09614-4_1

Brown, A. L. (1992). Design experiments: Theoretical and methodological challenges in creating complex interventions in classroom settings. *Journal of The Learning Sciences, 2*(2), 141-178.

Butler, S. (Ed.). (2017). *Macquarie Dictionary* (7th ed.). Macquarie Dictionary Publishers.

Carrejo, D. J. & Marshall, J. (2007). What is mathematical modelling? Exploring prospective teachers' use of experiments to connect mathematics to the study of motion. *Mathematics Education Research Journal, 19*(1), 45-76.

Christensen, J., Thönnessen, J., & Weber, B. (2020). Knowledge creation in reflective teaching and shared values in social education: A design for an international classroom. *Educatia 21 Journal, 19*, Article 02.

Clarke, D. & Cheeseman, J. (2000). Some insights from the first year of the early numeracy research project. Improving numeracy learning conference proceedings (pp. 6-10). ACER. Retrieved from https://research.acer.edu.au/research_conference_2000/2

Clarke, D. & Roche, A. (2009). Students' fraction comparison strategies as a window into robust understanding and possible pointers for instruction. *Educational Studies in Mathematics, 72*(1), 127-138.

Clarke, D., Mitchell, A., & Roche, A. (2005). Student one-to-one assessment interviews in mathematics: A powerful tool for teachers. In J. Mousley, L. Bragg, & C. Campbell (Eds.), *Mathematics: Celebrating achievement* (pp. 66-80). MAV.

Clarke, D., Roche, A., & Mitchell, A. (2008). Ten practical, research-based tips for making fractions come alive (and make sense) in the middle years. *Mathematics Teaching in the Middle School, 13(7*), 373-380.

Clarke, D., Roche, A., & Mitchell, A. (2011). One-to-one student interviews provide powerful insights and clear focus for the teaching of fractions in the middle years. In J. Way, & J. Bobis (Eds.), *Fractions: Teaching for understanding* (pp. 23-32). AAMT.

Collins, A. (1992). Toward a design science of education. In E. Scanlon & T. O'Shea (Eds.), *New directions in educational technology* (pp. 15-22). Springer-Verlag.

Department of Education and Training. (DET) (2016). *STEM in the Education State*. DET.

Department of Education and Training. (DET) (2020). *High Impact Teaching Strategies*. DET.

du Plessis, A. E. (2015). Effective education: Conceptualising the meaning of out-of-field teaching practices for teachers, teacher quality and school leaders. *International Journal of Educational Research, 72*(Suppl C), 89-102.

du Plessis, A. E. (2016). Leading teachers through the storm: Looking beyond the numbers and turning the implications of out-of-field teaching practices into positive challenges. *International Journal of Educational Research, 79*, 42-51.

Galleguillos, J. & Borba, M. C. (2018). Expansive movements in the development of mathematical modelling: Analysis from an activity theory perspective. *ZDM Mathematics Education 50*, 129-142. https://doi.org/10.1007/s11858-017-0903-3

Gravemeijer, K. & Cobb, P. (2006). Design research from the learning design perspective. In J. van den Akker, K. Gravemeijer, S. McKenney, & N. Nieveen (Eds.), *Educational design research* (pp. 17-51). Routledge.

Hobbs, L & Porsche R. (Eds.) (2022). *Out-of-field teaching across teaching disciplines and contexts*. Springer.

Hobbs, L. & Quinn, F. (2020). Out-of-field teachers as learners: Influences on teacher perceived capacity and enjoyment over time. *European Journal of Teacher Education, 44*(5), 627-651.
DOI: 10.1080/02619768.2020.1806230

Ingram, J., Burn, K., Fiddaman, J., Penfold, C., & Tope, C. (2021). The Influence of and interactions between different contexts in the learning and development of mathematics teacher educators. In M. Goos & K. Beswick (Eds.), *The learning and development of mathematics teacher educators: International perspectives and challenges* (pp. 263-280). Springer.

Kazemi, E., Gibbons, L.., Lewis, R., Fox, A., Hintz, A., Kelley-Petersen, M., & Balf, R. (2018). Math labs: Teachers, teacher educators, and school leaders learning together from their own students. *Journal of Mathematics Education Leadership, 19*(1), 23-36.

Kuhn, T. (1977). *The essential tension.* The University of Chicago Press.

Lovric, M. (2017). Tensions between mathematics and science disciplines: Creative opportunities to enrich teaching mathematics and science. In *Discussions on university science teaching, Proceedings of the Western Conference on Science Education*, 1(1). Retrieved from https://ojs.lib.uwo.ca/index.php/wcsedust/article/ view/3786

Lu, D. (13 Feb, 2022). Cracking the formula: How should Australia be teaching maths under the national curriculum? *The Guardian Australia.* Retrieved from https://www.theguardian.com/australia-news/2022/feb/13/cracking-the-formula-how-should-australia-be-teaching-maths-under-the-national-curriculum?s=09

Lünne, S., Schnell, S., & Biehler, R. (2021). Motivation of out-of-field teachers for participating in professional development courses in mathematics, *European Journal of Teacher Education, 44*(5), 688-705. DOI: 10.1080/02619768.2020.1793950

NEiTA & ACE. (2021). *NEiTA-ACE Teachers Report Card 2021.* Retrieved from https://www.austcolled.com.au/neita-ace-teachers-report-card/

Organisation for Economic Co-operation and Development. (OECD). (2018). *The future of education and skills Education 2030: The future we want.* OECD.

Park Rogers, M., Berry, A., Krainer, K., & Even, R. (2021). Finding common ground: A synthesis of science and mathematics teacher educators' experiences with professional growth. *International Journal of Science and Mathematics Education 19*, 167-S180. https://doi.org/10.1007/s10763-021-10188-9

Perryman, J. & Calvert, G. (2020). What motivates people to teach, and why do they leave? Accountability, performativity and teacher retention. *British Journal of Educational Studies, 68*(1), 3-23. DOI: 10.1080/00071005.2019.1589417

Price, A., Vale, C., Porsch, R., Rahayu, E., Faulkner, F., Ní Ríordáin, M., Crisan, C., & Luft, J.A. (2019). Teaching out-of-field internationally. In L. Hobbs, & G. Törner (Eds.), *Examining the phenomenon of "teaching out-of-field"* (pp. 53-83). Springer. https://doi-org/10.1007/978-981-13-3366-8_3

Risien, J. & Goldstein, B. E. (2021). Boundaries crossed and boundaries made: The productive tension between learning and influence in transformative networks. *Minerva, 59*, 539-563. https://doi.org/10.1007/s11024-021-09442-9

Shulman, L. (1987). Knowledge and teaching: Foundations of the new reform. *Harvard Educational Review, 57*(1), 1-22.

Smith, M. S. & Stein, M. K. (2018). *Five practices for orchestrating productive mathematics discussions* (2nd ed.). NCTM.

Stigler, J. W. & Hiebert, J. (1999). *The teaching gap: Best ideas from the world's teachers for improving in the classroom.* The Free Press.

Swan, M. (2006). *Collaborative learning in mathematics: A challenge to our beliefs and practices*. NRDC & NAICE.

Thomson, S., De Bortoli, L., & Underwood, C. (2017). *PISA 2015: Reporting Australia's results*. ACER. Retrieved from https://research.acer.edu.au/ozpisa/22

Thomson, S., Wernert, N., Buckley, S., Rodrigues, S., O'Grady, E., & Schmid, M. (2021). *TIMSS 2019 Australia. Volume II: School and classroom contexts for learning*. ACER. https://doi.org/10.37517/978-1-74286-615-4

Tytler, R., Hobbs, L., Brown, J., White, P., Campbell, C., Cripps Clark, J., Delaney, S., Herbert, S., Peters, A., Sawatzki, C., & Widjaja, W. (2021). Theory and practice relations in design-based research. In P. White, R. Tytler, J. Ferguson, & J. Cripps Clark (Eds.), *Methodological approaches to STEM education research* (Vol. 2., pp. 40-65). Cambridge Scholars.

Valsiner, J. (1997). *Culture and the development of children's action: A theory of human development* (2nd ed.). John Wiley.

Van Overschelde, J. P. (2022). Value-lost: The hidden cost of teacher misassignment. In: Hobbs, L., Porsch, R. (Eds.), *Out-of-field teaching across teaching disciplines and contexts*. Springer.

Walker, D. (2006). Toward productive design studies. In J. van den Akker, K. Gravemeijer, S. McKenney & N. Nieveen (Eds.), *Educational design resea*rch (pp. 8-13). Taylor & Francis.

Weldon, P. (2016). *Out-of-field teaching in secondary schools.* (Policy Insight 6). ACER.

Wood, T. (2002). What does it mean to teach mathematics differently? In B. Barton, K. C. Irwin, M. Pfannkuch & M. Thomas (Eds.), *Mathematics education in the South Pacific* (Proceedings of the 25th annual conference of MERGA, Vol. 1, pp. 61-67). MERGA.

Jill P. Brown (BSci(Ed), GC(Ma&MaEd), MEd, PhD, GCHE) - Deakin University, School of Education, 221 Burwood Highway, Burwood, 3125, Victoria, Australia. jill.brown@deakin.edu.au
Jill Brown is a senior lecturer in Mathematics Education at Deakin University. Following over two decades teaching secondary mathematics, she has now been involved in primary and secondary mathematics teacher education for almost as long. Her research has been published in prestigious journals and book chapters. She is interested in mathematical thinking including mathematical modelling and reasoning across all levels of schooling and content areas. She is also interested in the teaching and

learning of functions with a focus on multiple representations and digital technology use in secondary mathematics.
https://orcid.org/0000-0002-9202-3634

Carly Sawatzki (BA/BBus(Mgt), GradDipEd(Sec), MEd, PhD) - Deakin University, School of Education, 221 Burwood Highway, Burwood, 3125, Victoria, Australia. carly.sawatzki@deakin.edu.au
Carly Sawatzki is interested in how young people become financially capable within families, communities and schools. She has more than 15 years of experience in primary and secondary school teaching and pre-service teacher education across mathematics and humanities curriculum and pedagogy. Carly has published in prestigious international journals and led curriculum and research consultancies for Australian government agencies. She is regularly engaged by teacher associations to present professional learning and provides expert commentary for The Conversation, EduResearch Matters, EducationHQ and ABC Radio.
https://orcid.org/0000-0002-1463-8844

Wanty Widjaja (BEd (Hon), MEd, PhD, GCHE) - Deakin University, School of Education, 221 Burwood Highway, Burwood, 3125, Victoria, Australia. w.widjaja@deakin.edu.au
Wanty Widjaja is an Associate Professor in Mathematics Education at Deakin University. Her research activity aims to transform pedagogical approaches in mathematics classrooms as evident in the major foci: Lesson Study, mathematical reasoning and problem solving, and interdisciplinary approaches in teaching mathematics and science in primary and secondary schools.
https://orcid.org/0000-0002-7288-6088

CHAPTER 14

ANALYSING STUDENT GRAPHING IN MATHEMATICS AND SCIENCE: THE METHODOLOGICAL UTILITY OF A SEMIOTIC FRAMEWORK

JOHN CRIPPS CLARK
AND JOSEPH PAUL FERGUSON

Abstract

In STEM education and research the analysis of graphs arguably constitutes more of an art than a science due to the lack of a sound theoretical framework. Teachers of Mathematics and Science strive to better support their students to interpret and use graphs; to both create their own graphs and critique those of others. Science and Mathematics education researchers often lack a principled theoretical framework to critically analyse the nature and effectiveness of student-generated graphs as part of classroom teaching practice. Existing semiotic accounts have done much to expand our theoretical tools to explore student graphing. However, such approaches currently lack both the rigour and clarity required of a semiotic framework that offers a methodology to enable more meaningful and detailed exploration of student graphing. In this chapter, we outline the key aspects required of such a semiotic framework and propose that existing efforts to apply Charles Peirce's semiotic methodology to student graphing in Mathematics and Science need to be extended through a process of considered operationalisation. Doing so will open new insights into the nature of student graphing and enable teachers to teach graphing in ways that can better induct their students into the meaning-making practices of these disciplines.

Keywords: Graphing, mathematics, science, semiotic, Peirce

The world through graphs

As active citizens, we encounter and are required to both interpret and, less frequently, construct graphs. Currently, climate change and, more immediately, the COVID pandemic have highlighted the importance of graphical literacy, as we are bombarded with graphs (and the numbers they are based on) of infection, vaccinations, and deaths (Spiegelhalter & Masters, 2021). Graphing is recognised as a key competency in Science and Mathematics classrooms as reflected in curricula (see, for example, the latest version of the Australian National Curriculum from ACARA, 2021). Teachers of Mathematics and Science understand that graphing is important for their students. However, graphing often plays out in classrooms, in particular in Mathematics and Science, as a visual representational form that is detached from the phenomena of interest; existing as a 'thing-in-itself' or 'end-in-itself' as opposed to a 'means to an end' (i.e., to make and evaluate claims about particular phenomena) (Beichner, 1994; Leinhardt et al., 1990; McDermott et al., 1987; May, 2017). We don't mean 'thing-in-itself' in a Kantian sense but rather in the sense of treating graphs as isolated objects of study, as against developing an understanding of the relationships between the phenomenon and the distinctive components of the graph, through which meaning and use are established. This 'limited' realisation of graphing is not restricted to students and teachers, it also extends to education researchers exploring this practice in Science and Mathematics classrooms. There is a lack of appreciation for the importance of, and a lack of understanding of, the nature of graphing in Science and Mathematics among students, teachers, and researchers. But more importantly, for the purposes of this chapter, there is no satisfactory approach to evaluating and analysing students' graphing as a 'means to an end' (and not as a 'thing-in-itself'/'end-in-itself'). This is the methodological issue that we address in this chapter, as we articulate the problem and point to a solution using the semiotics of Charles Peirce that positions graphs as diagrams full of iconic potential. In doing so, we follow Leinhardt et al. (1990) in determining that investigations of graphs and graphing in Mathematics education, and we suggest also Science education, can be undertaken from three perspectives: tasks and the presentation of these tasks; the student and the development of their understanding; and the teacher and the nature of their instruction. We explore our methodological issue from all three perspectives.

Graphs in the world of teaching Science and Mathematics

Since, with Galileo, Science became profoundly intertwined with Mathematics, graphs have developed as powerful visual-linguistic tools to represent and interpret phenomena, and scientists and mathematicians have developed sophisticated conventions in the use and communication of graphs (Collins, 1993). Given how ubiquitous graphs are in Science and Mathematics, it is hardly surprising that they have become a prominent element of Science and Mathematics teaching. It is notable how ubiquitous graphing is in the Australian Curriculum (ACARA, 2021). It is most prominent in Mathematics, then Science, followed by Technologies. However, graphs and graphing occur in less obvious places such as the General Capabilities, including Literacy: "Students interpret still and moving images, graphs, … explain how analytical images such as figures, diagrams, tables, maps and graphs contribute to understanding of factual information." Even Chinese as a language to learn: "Locate, classify and organise relevant information, including analysing data in simple diagrams, tables and graphs, and re-present this information for known audiences". However, our focus in this chapter is Science and Mathematics, for which graphing forms a core competency of each discipline in distinct but closely aligned ways.

Research to date in Mathematics education (e.g., Ärlebäck & Doerr, 2018; de Beer et al., 2015; Leinhardt et al., 1990; Mulligan, 2015; Silber & Cai, 2021) and to a lesser extent in Science education (e.g., Glazer, 2011; Ring et al., 2019), often adopting a semiotic framing of some sort, has established key insights into the affordances of meaning making for students in undertaking graphing but also the significant challenges faced by students and their teachers to develop this skill. Students in Mathematics and Science not only find graphing and the interpretation of graphs challenging, but even when successful, they usually fail to translate and integrate this information with their existing knowledge and skills or link it to other representations. Although mastering the conventions of graphing is an important first step, students often struggle to be able to make the necessary judgement of the value of particular graphs to serve particular purposes or solve specific problems at hand ('means to an end'), even if they can enact the conventions of graphing ('thing in itself').

By often teaching graphs and graphing as a 'thing it itself' we focus on the features of the graph rather than their purpose (Beichner, 1994; McDermott et al., 1987). We establish, for example, rules for axes and labelling that, rather than being understood as communicative conventions, become the learning outcomes. Reading the graph becomes the objective (at 1 second the car is moving at 5 m/s) rather than a means of better interpreting and predicting the phenomenon that the graph is representing (the car is steadily accelerating until it reaches a steady speed and I predict that any cars that roll down a slope will gradually get faster until they reach their terminal velocity). Although the conventions of graphing are often discipline specific (as are the phenomena represented), the process of representation and interpretation crosses these boundaries and depends on the development of logical, critical, and creative thinking–key 21st Century skills (Tytler et al., 2020).

If we are to address these challenges of teaching the practices and capabilities of graphing we need a strong, clear, and comprehensive theoretical grounding for how we research graphs and graphing (which informs our methodology). Attending to this challenge necessitates developing a way for researchers to analyse students' graphs and graphing in a reliable and rigorous way (to honour graphing as a core practice) which grasps the structure and function of graphs. This attends to Peirce's famous pragmatic maxim: "Consider what effects, which might conceivably have practical bearings, we conceive the object of our conception to have. Then, our conception of those effects is the whole of our conception of the object" (Peirce, 1878/1992, p. 132). As Peircean scholar Cathy Legg points out (Ralón, 2022, a Pittsburgh School-culminating Peirce section, para.18), this maxim "seeks to understand every concept as a set of hypothetical conditionals pertaining to lived experience." Such an approach positions learning in Mathematics and Science as understanding and appreciating that graphs can be creatively productive and are not just reductively representational, which is foundational for critical visual literacy in these disciplines. Students need to be supported to realise that graphical representations can be created in certain ways so as to further certain agendas, in particular to engage with mathematical and scientific phenomena in meaningful ways. From a mathematical viewpoint, as Leinhardt et al. (1990) observe: "the way an individual interprets a graph often involves some level of algebraic construction in order to do the interpretation, whereas at other times it involves more direct visual comparison" (p. 53). In other words, mathematical and scientific graphs have meaning and indeed only exist because they enable us, which of course includes students, to do practical things (e.g., understand the relationship between distance and time) as lived experience (this is graphing as a 'means to an end'). If we adopt such an approach, we will be in a better position to work with teachers to support them and their students to appreciate graphs and graphing in Mathematics and Science within the context of authentic problems and data.

An example of graphing as representation of motion

The Modelling Motion resource, which was part of the reSolve project (Australian Academy of Sciences, 2022), is an example of an approach to graphing based on modelling a scientific and mathematical phenomenon as experienced by the students. In a series of lessons, Year 5-6 students are challenged with a question, then undertake practical activities focused on motion and force, part of which is the collaborative production of representations of speed, acceleration, and force in one and two dimensions, which students collaboratively interpret. In the first lesson, after a teacher-led class discussion of *what is speed* directed to the interpretation of speed as distance per unit time (as against the equally valid time per unit distance), students are challenged with the question: "Can you walk at a constant speed?" To answer this question, they work in teams to mark the position of one of their number every second (using the beat of a metronome) as they walk along a straight line across the classroom beside a paper streamer (Figure 14.1). The streamer is then divided into sections at the marks of the walker's position at each metronome beat. Each of these strips represents the distance travelled in a second (the speed). The strips are assembled (hopefully in order) to make a graph (Figure 14.2).

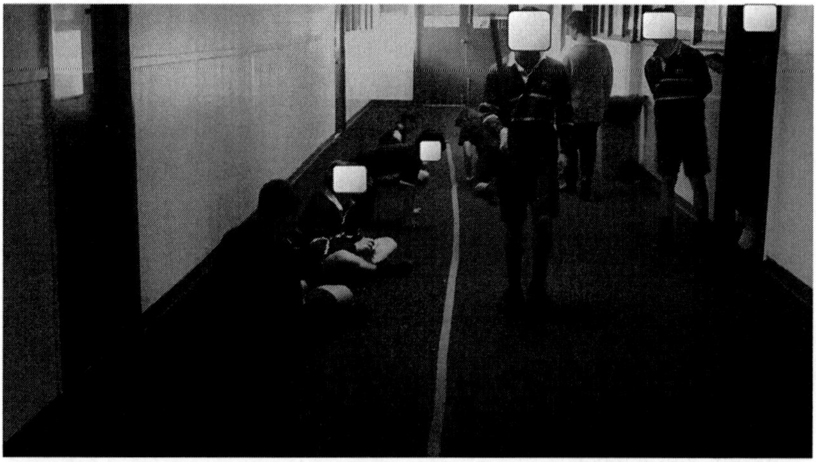

Figure 14.1: Collecting data for a student walking at a constant speed

The whole class analysis of these graphs affords the appreciation of a number of what are often somewhat arbitrarily taught knowledge and skills (Tytler et al., 2019):

- The communicative conventions of graphing: if all groups do not align the strips vertically, from left to right, starting at the x-axis, it is difficult to compare groups. If they do not have clear labelling, it is difficult to discuss;
- The difference between random and systematic errors: if a long strip is next to a very short strip, students will realise that they probably marked the wrong spot. Whereas a more random variation in height can be addressed by students thinking about how they could make their measurements more accurately;
- Students have direct experience of the phenomenon (and the graph is a direct representation of the phenomenon) and they can interpret the graph: shorter strips represent a slower speed, and longer strips represent a faster speed: shorter strips at the left or right hand side of the graph represent speeding up at the start of the walk or slowing down at the end.

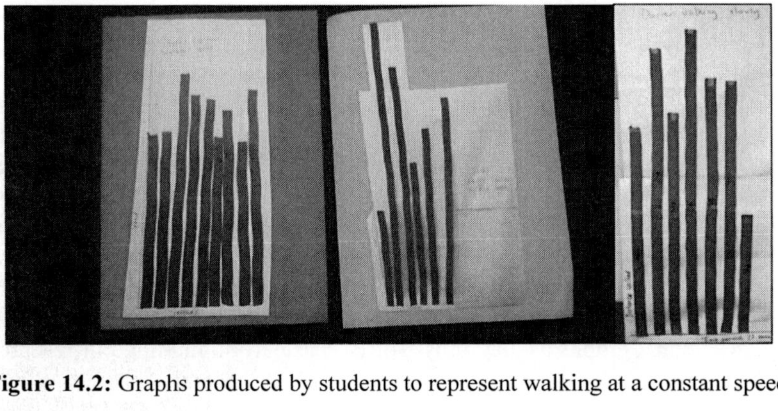

Figure 14.2: Graphs produced by students to represent walking at a constant speed

These strip graphs enable not only the development of the language conventions that enable graphing to become part of students' thinking in Mathematics and Science, but also provide a representation of powerful concepts that students would not otherwise be able to observe (Wainer, 1992), such as instantaneous speed or acceleration (the observation that the strip is longer when walking faster and the application of this representation to graph of speeding up or slowing down, i.e., acceleration). It's worth considering another Modelling Motion task; until students graph the motion of a ball rolling down a slope, they will generally describe the motion of the ball as slow at the top of the slope and fast at the bottom of the slope, with no concept of continuous change.

Peircean fits and starts: Gaps in theory

Much of the research to date exploring student graphing in Mathematics and Science education has been infused with a semiotic framing/focus to varying degrees. Semiotics is, very broadly defined, "the study of signs and sign-using behaviour" (Encyclopedia Britannica, 2020). We say "very broadly defined" because there are myriad ways to define and operationalise semiotics, with all well-regarded approaches rooted in either de Saussure's 'semiology' (a dyadic model of signs) or Peirce's 'semiotics' (a triadic model of signs). Semiotics has a rich history in Mathematics education (Presmeg et al., 2016) and Science education (Lemke, 1987); this is hardly surprising, considering the representationally-diverse and encounter-based nature of teaching and learning in these disciplines. While some researchers are very explicit about the semiotic nature of their undertakings, others' research is more implicit in its acknowledgement of semiotics. While this research has no doubt generated important insights into students' graphing practises, it has not yet explored graphs as a 'means to an end' and not as 'a-thing-in-itself'/'end-in-itself'. In this chapter, we will use Roth et al.'s (2005) work on graphical literacy and Duijzer et al.'s (2019) more recent work in the physics education context (with physics uniting Mathematics and Science) to make clear the methodological dangers of employing semiotic frameworks that are not firmly grounded in their theoretical/philosophical roots. From here we will (see following section) outline what a full embracing of semiotics, in particular Peirce's approach, can methodologically afford in exploring students' graphing in Mathematics and Science, drawing on May's (2017) Peircean contrasting of "graphs as images" with "graphs as diagrams" (p. 107) as part of aligning our research with Stjernfelt's (2007, p. xiv) "diagrammatology" agenda.

Roth et al. (2005) argue that the key challenge facing students in learning graphing specific to Mathematics and Science, and thus a significant challenge for teachers in their teaching of these graphing processes, is overcoming what they term "iconic confusions" (p. 21). Duijzer et al. (2019) make a similar proposal in identifying what they term "iconic reasoning" (p. 904) as detrimental to students progressing their induction into the graphing practises of physics. "Students committing these errors", Roth et al. (2005) propose–with 'error' here clearly referring to iconic meaning making–"take some feature in one domain (sign, referent) and, based on visual similarity, postulate the same visual feature in the other (referent, sign)" (p. 21). All of which amounts to students interpreting and constructing "a graph by its superficial characteristics" (Duijzer et al., 2019, p. 900). Some examples, involving student interpretation and construction of graphs, will make clearer the nature of this so-called error.

In the study by Duijzer et al. (2019), primary school students (ages 9-11) were required to undertake six distinct but related graphing motion tasks (involving the interpretation and construction of graphs) as part of a six-lesson physics sequence. Task 1 required students to interpret a distance (y-axis) and time (x-axis) graph of a motion event (Figure 14.3). They were also presented with pictorial illustrations (see p. 905) of three motion events (expressed in written form here): a) *a plane taking off;* b) *a person waiting at pedestrian lights to cross the road (the light is red);* c) *a person walking up stairs, with a flat landing between the first set of stairs and the second set of stairs*. If students interpreted the graph "on the basis of the shape of the graphical representation or superficial characteristics of the motion event" (p. 907) this was determined to be iconic reasoning. In regard to Task 1c, student iconic reasoning would be as follows: "*The staircase has almost the same shape as the graph*" (p. 907).

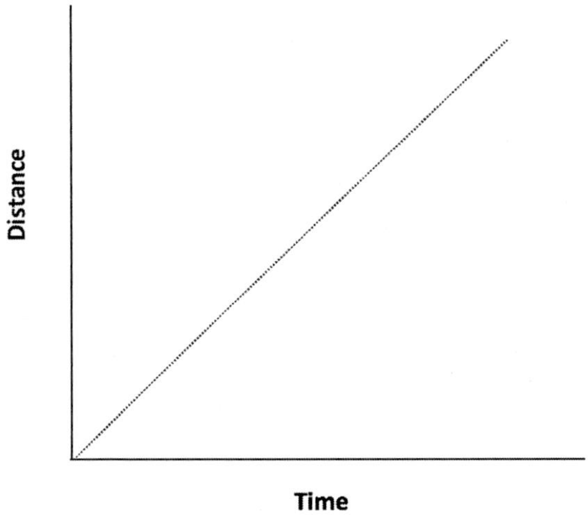

Figure 14.3: Recreation of graph prompt from Duijzer et al. (2019) for Task 1 (p. 905)

Task 4 posed students with the question: "*A person walks slowly from point A to point B and then fast back to point A again. Draw a graph that could fit this situation*" (Duijzer et al., 2019, p. 906). Students were provided with a distance (y-axis) and time (x-axis) template (i.e., grid paper) to complete the task. If students constructed their graph "on the basis of superficial characteristics of the description of the motion event" (p. 907) this was determined to be iconic reasoning (Figure 14.4).

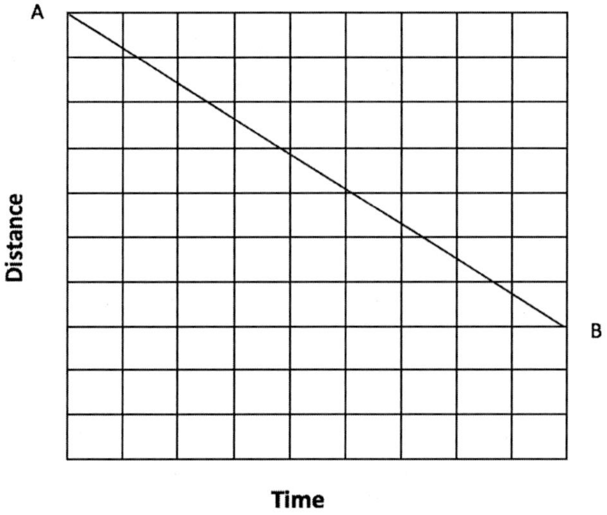

Figure 14.4: Recreation of student sample from Duijzer et al. (2019) for Task 4 (p. 907)

Therefore, according to this notion of iconicity employed by Duijzer et al. (2019) and grounded in Roth et al.'s (2005) reading of Peirce's semiotics, one of the main aims of teaching graphing in Science and Mathematics is to support students "to move beyond iconic interpretations of the graph" (Duijzer et al., 2019, p. 909). In addition to such pedagogical implications, if we use Roth et al.'s (2005) Peircean semiotic framework to analyse student graphing in Mathematics and Science, then iconic reasoning as an aspect of interpreting and constructing graphs is deemed as a degenerate form of reasoning, which has important methodological ramifications for researchers.

An expanded Peircean semiotic account of graphing: Developing theory

We propose, based on our own ongoing engagement with Peirce's work (Cripps Clark & Ferguson, 2021; Ferguson & Cripps Clark, 2021), that an approach such as that of Roth et al. (2005) locks us into a superficial reading of Peirce' semiotics that leads us to reduce Peirce's triadic model of meaning making (object, interpretant, representamen) and sign elements (icon, index, symbol) to a degenerate dyadic form. This is reflected in Roth et al.'s (2005) odd claim–odd from the Peircean perspective–that "the relationship between signifier and signified is always arbitrary" (p. 10), or in other words "the relation between signs and referents are arbitrary" (p. 16). But even more alarming is their suggestion that "the sign-referent relation is arbitrary even in the case of iconic similarity" (p. 20). This claim is fundamentally anti-Peircean in nature! Peirce repeatedly makes clear that it is only the symbol for which the relationship between object and representamen is arbitrary in nature, and he always frames this with reference to the interpretant (Bellucci, 2018). The distinctive character of the *symbol*, *index,* and *icon* is evident in Peirce's definitions of these sign elements:

> There are three kinds of signs. Firstly, there are likenesses, or icons; which serve to convey ideas of the things they represent simply by imitating them. Secondly, there are indications, or indices; which show something about things, on account of their being physically connected with them…Thirdly, there are symbols, or general signs, which have become associated with their meanings by usage. (Peirce, 1894/1998, p. 5).

It is important to keep in mind that Peirce's definition of signs evolved over the course of his life and his accounts varied in complexity (Bellucci, 2018). Here we present his definitions as stated in 'What is a sign?' as this is his most 'straightforward' definition of icon, index, and symbol that best serves our current needs. Bellucci (2018, p. 287) emphasises the distinction further: "…an icon refers to the object in virtue of its own qualities; an index refers to the object in virtue of being connected with it; a symbol refers to the object because it is interpreted as doing so."

If we return to the graphs generated by students to represent walking at a constant speed in Figure 14.2, then we can see examples of iconic, indexical, and symbolic sign making in action as Peirce prescribes. The headings of the x-axis and y-axis are mainly *symbolic* in nature as they constitute words which have agreed upon meaning due to communal and extended usage (i.e., graphing conventions, here showing us the meaning of the vertical and horizontal lines and how to interpret the meaning of the data plotted in relation to these lines). The individual pieces of streamer are mainly *indexical* in nature as they are physically connected with the specific walking events undertaken by the students. The arrangement of the streamers in the sequence of walking is mainly *iconic* in nature as it imitates the action of walking at a constant speed (or trying to walk at a constant speed) to reveal the structure of speed (distance per unit time or time per unit distance). We write 'mainly' here as almost all the signs we create and engage with, including in Mathematics and Science, are made up of different combinations of symbolic, indexical and iconic elements, with each sign being more or less symbolic, indexical, or iconic in nature. To encounter a 'pure' symbol, index, or icon is a rare event in our daily meaning making practices.

It seems that Roth et al. (2005) and those who follow their interpretation and application of Peircean semiotics to exploring graphing in Mathematics and Science education, including Djuizer et al. (2019), have misunderstood what Peirce meant when he said that icons "represent simply by imitating". As May (2017) makes clear, the iconic character of graphs in Mathematics and Science has not to do with the superficial resemblance of the graph with the phenomenon of interest, for example a motion event, but rather "similarity of relations" (p. 114). Therefore graphs resemble the "relational structures" (p. 114) of that which they represent–what we call the 'deep structure'–to such an extent that "experimenting with diagrams amounts to performing operations on the very objects of investigation that the diagram is supposed to represent" (Ambrosio & Campbell, 2017, p. 97). Peirce is very clear about this capacity of icons to afford discoveries:

> ...a great distinguishing property of the icon is that by the direct observation of it other truths concerning its object can be discovered than those which suffice to determine its construction...Given a conventional or other general sign of an object, to deduce any other truth than that which it explicitly signifies, it is necessary, in all cases, to replace that sign by an icon. (CP 2.279)[1]

Paolucci (2017) makes even clearer the error that Roth et al. (2005), and presumably those who follow their lead, make in misreading Peirce's account of iconicity:

---

[1] CP refers to *Collected Papers*. We use the standard protocol: CP x.y = *Collected Papers of Charles Sanders Peirce*, volume *x*, paragraph *y*.

> Iconicity has opposite features to the ones that have been attributed to it for a long time in the semiotic tradition. *It is not the iconic sign to be similar to the object, but it is the object to reveal through the sign new properties that were not involved in the construction of its sign.* The test for the iconicity of a sign rests in whether it is possible to manipulate the sign so that new information as to its object appears. (p. 75, italics in original)

By using an informed and philosophically comprehensive account of Peircean semiotics to frame student graphing in Mathematics and Science, in particular a semiotically accurate definition of the icon, May (2017) shows that the issue is not so much what Roth et al. (2005) call "iconic confusions" (p. 21), but rather a "graph-as-picture misconception" (p. 112). May goes on: "…students may attend to shape-features of graphs as if they were images rather than relational structures (diagrams)" (p. 109) and in so doing treat "graphs-as-images" instead of "graphs-as-diagrams". In this way:

> …the 'graph-as-picture' misconception is not generally based in direct correspondence errors but in the cognitive consequences of a desymbolization of diagrams to images. It is, in other words, a typological error of a semiotic nature. (May, 2017, p. 114)

This alternative framing of iconicity (alternate to the commonly used framing of Roth et al., 2005) and in particular students' struggles to graph in ways appropriate to Mathematics and Science, has important ramifications for teachers' graphing pedagogy and researchers' graphing methodology; iconic reasoning is no longer degenerate but rather the driver of productive and creative reasoning as/through diagrams. It is not a matter of teachers supporting students "to move beyond iconic interpretations of the graph" (Duijzer et al., 2019, p. 909), but rather teachers ought to support students 'to fully inhabit iconic interpretations of the graph'; to move beyond superficial resemblance to deep structure resemblance. Similarly, researchers ought to analyse students' diagrams as iconic in terms of the degree to which they resemble the relational structures of the phenomenon of interest. In this way, if a student interprets and/or constructs a graph in Mathematics or Science based on superficial characteristics of the motion event, this is not a problem because it is iconic reasoning, on the contrary it is a problem because it is not iconic reasoning in the proper Peircean sense.

The response of teachers and researchers should not be to identify these as iconic confusions, but rather as opportunities to support students to engage with graphs as diagrams in the sense of appreciating and understanding the distinctive iconic, indexical, and symbolic character of these representations to afford meaning making. It is typological errors of a semiotic nature that teachers, and we as researchers, ought to address in students' graphing in Mathematics and Science, not direct correspondence errors that ignore the fundamental nature of Peircean iconicity. We of course mean 'error' here in the non-judgemental scientific and mathematical sense of not aligning with accepted disciplinary practices (not in the judgemental sense of students doing the 'wrong' thing).

Towards a semiotics of graphing – Diagrammatology for the classroom

We suggest that there is a methodological opportunity, indeed necessity, to expand the theoretical/philosophical underpinnings of the semiotics employed in research exploring students' graphing in Mathematics and Science. We need to continue developing and implementing more sophisticated understandings of Peirce's semiotics, as exemplified by May (2017). There is also a need for a contribution of more sophisticated semiotic approaches that make accessible to teachers and their students the structure and function of the components of graphs and elements of graphing and their relationships–in other words the logic of graphing–in particular the iconic nature of such diagrammatic reasoning. In doing so, we propose that Roth et al.'s (2005) "critical graphicacy" (p. 1) needs to be reconceptualised as 'critical graphical semioticity' as opposed to 'critical graphical literacy'. In other words, graphing needs to be positioned as first and foremost a semiotic undertaking (i.e., signs) of Peircean triadic nature as opposed to a linguistic undertaking (i.e., literacy, the written and spoken word) of a Sausserian dyadic nature. Dyadic semiotics doesn't allow for these deeper levels of relational similarity because the interpretant is needed for this flexibility of meaning associations. Thus, we strive towards a *semiosis of graphing* for Mathematics and Science education, which will be a form of Stjernfelt's (2007) "diagrammatology" (p. xiv).

However, robust theoretical framings are not sufficient; we need to also ensure that in our research we attend to the products and processes of graphing, for it is through making graphing meaningful that we can motivate and engage students (Daniels, 2012) and it is only through entangling understanding in process and product that students can critically apply graphing in their social situation of development (Davydov, 2008). We must extend May's (2017) work by fully operationalising Peirce's taxonomy of signs for application to the 'realities' of the Science and Mathematics classroom contexts, in order to unpack in semiotic detail the nature of students' graphing. This is no easy task considering the apparent 'gap' between Peirce's idealism–paired as it is with his realism–and the 'messiness' of the classroom context and the data that we as researchers generate when immersed in such contexts. But Stjernfelt (2007, 2014) has shown us that such operationalisation of Peirce's ideas is possible if we engage with Peirce's full philosophy and appreciate and honour the generality of his ideas. Peirce is a realist–"I am myself a scholastic realist of a somewhat extreme stripe" (CP, 5.470)–in that he vehemently opposes nominalism and so believes in the existence of universals, reflected in his notion of Thirdness (e.g., habits, laws). We begin to address this significant methodological challenge in an upcoming paper with Peircean scholar Cathy Legg, who has shown the value of Peirce's ideas, in particular those pertaining to diagrams (Legg, 2017). This will enable us to productively frame our research of teaching practices in graphing. We propose a need for Science and Mathematics education researchers to work more closely and more often with our colleagues/friends in philosophy, particularly with the tradition of pragmatism. In terms of how our research ought to impact teacher practice–an important methodological issue in its own right (Cripps Clark & Ferguson, 2021)–we need to work with teachers so they can notice and support diagrammatic reasoning, such as that involved in graphing, in their students. In the spirit of Peirce the idealist and lover of norms, we must not settle for what graphing currently is in Science and Mathematics education, but *realise it for what it ought to be*.

References

Australian Academy of Sciences. (2022). *Modelling motion*. Retrieved from https://www.resolve.edu.au/modelling-motion

Australian Curriculum, Assessment and Reporting Authority (ACARA). (2021). *Australian Curriculum – Version 8.4*. Australian Curriculum, Assessment and Reporting Authority.

Ärlebäck, J. B. & Doerr, H. M. (2018). Students' interpretations and reasoning about phenomena with negative rates of change throughout a model development sequence. *ZDM Mathematics Education, 50*(1-2), 187-200.

Ambrosio, C. & Campbell, C. (2017). The chemistry of relations: Peirce, perspicuous representations, and experiments with diagrams. In K. A. Hull & R. K. Atkins (Eds.), *Peirce on perception and reasoning: From icons to logic*, (pp. 86-106). New York, NY: Taylor & Francis.

Beichner, R. J. (1994). Testing student interpretation of kinematics graphs. *American Journal of Physics*, *62*(8), 750-762.

Bellucci, F. (2018). Peirce's speculative grammar – Logic as semiotics. Routledge.
Collins, B. B. (1993). Data visualization – Has it all been seen before? In R. A. Earnshaw & D. F. Watson (Eds.), *Animation and scientific visualization: Tools and applications*. Academic Press.
Cripps Clark, J. & Ferguson, J. P. (2021). Researching together in 21st Century Australian universities: A pragmatist accounting of collaborative research in science education. In P. White, R. Tytler, J. Ferguson & J. Cripps Clark (Eds.), *Methodological approaches to STEM education research: Volume 2* (pp. 302-319). Cambridge Scholars Publishing.
Daniels, H. (2012). Changing situations and motives. In M. Hedegaard, A. Edwards, & M. Fleer (Eds.), *Motives in children's development: Cultural-historical approaches* (pp. 109-128). Cambridge University Press.
Davydov, V. V. (2008). *Problems of developmental instruction: A theoretical and experimental psychological study* (P. Moxhay, Trans.). Nova Science Publishers.
Duijzer, C., Van den Heuvel-Panhuizen, M., Veldhuis, M., & Doorman, M. (2019). Supporting primary school students' reasoning about motion graphs through physical experiences. *ZDM Mathematics Education 51*, 899–913.
de Beer, H., Gravemeijer, K., & van Eijck, M. (2015). Discrete and continuous reasoning about change in primary school classrooms. *ZDM Mathematics Education, 47*(6), 981-996.
Encyclopaedia Britannica (2020). "Semiotics". Retrieved from https://www.britannica.com/science/semiotics
Ferguson, J. & Cripps Clark, J. (2021). Where the truth lies: Peirce through the lens of the third man. In N. Barnes & A. Bedford (Eds.), *Unlocking social theory with popular culture: Remixing theoretical influencers* (pp. 33-46). Springer.
Glazer, N. (2011). Challenges with graph interpretation: A review of the literature. *Studies in Science Education, 47*(2), 183-210.
Legg, C. (2017). Diagrammatic teaching: The role of iconic signs in meaningful pedagogy. In I. Semetsky (Ed.), *Edusemiotics – A handbook*, (pp. 29-45). Springer.
Lemke, J. L. (1987). Social semiotics and science education. *The American Journal of Semiotics, 5*(2), 217-232.
Leinhardt, G., Zaslavsky, O., & Stein, M. K. (1990). Functions, graphs, and graphing: Tasks, learning, and teaching. *Review of Educational Research*, *60*(1), 1-64.

McDermott, L. C., Rosenquist, M. L., & Van Zee, E. H. (1987). Student difficulties in connecting graphs and physics: Examples from kinematics. *American Journal of Physics*, *55*(6), 503-513.

May, M. (2017). Graphs as images vs. graphs as diagrams: A Problem at the intersection of semiotics and didactics. In K. A. Hull & R. K. Atkins (Eds.), *Peirce on perception and reasoning: From icons to logic*, (pp. 107-118). Taylor & Francis.

Mulligan, J. (2015). Moving beyond basic numeracy: Data modeling in the early years of schooling. *ZDM Mathematics Education, 47*(4), 653-663.

Presmeg, N., Radford, L., Roth, W-M., & Kadunz, G. (2016). *Semiotics in mathematics education*. Springer.

Peirce, C. S. (1878/1992). How to make our ideas clear. In N. Houser & C. Kloesel (Eds.), *The essential Peirce — Selected philosophical writings (1867– 1893) (Vol. 1)*. Indiana University Press.

Peirce, C. S. (1894/1998). What is a sign? In N. Houser, A. De Tienne, J. R. Eller, C. L. Clark, A. C. Lewis, & D. B. Davis (Eds.), *The essential Peirce—Selected philosophical writings (1893–1913)* (Vol. 2, pp. 4–10). Indiana University Press.

Peirce, C. S. (1931-1935). *Collected Papers*, C. Hartsthorne & P. Weiss (Eds.). (Vol 2). Harvard University Press.

Peirce, C. S. (1931-1935). *Collected Papers*, C. Hartsthorne & P. Weiss (Eds.). (Vol 5). Harvard University Press.

Paolucci, C. (2017). Semiotics, schemata, diagrams, and graphs: A new form of diagrammatic Kantism by Peirce. In K. A. Hull & R. K. Atkins (Eds.), *Peirce on perception and reasoning: From icons to logic* (pp. 74-85). Taylor & Francis.

Ralón, L. (2022, February 23). *Interview with Cathy Legg*. Figure/Ground. Retrieved from http://figureground.org/a-conversation-with-cathy-legg/

Ring, M., Brahm, T., & Randler, C. (2019). Do difficulty levels matter for graphical literacy? A performance assessment study with authentic graphs. *International Journal of Science Education, 41*(13), 1787-1804.

Roth, W. M., Pozzer-Ardenghi, L., & Han, J. H. (2005). *Critical graphicacy - Understanding visual representation practices in school science*. Springer.

Silber, S. & Cai, J. (2021). Exploring underprepared undergraduate students' mathematical problem posing. *ZDM Mathematics Education, 53*(4), 877-889.

Spiegelhalter, D. & Masters, A. (2021). Covid by numbers: Making sense of the pandemic with data. Penguin.

Stjernfelt, F. (2007). *Diagrammatology*. Springer.

Stjernfelt, F. (2014). *Natural propositions: The actuality of Peirce's doctrine of dicisigns.* Springer.

Tytler, R., Williams, G., Hobbs, L., & Anderson, J. (2019). Challenges and opportunities for a STEM interdisciplinary agenda. In B. Doig, J. Williams, D. Swanson, R. Borromeo Ferri & P. Drake (Eds,) *Interdisciplinary mathematics education: The state of the art and beyond* (pp. 51-81). Springer.

Tytler, R., Prain, V., Ferguson, J., & Cripps Clark, J. (2020). Methodological challenges in researching, teaching and learning student capabilities: The "creativity" case. In P. White, R. Tytler, J. Ferguson, & J. Cripps Clark (Eds.), *Methodological approaches to STEM education research: Volume 1* (pp. 1-20). Cambridge Scholars.

Wainer, H. (1992). Understanding graphs and tables. *Educational Researcher*, *21*(1), 14-23.

Acknowledgements

Thank you to Cathy Legg for guiding us on the true Peircean path.

John Cripps Clark (BSc, DipEd, MAppSc, PhD) - Deakin University, School of Education, Faculty of Arts & Education, 221 Burwood Highway, Burwood, VIC 3125, Australia. john.crippsclark@deakin.edu.au
John is a science and science education teacher and theorises and researches STEM education, online learning, cultural-historical theory, visual-linguistic reasoning, the pedagogy of judging the accuracy and reliability of information, school gardens, overseas professional experience, science games, Pragmatism, and film. For the last twenty years, John has run a cultural-historical & activity research reading group with members across the Asia-Pacific region. He is a coordinator for the Australian Association for Research in Education special interest group: Cultural-Historical & Activity Theory and has organised four Asia-Pacific ISCAR-AARE Summer Schools.
https://orcid.org/0000-0002-0335-6964

Joseph Paul Ferguson (BSc, BA(Hons), MTeach, PhD) - Deakin University, School of Education, Faculty of Arts & Education, 221 Burwood Highway, Burwood, VIC 3125, Australia.
joe.ferguson@deakin.edu.au
Joseph is an educational researcher and teacher educator interested in investigating the various manifestations of reasoning in the science classroom, in particular creative reasoning. Through an exploration of the philosophies and theories that underpin inquiry practices and the associated methodologies that enable this research, Joseph seeks to add to efforts to clarify what it means for teachers and students to enact inquiry in the classroom. Joseph is particularly interested in the affordances of video-based methodologies, informed by film theory and philosophy, to maximise the potential of such research to support teachers and students in their teaching and learning of science.
https://orcid.org/0000-0002-0971-3256

CHAPTER 15

COSMIC AESTHETICS: HARMONISING VARYING THEORETICAL PERSPECTIVES IN A THESIS BY PUBLICATION

SAEED SALIMPOUR, RUSSELL TYTLER, MICHAEL FITZGERALD AND URBAN ERIKSSON

Abstract

Cosmology, as a field of science, represents a reasoned pursuit to understand the fundamental mechanisms of the Universe, with the aim of charting the evolutionary timeline of the Universe. These mechanisms interact to manifest the harmony and order in the Universe on the largest scales. Cosmology, although a precision observational science, is richly layered with deep roots in philosophy, religion, mythology and culture. More importantly, the Cosmos and, by extension, cosmology encapsulates a certain innate mystery and awe, together with a spectrum of aesthetic experiences from those grounded in the everyday to those grounded in the discipline of cosmology itself. This spectrum of aesthetics is an important part of the teaching and learning of cosmology; owing to the nature of the field where there is an interplay between declarative knowledge, consensus ideas and speculation. This chapter unpacks a personal perspective on the challenges that arise from the methodological approach of bringing together various theoretical and philosophical perspectives. The aim of combining these multiple perspectives being to connect aesthetic experiences, meaning-making, conceptions, reasoning and epistemic practices of the field of cosmology into a coherent and harmonious line of inquiry in a thesis by publication focusing on cosmology education at high school.

Keywords: Cosmology education, aesthetics, methodology, interdisciplinary

Introduction

"Space: the final frontier.

These are the voyages of the starship Enterprise.

Its continuing mission: to explore strange new worlds.

To seek out new life and new civilizations.

To boldly go where no one has gone before!"

(Star Trek: The Next Generation, 1987)

The above monologue captures the essence of a PhD journey, with the caveat that although a PhD may be limited to a defined time, the pursuit of knowledge is a 'continuing mission'. But then again, is a PhD ever finished?

My pathway through the thesis has aspects of a self-study where each of the individual research projects reported in the seven papers contributed to my growing insights into the ways in which aesthetics, conceptions, reasoning and representations are manifested in cosmology and cosmology education. In this journey, my supervisory team acted as important critical friends which allowed perspectives to be refined and unpacked through rich discussions. Therefore, this chapter includes them as authors even though, as a story of a personal journey, it is written in the single voice of the first author.

Coming from a background in science, visual arts and education, my goal has always been to work at the interface of these three (Figure 15.1) and harmonise some of the philosophical, theoretical and methodological perspectives that underpin each of these fields. This diversity of perspectives presents a challenge to not only frame a thesis, but also to characterise the methodological foundations.

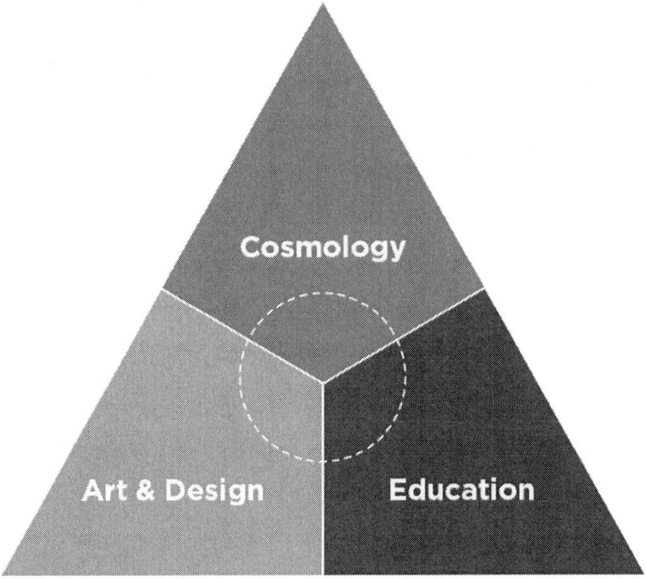

Figure 15.1: My thesis and philosophical lens is informed by working at the interface of three fields: Cosmology, Art & Design, and Education

The relationship between knowledge and aesthetics has been one of the driving forces in my varied career. From my own philosophical lens, which over the years has had various influences, knowledge is much more than content and skills, rather it is about beauty, curiosity, exploration, connections, relationships and interactions. Knowledge in Latin is *Scientia*, which is the root of the word Science. Therefore, Science in my opinion is not a field or a collection of sub-fields, rather it is the pursuit of knowledge, perhaps science as opposed to Science (Salimpour, 2021). My perspective is that knowledge in science is not a collection of disparate facts, theories and laws; rather science is an interaction of exploration, play, facts, observations, theories, laws, beauty and innate aesthetic experiences. All of which connect to develop a factual narrative about one of the greatest mysteries–the evolution of the Cosmos.

Taken together the above notions of beauty, play, knowledge, curiosity, exploration, connections, interactions and aesthetic experiences all formed vital aspects within my thesis and inspired various lines of inquiry. The challenge was bringing these together in a way that builds a coherent narrative exploring cosmology education in high school.

Given the focus of my thesis on cosmology education, the next section in this chapter aims to provide the reader with a brief introduction to cosmology. It should be emphasised that this section is meant to be a snapshot rather than an in-depth look at the field of cosmology, or an introductory chapter on cosmology.

Cosmology

Cosmology is considered to be the oldest science (North, 2008), and for millennia the beauty and mystery of the Cosmos has piqued the curiosity of all sentient beings. I can count on one hand the number of people that in my experience have not shown any interest in the Universe, or the topics of astronomy. Cosmology has roots in philosophy, culture and religion, and as such understanding the nature of the Universe can be explored from various perspectives and methodological approaches, each contributing a different piece to the puzzle.

Cosmology comes from the Greek words *kosmos* (order, harmony), and *logos* (thought, reason) (Figure 15.2). At its core the science of cosmology aims to provide a reasoned account of the underlying physical mechanisms that provide order in the Universe on the largest scale. The aim is to understand and characterise the evolutionary timeline of the Universe, and ultimately predict its future evolution (e.g., Peebles et al., 1994; Peebles, 1993).

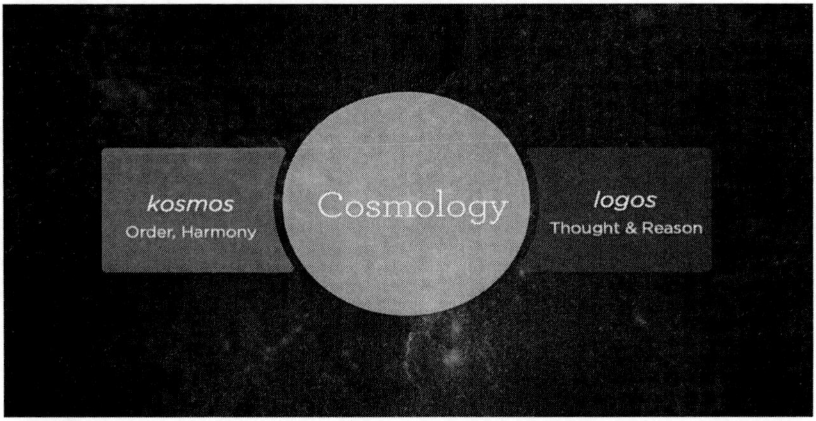

Figure 15.2: The origins of the word Cosmology from the words *kosmos* (order, harmony), and *logos* (thought, reason)

Providing an overview of the current consensus views in cosmology is beyond the scope of this chapter; however, a brief summary is perhaps useful to provide context. As of this writing, the ‘consensus’ framework or ‘concordance model’ in Cosmology is the Lambda Cold Dark Matter (LambdaCDM, written as ΛCDM) (e.g., Aghanim et al., 2020; Bull et al., 2016). In this framework, Lambda (Λ) signifies the Cosmological Constant, attributed to Dark Energy, the phenomenon causing the Universe to undergo accelerated expansion on the largest scales in the order of 100Mpc (~330 million light years) (Perlmutter et al., 1999; Riess et al., 1998; Schmidt et al., 1998). The word ‘cold’ refers to the motion of the Dark Matter particles, the nature of which is still a mystery and the subject of intense research (e.g., Bertone & Hooper, 2018; Bertone & Tait, 2018; Feng, 2010; Frenk & White, 2012). The ΛCDM model is based on several lines of observation and theoretical work, grounded in the mathematics of General Relativity (Einstein, 1917). Based on the ΛCDM model, the Universe on the largest scales has a flat geometry that is undergoing accelerated expansion, which is attributed to Dark Energy (e.g., Riess et al., 1998). The composition of the Universe is dominated by Dark Energy (~68%) and Dark Matter (~27%) with normal ‘everyday’ stuff making up a tiny fraction (~5%) (Aghanim et al., 2020).

One of the sources of observational data that has contributed to understanding the Universe on the largest scales is the Cosmic Microwave Background (CMB) Radiation–the remnant glow from the era when the Universe was around 380,000 years old (e.g., Ade et al., 2011; Penzias & Wilson, 1965) (Figure 15.3). The data from the CMB allows scientists to deduce various key measurements about the nature of the Universe, for example, its composition, geometry, overall density, and age (e.g., Ade et al., 2016; Aghanim et al., 2020).

Figure 15.3: The CMB as measured and visualised by the Planck space-based satellite. This represents the glow left over from when the Universe was 380,00 years old. Image credit: ESA/Planck Collaboration.

We have yet to develop a complete picture of the evolutionary timeline of the Universe; both into the distant past and into the future. This is owing to the enormous temporal and spatial scales involved and the finite speed of light, which governs how far back in time we can observe with our instruments–*lookback time*. Therefore, in this era of precision observational cosmology there are still rich debates about the intricacies and nuances of the Universe (e.g., Di Valentino et al., 2020; Efstathiou & Gratton, 2020; Handley, 2019; Park & Ratra, 2019). Equally vital is our limited understanding of the missing pieces to the puzzle, namely: the nature of Dark Matter, Dark Energy, refinements of the key parameters in the standard model of cosmology, aspects of the nature of inflation, and a few others. The fascinating journey of scientific discovery is not the known unknowns, rather the unknown unknowns.

Given the richness of the field, the grandeur of exploration, open questions and missing pieces of the puzzle, as educators we owe it to the future generation to engage them in the aesthetic experiences of building knowledge and curiosity.

Foundations

Lines of Inquiry

Exploration and the pursuit of knowledge at their core are driven by some fundamental overarching lines of inquiry. In the context of my thesis, these are best encompassed by two seminal works: (i) *Critique of Pure Reason* (Kant, 1781), and (ii) the 19th century painting "*D'où venons-nous ? Que sommes-nous ? Où allons-nous ?*" (Gauguin, 1897). In *Critique of Pure Reason,* Kant poses three questions: *"What can I know?", "What should I do?", and "What may I hope?".* The title of the painting by Gauguin translates to: *"Where Do We Come From? What Are We? Where Are We Going?".* Taken together these six questions provided the inspiration for developing the research questions in my thesis and lay the foundations for the several lines of inquiry that made up my thesis (Salimpour, 2021).

Kant's questions provided the overarching layer, his first question *"What can I know?"* sets the stage for the research questions that drive my thesis; not the research questions themselves but rather the essence of the exploration. The second question *"What should I do?"* perhaps translates well to the methodology and methods of my thesis; therefore, not only what to do, but how it can be done. The third question *"What may I hope?"* captures my aim as a PhD candidate and growing researcher, coupled with the aim of my research.

The questions in the title of the painting by Gauguin although aimed at representing the nature of being and life from a deep philosophical lens, inspired the three overarching research themes that formed part of the framework for my thesis. To truly explore any line of inquiry it is vital to unpack and characterise all that has come before, the landscape if you will. Gauguin's first question–*"Where Do We Come From?"*–encompasses this aspect of my thesis, wherein, my aim was to paint a picture of the landscape of cosmology education looking at aspects such a curriculum, research and the representations in cosmology. Student learning was always at the heart of my research, and as such it was necessary to understand student perspectives, conceptions and reasoning about topics in cosmology that were present both in curricula and beyond the classroom. This was inspired by Gauguin's second question "*What Are We?*", in the context of my thesis it related to what students know and how they reason. The final question "*Where Are We Going?*" from a broad perspective is answered by the concept of "re-imagining of science education" (Tytler, 2007). In the context of my thesis, this relates to the tangible resources that were developed as part of my research to support teachers in bringing the grandeur of cosmology into the classroom, in a way that echoes contemporary science.

Having laid the framework for the lines of inquiry, the next section tackles the first methodological challenge, which involved bringing together the theoretical perspectives from aesthetics, social semiotics, and conceptions and reasoning.

Aesthetics

The construct of aesthetics is richly layered and is traditionally associated with art and beauty (e.g., Gardner, 1996; Hegel, 1835; Welsch, 1997). However, aesthetics is much more than that, it encompasses various aspects of experience and knowledge (e.g., Wickman, 2006). For example, the Persian philosopher Elahi Ghomshei conceptualises the construct of aesthetics in terms of 'flight', and characterises three types of flight: *Flight to Knowledge*, *Flight to Beauty (Aesthetics)*, and *Flight to Goodness*. Ghomshei's notion of flight is one where the individual is transported cognitively to a different state of consciousness and by extension having a different experience. Ghomshei's notion of flight should not be confused with Csikszentmihalyi's flow (Csikszentmihalyi, 1975), although they share some aspects in terms of both cognitive and experiential shifts.

Ghomshei unpacks beauty from a perspective enabling him to characterise it from both the lens of visual appeal and that of experience. This is also echoed by Girod et al. (2003), in the context of learning science, where "science learning is something to be swept-up in, yielded to, and experienced" (pp. 575-576).

In the Persian language there are several words for beauty, and more importantly aesthetics is closely related to knowledge. Therefore, in a sense building knowledge, or rather the pursuit of knowledge, has associated with it an experience which can be conceptualised in terms of beauty, but more holistically in terms of aesthetics. Murray Gell-Mann puts it eloquently "…in fundamental physics a beautiful or elegant theory is more likely to be right than a theory that is inelegant", "…and we published before knowing that, because we figured it was so beautiful, it's gotta be right!" (Gell-Mann, 2007).

My philosophical lens places a premium on the connections and relationships between fields of inquiry, as a result aesthetics formed an integral part of my thesis (Salimpour, 2021). My aim was not only to provide a theoretical framework that brings aesthetics into cosmology education, but rather incorporates the various philosophical underpinnings into the resources for teachers. This was done through different approaches in the papers that constituted my thesis, so that the aesthetic focus appears through different aspects of disciplinary knowledge, representations, student learning, and teacher appreciation of this dimension. This would ultimately enable teachers to manifest the aesthetic experiences (Wickman, 2006) associated with the discipline and the awe of cosmology in their classroom (Salimpour, Tytler, Eriksson, et al., 2021). This goal proved to be methodologically challenging because of the varying lenses that aesthetics has been explored (e.g., Dewey, 1897; Gardner, 1996; Hegel, 1835; Shusterman, 2000; Wickman, 2006). In the context of the classroom, aesthetics is often associated with visual appeal, my aim was to go beyond that to incorporate the disciplinary aesthetic experiences into the teaching and learning of cosmology (Salimpour, Tytler, Eriksson, et al., 2021). This was in no way disregarding the rich landscape of theoretical research into aesthetics, rather it was about drawing on previous work to enable me to achieve my goal of creating connections and relationships.

Social Semiotics

Semiotics, from the Greek *sémeion*, is in essence concerned with the study of signs and meaning. Eco (1976) describes semiotics as being "…concerned with everything that can be taken as a sign" (p. 7). Therefore, a sign encompasses a range of representational modes, which includes words, images, drawings, paintings, objects, gestures, each of which carries meanings that can be interpreted by an observer or social group. The theoretical underpinning of semiotics from a linguistic perspective are found in the works of Ferdinand de Saussure (Saussure, 1916/2013), and from a pragmatist perspective in the work Charles Peirce (Peirce, 1992, 1998).

Social Semiotics draws on the field of semiotics and specifically the work of Halliday on human meaning-making (*semiosis*), in the context of language (Halliday, 1978). At its core social semiotics is concerned with how semiotic systems (language, representations, gestures) are used in social meaning-making (e.g., Airey & Linder, 2017; Hodge & Kress, 1988; Kress, 2010; Lemke, 1987). The creation and interpretation of signs is a vital part of society and social interactions; therefore, it is only natural to state that representations created within a field or discipline have meanings for that community. Therefore, one aspect of education is enabling students to unpack the disciplinary information both visible and hidden (Airey, 2015) encompassed in representations and concepts in cosmology.

The concepts in cosmology involve intangible spatial and temporal relations, which can only be characterised and unpacked through various representations (e.g., mathematical parametrisations, diagrams, visualisations of data; computer simulations). Therefore, it is vital to understand not only the use of representations in cosmology education, but also how these representations are created–their innate anatomy (Salimpour, Tytler, Eriksson, et al., 2021).

One of the methodological challenges, which naturally became one of the aims of my thesis was to determine how notions of aesthetics (e.g., Wickman, 2006) and the construction of representations (Tytler et al., 2013) could manifest within the broader framework of social semiotics. From my perspective, meaning-making and the reading of signs involves some level of aesthetic experience, and my aim was to characterise that within social semiotics, and ultimately within the pedagogical approaches to teaching cosmology. To address this challenge an in-depth qualitative review of all representations in cosmology and cosmology education was conducted, in order to unpack the fundamental characteristics of the representations. This analysis through the categorisation of representations and iterative coding of the various characteristics allowed the aesthetic dimensions to emerge. The key approach to overcome the challenge was to take a methodologically agnostic position, and let the data guide the analysis. This allowed various aesthetic dimensions of the representations to emerge, which were grouped into the everyday aesthetic, art aesthetic and disciplinary aesthetic, representing the awe-inspiring, colourful nature of the universe, and the experiences associated with disciplinary rigour and discernment. The aesthetic dimensions were strongly embedded in the representational framing. Furthermore, each representation was looked at by myself given my background in cosmology and visual arts, and iterative discussions with the supervisory team with regards to the emergent codes.

The outcome from this analysis was the development of a qualitative framework for analysing representations in astronomy–anatomy of representations (AoR) (Salimpour, Tytler, Eriksson, et al., 2021). Most importantly, the AoR allowed aesthetics, disciplinary discernment, affordances and representation construction to be brought into the overarching framework of social semiotics (Salimpour, 2021).

Conceptions and reasoning

The notions of logic, reason and harmony are vital to any exploratory endeavour of which science is one avenue. In the context of science education one of the aims is developing student disciplinary knowledge and discernment (Airey & Eriksson, 2019; Eriksson, 2019). One aspect of this is enabling students to become familiar with the reasoning approaches that allows them to understand the various concepts in science. In cosmology education, this capability is vital given the complex spacetime relations, and intangibility of the underlying concepts. Research in the field of astronomy education has revealed a range of persistent alternative conceptions students possess (e.g., Bailey et al., 2012; Comins, 2001; Danaia & McKinnon, 2007; Miller & Brewer, 2010; Slater et al., 2018).

The literature related to student conceptions is vast, having roots in the work of Piaget (1950/2001). Conceptions and reasoning can be looked at from a developmental lens; however, the work of Driver and Easley (1978), perhaps best resonated with what I aimed to achieve in my thesis. To truly unpack and characterise student conceptions that are contrary to consensus views in science it is necessary to understand how students reason (Salimpour, Tytler, et al., 2020). Driver and Easley (1978) highlight two approaches that are often used in student conceptions research: *nomothetic*, where student answers are evaluated based on some predetermined scale or consensus views; and *ideographic*, where student answers are evaluated based "on their own terms" (p. 62). My aim was to draw on both these approaches to unpack both student conceptions and more importantly the underlying reasoning for the conceptions students had, irrespective of whether they aligned to consensus views. In terms of pedagogy, it is vital to understand the underlying reasoning patterns before designing instructional resources and approaches. This involved looking at alternative conceptions from the standpoint lens that they can be a productive basis to bring about conceptual growth (e.g., Smith et al., 1994; Suchman, 1962). Therefore, the methodological approaches I used to gather and analyse data had to be flexible in order to capture the richness of underlying information. This was achieved by creating open-ended surveys, that allowed students to express their reasoning (Salimpour, Tytler, et al., 2020; Salimpour, Tytler, Fitzgerald, et al., 2021), rather than having consensus multiple-choice answers that would not be able to unpack the depth and richness in student reasoning. The methodological challenges associated with the analysis of open-ended survey responses is explained in Salimpour et al. (2020), where it was vital to situate the analysis on the interface of the positivist-interpretivist divide. This approach allows the researcher to draw "on various paradigms as the data is explored *in vivo*. The data in turn guides the methodological framework as patterns and exceptions emerge" (p. 221).

Diversity of methods

Pick up any textbook on research, and the reader is presented with a range of methodologies, each driven by different philosophical perspectives (e.g., Tuck et al., 2014). In the context of social science research, the methodology and methods are multifaceted, compared to fundamental science, where the underlying laws of nature and the phenomenon being explored help situate the methodology and methods (Salimpour, 2021).

The nature of my thesis with its grand sweep aim of exploring the 'what', 'why' and 'how' of cosmology education meant that a range of methodologies had to be considered and brought together to answer both the overarching and fine-grained lines of inquiry. This meant that a mixed-methods approach was a natural way forward. A mixed-methods approach "involves a symbiotic, harmonious interaction of both qualitative and quantitative methodologies and methods" (Salimpour, 2021, p.96). This interaction in the context of my thesis occurred not only within each of the individual publications but also across each of the overarching themes.

One of my aims was to go beyond merely dots on graph, numbers and statistics, to unpacking the essence of what those dots on a graph meant, and how those numbers came about. This was inspired by Cobb et al. (2003), who stated "...complexity is the hallmark of educational settings" (p. 9). This is because teaching and learning involves "multiple elements interacting in a myriad of different ways" (Salimpour, 2021, p.96). My aim was not only exploring cognitive processes, and content knowledge, rather, also exploring the interaction between the reality of cosmology and how it is (and can be) framed in schools. Some of the methodological challenges that arise from this interaction is to what extent the consensus ideas from cosmology research need to be brought into the classroom, keeping in mind the context and content of the curriculum (Salimpour, Bartlett, et al., 2020). It would be nigh impossible to delve deeply into all the concepts in cosmology within the limited time of a typical school classroom.

Even though I had my own philosophical and scientific lenses, as a researcher I had to approach the lines of inquiry from a methodologically agnostic lens, and let the data guide the methodology. For example, when analysing the range of representations in cosmology and cosmology education, the representations themselves guided the analysis. Another example is when analysing the open-ended responses from students, their responses and reasoning guided the emergence of themes.

Bringing it together

One of the key characteristics of a thesis by publication in general is creating a coherent narrative that brings together both the line(s) of inquiry and the different publications that form the key research output. Therefore, it is vital to think about each of the publications not in isolation but as part of a grand sweep narrative. This required some sort of overarching framework to help 'package' the various components of the thesis. Keeping in mind the six questions that inspired the lines of inquiry at the start, these were refined into three overarching themes: (i) The Cosmic Landscape; (ii) Beyond Conceptions; and (iii) Cosmology in the Classroom (Figure 15.4).

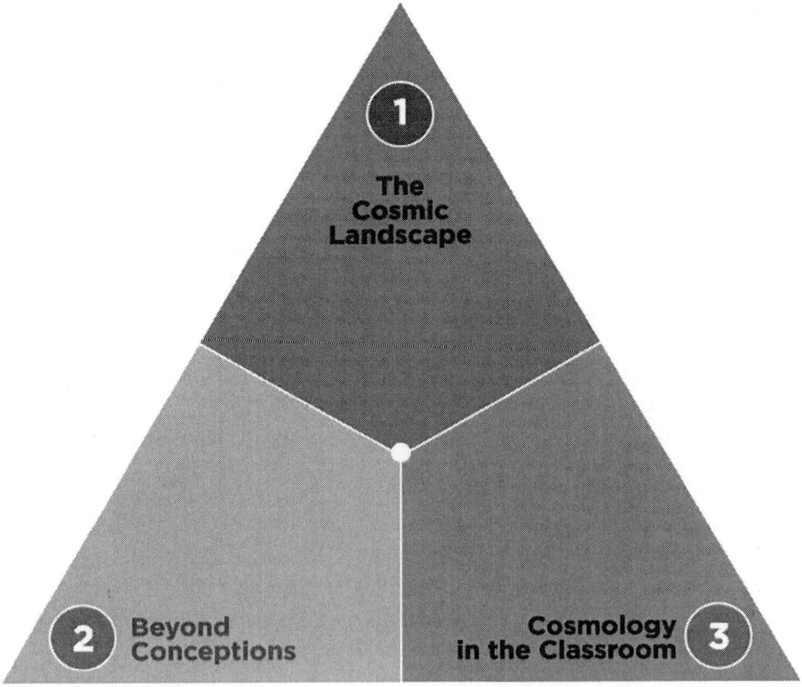

Figure 15.4: The three overarching themes that form the backbone of my thesis, inspired by questions posed by Kant and Gauguin (Salimpour, 2021)

The conceptual framework of my thesis is in essence an interaction between various theoretical perspectives (Figure 15.5). The main stage is Contemporary Science Education, which encompasses theoretical perspectives from the Nature of Science (e.g., Clough, 2017; Duschl, 2008; Lederman, 1992) and Science-as-Practice (Lehrer & Schauble, 2007). The disciplinary stage in my thesis is cosmology, and it is here where I draw on the three main theoretical perspectives of aesthetics, social semiotics, and conceptions and reasoning, to explore student learning in cosmology. Although the layers in the conceptual framework are demarcated with white lines (Figure 15.5), this is done for clarity, but in reality, these layers have to interact to ensure the pedagogy is coherent. The interaction of the layers is owing to the fact that the notions captured in this framework have underpinnings in both science education, but also from the disciplinary lens of cosmology. One instance of this interaction is the evolution of the epistemic practices in cosmology driven by, for example, big data, precision observational methods, and large-scale collaborations, all of which feedback into the notion of contemporary science and science education.

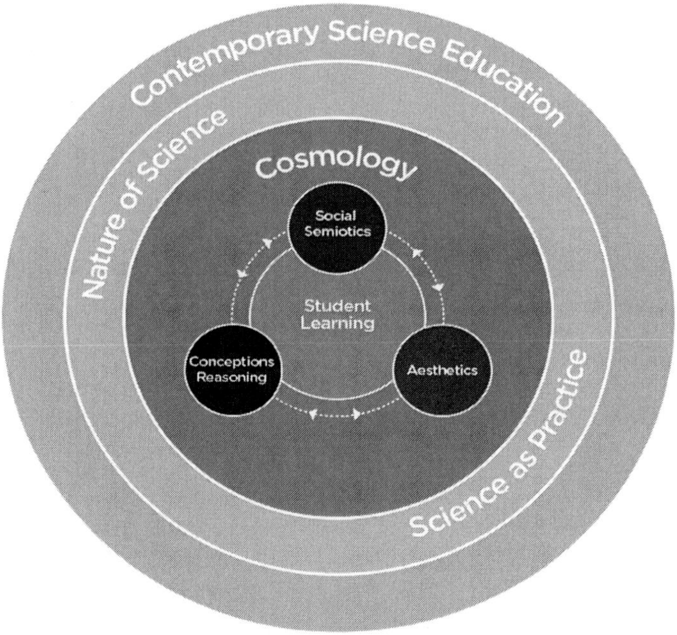

Figure 15.5: The conceptual framework of my thesis is a layered interaction of various theoretical lenses, at the core of which is student learning (Salimpour, 2021)

This conceptual framework is situated at the intersection of the three overarching themes defining this thesis (Figure 15.6). The overarching themes help contribute to the various theoretical perspectives and help guide the lines of inquiry within each of the themes. Although it could be argued that contemporary science education should encompass the three themes; however, that would mean that the landscape of cosmology education would be incomplete, in a sense falling short of answering Gauguin's question *"Where Do We Come From?"*.

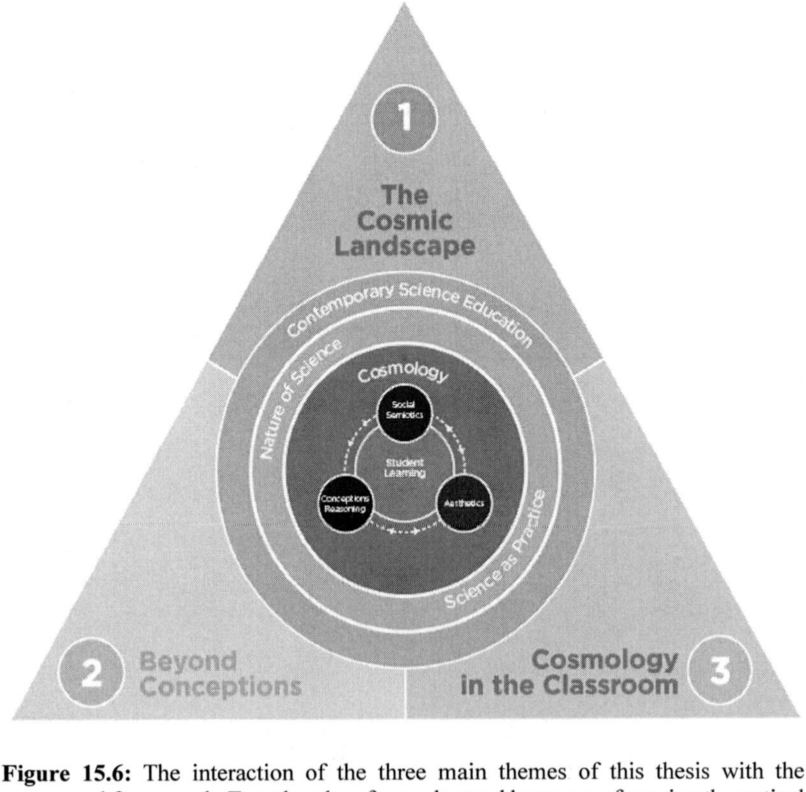

Figure 15.6: The interaction of the three main themes of this thesis with the conceptual framework. Together they form a layered harmony of varying theoretical perspectives.

Discussion

The scope of my thesis encompassed a broad sweep exploration through the grandeur of the cosmos both as a discipline and its 'everyday' awe. In addition, my thesis encompassed a vision for bringing these ideas into the classroom. Although this vision was befitting the subject, the challenge was finding an avenue to encompass this grand vision within the requirements of academic rigour that a thesis requires. This meant that there was a tension between developing a thesis that captured my passions and perspectives, but also kept true to the methodological traditions of a doctoral research. Transforming and reshaping this tension into a 'cosmic harmony' was made possible via two aspects: a thesis by publication structure, and the notion of aesthetics, these are unpacked below.

A thesis by publication allowed me (with the help of my supervisory team) to use the individual publications to situate my thesis within the rigours of a doctoral research. Each of the papers represented different theoretical and methodological perspectives, and the research within each of the individual studies used robust analysis methods. In pursuing the interlocking foci for the different papers (curriculum, student reasoning, representations used in cosmology texts, and a model for utilising cosmology research data in classrooms) I was able to refine a grand vision that cohered around these elements such that my thesis was able to focus on this narrative in ways that would be difficult for a normal thesis structure. This allowed for a meta-perspective to be unpacked. Creating a narrative that tied together the theoretical and methodological diversity. The thesis had the task of bringing together the individual papers in such a way as to make apparent the wider logic of my vision for cosmology education.

The second aspect was the notion of aesthetics, which in a sense became the 'glue that binds' my different foci, theoretical perspectives, and methodologies. This helped bring together the various themes in my thesis and my wider research agenda. Although aesthetics played some part in the individual papers it could not be explored fully in these because of demands of rigour and brevity. However, the coherent focus on aesthetics was emergent as these strands and their interactions were pulled together and refined. In a sense the analysis of representations provided the foundations for framing the various dimensions of aesthetics in my thesis.

Conclusion

This chapter aimed to capture in part the process of developing a thesis by publication and bringing together the various theoretical perspectives that drove the lines of inquiry in the three themes underpinning my thesis. My own philosophical perspectives, shaped by working at the interface of cosmology, visual arts, and education, informed and inspired how I approached the methodological aspects of my thesis and the various publications. My personal perspectives were not developed in isolation, and several rich discussions with my co-authors helped refine and unpack the underlying theoretical frameworks. The aim was to find ways of bringing together the methodological approaches that were suited to exploring the overarching and fine-grained lines of inquiry. Although a mixed-methods approach seemed to be a natural pathway, the various publications each had their own (sometimes multiple) methodologies. Therefore, as a researcher I had to take a methodologically agnostic approach in order to tap into the rich potential of diversity. Perhaps this way of thinking was inspired by the Vulcan IDIC "*Infinite Diversity in Infinite Combinations. The ways our differences combine to create meaning and beauty.*" (Star Trek: The Original Series, Episode: Is There in Truth No Beauty?). These two statements capture my aim of harmonising the varying theoretical, philosophical and methodological perspectives over a range of publications into a thesis. Therefore, it was not only about bringing together those aspects of methodologies that were similar, but also the differences were vital in allowing me to create the connections needed to bring things together.

References

Ade, P. A. R., Aghanim, N., Arnaud, M., Ashdown, M., Aumont, J., Baccigalupi, C., Baker, M., Balbi, A., Banday, A. J., Barreiro, R. B., Bartlett, J. G., Battaner, E., Benabed, K., Bennett, K., Beno t, A., Bernard, J. P., Bersanelli, M., Bhatia, R., Bock, J. J., … Zonca, A. (2011). Planck early results. I. The Planckmission. *A&A*, *536*, A1. https://doi.org/10.1051/0004-6361/201116464

Ade, P. A. R., Aghanim, N., Arnaud, M., Ashdown, M., Aumont, J., Baccigalupi, C., Banday, A. J., Barreiro, R. B., Bartlett, J. G., Bartolo, N., Battaner, E., Battye, R., Benabed, K., Benoît, A., Benoit-Lévy, A., Bernard, J.-P., Bersanelli, M., Bielewicz, P., Bock, J. J., … Zonca, A. (2016). Planck 2015 results. XIII. Cosmological parameters. *Astronomy and Astrophysics*, *594*, A13. https://doi.org/10.1051/0004-6361/201525830

Aghanim, N., Akrami, Y., Ashdown, M., Aumont, J., Baccigalupi, C., Ballardini, M., Banday, A. J., Barreiro, R. B., Bartolo, N., Basak, S., Battye, R., Benabed, K., Bernard, J.-P., Bersanelli, M., Bielewicz, P., Bock, J. J., Bond, J. R., Borrill, J., Bouchet, F. R., … Zonca, A. (2020). Planck 2018 results. VI. Cosmological parameters. *Astronomy and Astrophysics*, *641*, A6. https://doi.org/10.1051/0004-6361/201833910

Airey, J. (2015). *Social semiotics in higher education: Examples from teaching and learning in undergraduate physics*. 103. http://urn.kb.se/resolve?urn=urn:nbn:se:uu:diva-266049

Airey, J., & Eriksson, U. (2019). Unpacking the Hertzsprung-Russell Diagram: A social semiotic analysis of the disciplinary and pedagogical affordances of a central resource in astronomy. *Designs for Learning*, *11*(1), 99-107. https://doi.org/10.16993/dfl.137

Airey, J. & Linder, C. (2017). Social semiotics in university physics education. In D. F. Treagust, R. Duit, & H. E. Fischer (Eds.), *Multiple representations in physics education* (pp. 95-122). Cham, Switzerland: Springer. https://doi.org/10.1007/978-3-319-58914-5_5

Bailey, J. M., Coble, K., Cochran, G., Larrieu, D., Sanchez, R., & Cominsky, L. R. (2012). A multi-institutional investigation of students? Preinstructional ideas about cosmology. *Astronomy Education Review*, *11*(1). https://doi.org/10.3847/AER2012029

Bertone, G. & Hooper, D. (2018). History of dark matter. *Reviews of Modern Physics*, *90*(4), 045002. https://doi.org/10.1103/RevModPhys.90.045002

Bertone, G. & Tait, T. M. P. (2018). A new era in the search for dark matter. *Nature*, *562*(7725), 51-56. https://doi.org/10.1038/s41586-018-0542-z

Bull, P., Akrami, Y., Adamek, J., Baker, T., Bellini, E., Beltrán Jiménez, J., Bentivegna, E., Camera, S., Clesse, S., Davis, J. H., Di Dio, E., Enander, J., Heavens, A., Heisenberg, L., Hu, B., Llinares, C., Maartens, R., Mörtsell, E., Nadathur, S., … Winther, H. A. (2016). Beyond ΛCDM: Problems, solutions, and the road ahead. *Physics of the Dark Universe*, *12*, 56-99. https://doi.org/10.1016/j.dark.2016.02.001

Clough, M. P. (2017). History and nature of science in science education. In K. S. Taber & B. Akpan (Eds.), *Science education: An international course companion* (pp. 39-51). SensePublishers. https://doi.org/10.1007/978-94-6300-749-8_3

Cobb, P., Confrey, J., diSessa, A., Lehrer, R., & Schauble, L. (2003). Design experiments in educational research. *Educational Researcher*, *32*(1), 9–13. https://doi.org/10.3102/0013189X032001009

Comins, N. F. (2001). *Sources of misconceptions in astronomy*. 38.

Csikszentmihalyi, M. (1975). *Beyond boredom and anxiety* (MELB; First edition.). Jossey-Bass Publishers. https://search.ebscohost.com/login.aspx?direct=true&AuthType=sso&db=cat00097a&AN=deakin.b1403306&authtype=sso&custid=deakin&site=eds-live&scope=site

Danaia, L. & McKinnon, D. H. (2007). Common alternative astronomical conceptions encountered in junior secondary science classes: Why is this so? *Astronomy Education Review*, *6*, 32-53.

Dewey, J. (1897). The aesthetic element in education. In J. A. Boydston (Ed.), *The early works of John Dewey, Volume 5, 1882—1898: Early essays, 1895-1898 (republished 1972)*. Southern Illinois University Press.

Di Valentino, E., Melchiorri, A., & Silk, J. (2020). Planck evidence for a closed Universe and a possible crisis for cosmology. *Nature Astronomy*, *4*, 196-203. https://doi.org/10.1038/s41550-019-0906-9

Driver, R. & Easley, J. (1978). Pupils and paradigms: A review of literature related to concept development in adolescent science students. *Studies in Science Education*, *5*(1), 61-84. https://doi.org/10.1080/03057267808559857

Duschl, R. (2008). Science education in three-part harmony: Balancing conceptual, epistemic, and social learning goals. *Review of Research in Education*, *32*(1), 268-291. https://doi.org/10.3102/0091732X07309371

Eco, U. (1976). *A theory of semiotics*. Indiana University Press.

Efstathiou, G. & Gratton, S. (2020). The evidence for a spatially flat universe. *Monthly Notices of the Royal Astronomical Society*, *496*, L91-L95. https://doi.org/10.1093/mnrasl/slaa093

Einstein, A. (1917). Kosmologische Betrachtungen zur allgemeinen Relativitätstheorie. *Sitzungsberichte Der Königlich Preußischen Akademie Der Wissenschaften (Berlin), Seite 142-152.* http://adsabs.harvard.edu/abs/1917SPAW.......142E

Eriksson, U. (2019). Disciplinary discernment: Reading the sky in astronomy education. *Physical Review Physics Education Research*, *15*(1), 010133. https://doi.org/10.1103/PhysRevPhysEducRes.15.010133

Feng, J. L. (2010). Dark matter candidates from particle physics and methods of detection. *Annual Review of Astronomy and Astrophysics*, *48*(1), 495-545. https://doi.org/10.1146/annurev-astro-082708-101659

Frenk, C. s., & White, S. d. m. (2012). Dark matter and cosmic structure. *Annalen Der Physik*, *524*(9-10), 507-534. https://doi.org/10.1002/andp.201200212

Gardner, S. (1996). Aesthetics. In E. Bunnin & E. P. Tsui-James (Eds.), *The Blackwell companion to philosophy* (pp. 229-256). Oxford, UK: Blackwell.
Gell-Mann, M. (2007). *Beauty, truth and ... Physics?* Retrieved from https://www.ted.com/talks/murray_gell_mann_beauty_truth_and_physics
Girod, M., Rau, C., & Schepige, A. (2003). Appreciating the beauty of science ideas: Teaching for aesthetic understanding. *Science Education*, *87*(4), 574-587. https://doi.org/10.1002/sce.1054
Halliday, M. A. K. (1978). *Language as social semiotic: The social interpretation of language and meaning*. Edward Arnold.
Handley, W. (2019). Curvature tension: Evidence for a closed universe. *ArXiv E-Prints*, *1908*, arXiv:1908.09139.
Hegel, G. W. F. (1835). *Introductory lectures on aesthetics* (T. Duncan, Trans.). Penguin.
Hodge, R. & Kress, G. (1988). *Social semiotics*. Cornell University Press.
Kant, I. (1781). *The critique of pure reason* (J. M. D. Meiklejohn, Trans.). Infomotions, Inc.
Kress, G. R. (2010). *Multimodality: A social semiotic approach to contemporary communication*. Routledge.
Lederman, N. G. (1992). Students' and teachers' conceptions of the nature of science: A review of the research. *Journal of Research in Science Teaching*, *29*(4), 331-359. https://doi.org/10.1002/tea.3660290404
Lehrer, R. & Schauble, L. (2007). Scientific thinking and science literacy. In *Handbook of child psychology*. American Cancer Society. https://doi.org/10.1002/9780470147658.chpsy0405
Lemke, J. L. (1987). Social semiotics and science education. *The American Journal of Semiotics*, *5*(2), 217–232.
Miller, B. W. & Brewer, W. F. (2010). Misconceptions of astronomical distances. *International Journal of Science Education*, *32*(12), 1549-1560. https://doi.org/10.1080/09500690903144099
North, J. (2008). *Cosmos: An illustrated history of Astronomy and Cosmology*. University of Chicago Press. http://press.uchicago.edu/ucp/books/book/chicago/C/bo5561644.html
Park, C.-G. & Ratra, B. (2019). Using the Tilted flat-ΛCDM and the Untilted Non-flat ΛCDM Inflation models to measure cosmological parameters from a compilation of observational data. *The Astrophysical Journal*, *882*, 158. https://doi.org/10.3847/1538-4357/ab3641
Peebles, P. J. E. (1993). *Principles of physical cosmology*. Princeton University Press. Retrieved from http://press.princeton.edu/titles/5263.html

Peebles, P. J., Schramm, D. N., Turner, E. L., & Kron, R. G. (1994). The evolution of the universe. *Scientific American, 271*(4), 52-57. https://doi.org/10.1038/scientificamerican1094-52

Peirce, C. S. (1992). *The essential Peirce, volume 1: Selected philosophical writings, (1867–1893)*. Indiana University Press.

Peirce, C. S. (1998). *The essential Peirce, volume 2: Selected philosophical writings, 1893-1913* (P. E. Project, Ed.). Indiana University Press.

Penzias, A. A. & Wilson, R. W. (1965). A measurement of excess antenna temperature at 4080 Mc/s. *The Astrophysical Journal*, *142*, 419-421.

Perlmutter, S., Aldering, G., Goldhaber, G., Knop, R. A., Nugent, P., Castro, P. G., Deustua, S., Fabbro, S., Goobar, A., Groom, D. E., Hook, I. M., Kim, A. G., Kim, M. Y., Lee, J. C., Nunes, N. J., Pain, R., Pennypacker, C. R., Quimby, R., Lidman, C., … Project, T. S. C. (1999). Measurements of Ω and Λ from 42 High-Redshift Supernovae. *The Astrophysical Journal*, *517*(2), 565-586. https://doi.org/10.1086/307221

Piaget, J. (2001). *The psychology of intelligence* (New edition.). Routledge. (Original work published 1950)

Riess, A. G., Filippenko, A. V., Challis, P., Clocchiatti, A., Diercks, A., Garnavich, P. M., Gilliland, R. L., Hogan, C. J., Jha, S., Kirshner, R. P., Leibundgut, B., Phillips, M. M., Reiss, D., Schmidt, B. P., Schommer, R. A., Smith, R. C., Spyromilio, J., Stubbs, C., Suntzeff, N. B., & Tonry, J. (1998). Observational evidence from supernovae for an accelerating universe and a cosmological constant. *The Astronomical Journal*, *116*(3), 1009-1038. https://doi.org/10.1086/300499

Salimpour, S. (2021). *Visualising the cosmos: Teaching cosmology in high school in the era of big data* [Doctoral Thesis]. Deakin University.

Salimpour, S., Bartlett, S., Fitzgerald, M. T., McKinnon, D. II., Cutts, K. R., James, C. R., Miller, S., Danaia, L., Hollow, R. P., Cabezon, S., Faye, M., Tomita, A., Max, C., de Korte, M., Baudouin, C., Birkenbauma, D., Kallery, M., Anjos, S., Wu, Q., … Ortiz-Gil, A. (2020). The gateway science: A review of astronomy in the OECD school curricula, including China and South Africa. *Research in Science Education*. https://doi.org/10.1007/s11165-020-09922-0

Salimpour, S., Tytler, R., Eriksson, U., & Fitzgerald, M. (2021). Cosmos visualized: Development of a qualitative framework for analyzing representations in cosmology education. *Physical Review Physics Education Research*, *17*(1), 013104. https://doi.org/10.1103/PhysRevPhysEducRes.17.013104

Salimpour, S., Tytler, R., & Fitzgerald, M. T. (2020). Exploring the cosmos: The challenge of identifying patterns and conceptual progressions from student survey responses in cosmology. In R. Tytler, P. White, J.

Ferguson, & J. C. Clark (Eds.), *Methodological Approaches to STEM Education Research* (Vol. 1). Cambridge University Press.

Salimpour, S., Tytler, R., Fitzgerald, M. T., & Eriksson, U. (2021). Is the universe infinite? Exploring high student conceptions of cosmology concepts using open-ended surveys. *In Review*.

Saussure, F. de. (2013). *Course in general linguistics* (R. Harris, Trans.). Bloomsbury Academic, An imprint of Bloomsbury Publishing Plc. (Original work published 1916)

Schmidt, B. P., Suntzeff, N. B., Phillips, M. M., Schommer, R. A., Clocchiatti, A., Kirshner, R. P., Garnavich, P., Challis, P., Leibundgut, B., Spyromilio, J., Riess, A. G., Filippenko, A. V., Hamuy, M., Smith, R. C., Hogan, C., Stubbs, C., Diercks, A., Reiss, D., Gilliland, R., … Ciardullo, R. (1998). The High-Z Supernova search: Measuring cosmic deceleration and global curvature of the universe using Type Ia Supernovae*. *The Astrophysical Journal*, *507*(1), 46. https://doi.org/10.1086/306308

Shusterman, R. (2000). *Pragmatist aesthetics: Living beauty, Rethinking art*. Rowman & Littlefield.

Slater, E., Morris, J., & Mckinnon, D. H. (2018). Astronomy alternative conceptions in pre-adolescent students in Western Australia. *International Journal of Science Education*, *0*(0), 1-23. https://doi.org/10.1080/09500693.2018.1522014

Smith, J. P., diSessa, A. A., & Roschelle, J. (1994). Misconceptions reconceived: A constructivist analysis of knowledge in transition. *Journal of the Learning Sciences*, *3*(2), 115-163. https://doi.org/10.1207/s15327809jls0302_1

Suchman, J. R. (1962). Creative thinking and conceptual growth. *Gifted Child Quarterly*, *6*(3), 95-99. https://doi.org/10.1177/001698626200600309

Tuck, E., McKenzie, M., & McKenzie, M. (2014). *Place in research: Theory, methodology, and methods*. Taylor & Francis Group. Retrieved from http://ebookcentral.proquest.com/lib/deakin/detail.action?docID=1762000

Tytler, R. (2007). *Re-imagining science education: Engaging students in science for Australia's future*.

Tytler, R., Prain, V., Hubber, P., & Waldrip, B. (Eds.). (2013). *Constructing representations to learn in science*. SensePublishers. https://doi.org/10.1007/978-94-6209-203-7

Welsch, W. (1997). *Undoing aesthetics*. SAGE Publications.

Wickman, P.-O. (2006). *Aesthetic experience in science education: Learning and meaning-making as situated talk and action*. Routledge.

Acknowledgements

Saeed Salimpour would like to thank his supervisory team for their immense support before, during and after the PhD journey, the teachers and students at the various schools who participated in the various studies that formed part of this thesis, the reviewers of the various publications, and the myriad collegial discussions with those outside the supervisory team.

Saeed Salimpour (BArt/Design, MSc (Astrophysics), MTeach (Physics/Visual Arts), PhD) - Max Planck Institute for Astronomy (IAU Office of Astronomy for Education), Heidelberg, Germany; Deakin University, School of Education, 221 Burwood Highway, Burwood, 3125, Victoria, Australia. astrophysics@saeedsalimpour.com
Saeed Salimpour is a Postdoctoral researcher in astronomy education at the International Astronomical Union Office of Astronomy for Education. He has a background in Cosmology, Visual Art/Design, and Education. His research interest crosses disciplinary boundaries and includes Science Education (with a focus on Cosmology), Creativity, Representation, Aesthetics, Big Data Visualisation, Learning Progressions, Concept Inventories, VR/AR, Curriculum Development, Indigenous/Cultural Astronomy, and Student Research in Astronomy. He has worked with international collaborations in astronomy education research; implemented research-based astronomy projects involving colour imaging and exoplanet science for middle and secondary school students; conducted teacher training workshops; taught high school physics; lectured at university; exhibited in art exhibitions; and given public talks on various topics about the Universe. His goal is to bring the science and beauty of Cosmos to everyone, whilst working at the interface of Science, Art and Education. He believes in "being curious for the sake of finding things out!"
https://orcid.org/0000-0002-0387-3152

Russell Tytler (BSc (Hons), BEd, MSc, MEd, PhD, FASSA) - Deakin University, School of Education, 221 Burwood Highway, Burwood, 3125, Victoria, Australia. russell.tytler@deakin.edu.au
Russell Tytler is Alfred Deakin Professor and Chair in Science Education at Deakin University, Melbourne. He has researched and written extensively on student learning and reasoning in science. His interest in the role of

representation as a multimodal language for reasoning in science extends to pedagogy and teacher learning. He researches and writes on student engagement with science and mathematics, aesthetics and identity in learning, school-community partnerships, and STEM curriculum policy and practice. He is widely published and has been chief investigator on a range of Australian Research Council and other research projects.
https://orcid.org/0000-0003-0161-7240

Michael Fitzgerald (BSc (Hons), MEd, PhD) - Las Cumbres Observatory, Goleta, California, USA; Deakin University, School of Education, 221 Burwood Highway, Burwood, 3125, Victoria, Australia.
psyfitz@gmail.com
Michael Fitzgerald is a Project Astronomer with Las Cumbres Observatory. He leads the development of the Photon Ranch Education Course. He has had a long interest in using robotic telescopes for education uses as well as student research projects. His particular interests are in stars that pulse periodically, stars that group together and how astronomy can be made interesting and inspiring for everybody. Michael grew up in Tasmania and now lives in the leafy north-eastern suburbs of Melbourne.
https://orcid.org/0000-0001-6554-1826

Urban Eriksson (BaSc, MEdu, PhLic, PhD, Docent) - Lund University, Sweden. urban.eriksson@fysik.lu.se
Urban Eriksson is an Associate Professor in Physics, specialising in Astronomy Education Research (AER) at Lund University, and Director for the National Resource Center for Physics Education. He is also Head of Department for Educational Sciences at Lund University. He has a background in astrometry, extensive experience in AER and teaching astronomy at university. He is Vice President of the International Astronomical Union (IAU)–Commission C1 Astronomy Education and Development, and editor of the Astronomy Education Journal. His research revolves around novice and experts learning astronomy via disciplinary discernment from semiotic resources used in astronomy, including extrapolating three-dimensionality from 1D and 2D resources, such as astronomical imagery and simulations presented on multimodal formats.
https://orcid.org/0000-0001-6638-1246

CHAPTER 16

APPLYING A SCOPING REVIEW METHODOLOGY TO INVESTIGATE ADAPTIVE EXPERTISE: INSIGHTS AND CHALLENGES FOR RESEARCH IN STEM EDUCATION

LIHUA XU, KENNEDY CHAN, AMANDA BERRY, JOSEPH PAUL FERGUSON, GAHYOUNG KIM, JAN VAN DRIEL, COLLEEN VALE AND WANTY WIDJAJA

Abstract

Adaptive experts are considered to be those who are not only able to perform procedural skills efficiently, but also apply meaningfully learned procedures flexibly and creatively in new contexts. Therefore, adaptive expertise encompasses both efficiency and innovation. The notion of adaptive expertise is not new, but its characteristics and development remain under-researched in teacher education with limited empirical evidence about the characteristics of adaptive teachers or their impact on student learning. Given this relatively new field in teacher education, we sought to determine a review methodology that would help us to examine the types of available evidence, clarify key concepts and characteristics associated with the construct of adaptive expertise in different education fields, and to identify research gaps in the existing literature. Scoping review is a relatively new qualitative research synthesis approach that is suited to map the literature on a given topic across diverse and evolving fields. In this chapter, we discuss four methodological issues that emerged during the research process of deciding on and undertaking a review methodology, and the opportunities and challenges we encountered along the way. This chapter offers some insights into research on teacher adaptive expertise that might be useful to inform future research in STEM education, and draws some implications for scoping review as a continually developing methodology in education research that need further consideration and refinement.

Keywords: Adaptive expertise, research synthesis, scoping review, mathematics, science

Introduction

In the past few decades, research synthesis has been widely employed in a variety of disciplines in the social sciences, ranging from psychology to education. The purpose of research synthesis is to "produce new knowledge by making explicit connections and tensions between individual study reports that were not visible before" (Suri, 2011, p. 63). Research synthesis can take on different forms. For example, quantitative research synthesis such as *statistical research synthesis* emphasises statistical integration of primary research reports, such as meta-analysis and best-evidence synthesis (Suri & Clarke, 2009). In comparison, qualitative research synthesis such as *systematic reviews* emphasise a priori protocols, comprehensive searches, transparency to reduce biases, and the involvement of stakeholders (e.g., target groups, industry representatives) in the review process (Suri & Clarke, 2009). *Scoping reviews*, as a relatively new and rapidly evolving qualitative research synthesis approach (Tricco et al., 2016), are "exploratory projects that systematically map the literature available on a topic, identifying key concepts, theories, sources of evidence and gaps in the research" (Grimshaw, 2020, p. 34). Each type of research synthesis is underpinned by different assumptions about the nature of knowledge and how the knowledge ought to be integrated and synthesised to move the field forward.

Constructs such as adaptive expertise have been widely employed in a range of research fields in the social sciences, including education. When the same construct is applied across multiple fields and disciplines, meanings can differ, including some ambiguity in how the construct is defined and operationalised. However, the way in which researchers work with constructs that are broadly or ambiguously defined across diverse fields is often implicit in research reports. This situation presents significant challenges for undertaking research synthesis. In particular, education as a field of research has been criticised for its fragmented nature in that researchers tend not to build on each other's work as much as they ought, in order to advance the field and to inform practice and policy at scale (Burkhardt & Schoenfeld, 2003). Therefore, research synthesis that draws together different conceptualisations and empirical evidence related to the same construct, such as adaptive expertise, is an important step in advancing the field of education research.

In this chapter, we argue that it is important for researchers to make explicit the methodological decisions and dilemmas they encounter during the research synthesis process. These accounts can complement generic guidelines on how to conduct research synthesis (Suri, 2011) and existing guidance regarding particular issues during the process of research synthesis (e.g., ethical consideration (Suri, 2020), purposeful sampling (Suri, 2011)). To illustrate this methodological decision making process, we employ a current research project on adaptive expertise in the context of interdisciplinary mathematics and science education. We present and discuss four interrelated methodological issues and challenges that emerged during the early stages of conducting a scoping review. This chapter will also draw some implications for future research employing similar methodologies in STEM education.

Research background and context

Our research project, 'Primary teachers' adaptive expertise in interdisciplinary maths and science', funded by the Australia Research Council (2021-2023), aims to support Australian primary teachers to teach interdisciplinary mathematics and science by collaborative co-design involving in-service primary teachers and the research team. In Australia, primary teachers are usually trained as generalists who often lack understanding of how to connect mathematics and science in meaningful ways to promote students' interdisciplinary learning (Timms et al., 2018). In the project, we draw on the construct of teachers' adaptive expertise, which is widely considered a gold standard for professionalism in teaching (Hammerness et al., 2005; Von Esch & Kavanagh, 2018). The notion of adaptive expertise was first proposed by Hatano and Iganaki (1984) to distinguish between two different types of expertise: *routine* and *adaptive*. Adaptive experts are "those who not only perform procedural skills efficiently but also understand the meaning of the skills and nature of their object" (Hatano & Inagaki, 1984, p. 28). Teachers with adaptive expertise are ready to invent new strategies to use with students "for whom the existing routines are not enabling success" (Hammerness et al., 2005, p.

360). To provide a strong theoretical basis for investigating adaptive expertise, we embarked on an initial scan of the literature to clarify key concepts and characteristics associated with the construct, and to identify research gaps in the existing literature. We quickly settled on employing qualitative research synthesis, as our initial foray into the existing literature revealed few quantitative studies on adaptive expertise. Our next decision related to the type of qualitative review to employ. In the next section, we explain the various methodological issues that we encountered, related to:

1. Selecting a review approach
2. Identifying reviewing processes
3. Applying sampling strategies
4. Working with multiple, overlapping constructs

Methodological issue 1: Selecting a review approach

As we continued to explore the literature, it became apparent that adaptive expertise was a growing area, across a number of education fields. At the same time, we located only a few empirical studies on adaptive expertise in teacher education. Thus, the first methodological issue that we faced was *whether to undertake a systematic review or a scoping review*. Systematic review is the most widely-used research synthesis method. However, scoping review is rapidly emerging as an appealing methodology for those researchers looking for a more flexible research synthesis approach that is suitable to apply to diverse and evolving literature landscapes (Tricco et al., 2016). The two approaches are similar in both scale and rigour, with both of these research synthesis methodologies constituting comprehensive and systematic approaches to reviewing the literature on a chosen topic (Hanneke et al., 2017). At a broad level, "if the intention of the review is to inform clinical decision making…then a systematic review is more appropriate" while "scoping reviews…are more appropriate to assess and understand the extent of the knowledge in an emerging field or to identify, map, report, or discuss the characteristics or concepts in that field" (Peters et al., 2020, p. 5). More specifically, systematic reviews seek to find, critique and synthesise all available evidence for a precise and focused question to determine the effectiveness of interventions, or to prepare for a meta-analysis, to support decision making and development of best practice, and to make recommendations for further research. Systematic reviews are also exhaustive, transparent and must be reproducible, to enable updating and verification (Tricco et al., 2018). In contrast, scoping reviews are broader in scope, and seek to explore and map overall themes and gaps in the available research literature on a broad topic; they may be used to scope further research and determine the possible usefulness of a future systematic review; and involve searching that is systematic and transparent, but with criteria modified as the review proceeds (Tricco et al., 2018). Table 16.1 summarises the different nature and purposes of systematic reviews and scoping reviews which we carefully considered in deciding how to best proceed.

Table 16.1: The difference between systematic reviews and scoping reviews based on Peters et al. (2020) and Arksey and O'Malley (2005)

	Systematic review	**Scoping review**
Title, research question, inclusion criteria	Precise question and prior clear inclusion criteria based on PICO mnemonic (population, intervention, comparator and outcome) to determine effectiveness of intervention e.g., *What is the effectiveness of Nissen fundoplication (intervention) in comparison to anterior partial fundoplication (90 degree, 120 degree and 180 degree) and posterior 270 degree fundoplication (comparator) in terms of symptom control of gastro-esophageal reflux disease (population), and what are the side effects of these surgical interventions (outcome)?* (Marshall-Webb et al., 2018)	Broad question and editable inclusion criteria based on PCC mnemonic (population, concept and context) to overview or map of concept e.g., *What quality-of-life questionnaires (concept) are available for paediatric patients (population) following tonsillectomies with or without adenoidectomies for chronic infections or sleep disordered breathing?* (Kao et al., 2017)
Identifying studies	Searching all the evidence to minimise bias with strict inclusion criteria	Searching available evidence with expansive inclusion criteria

Synthesis findings	Synthesis of the evidence with critical appraisal for assessing the quality of evidence	Overview or a map of the available evidence without critical appraisal as not necessary to assess the quality of evidence

As discussed above, systematic reviews follow a linear process by which the research question and problem is clearly defined, followed by the identification of inclusion and exclusion criteria to enable a comprehensive search of all relevant literature to be evaluated and synthesised as part of the research process. However, this linear process did not suit our research intention to clarify the meanings and characteristics of the construct of adaptive expertise and how it has been employed in empirical studies across a number of research areas in the field of education. The scoping review, with its flexible and iterative approach to searching and reviewing, seemed better aligned with our overall research purpose. It allowed us to start with a broadly framed research question *'How is adaptive expertise conceptualised and operationalised?'* that encompasses a number of research areas and which could then be further refined as part of the scoping review process to focus on empirical studies.

Methodological issue 2: Identifying reviewing processes

Following our decision to conduct a scoping review, we chose to follow the guidelines of Arksey and O'Malley (2005) because this is the most widely used and accepted model. They outline six foundational stages involved in a scoping review:

1. Identifying the research question
2. Identifying relevant studies
3. Study selection
4. Charing the data (i.e., equivalent of data extraction in systematic review)
5. Collating summarising and reporting the results
6. A consultation exercise involving key stakeholders (optional, but highly recommended).

As the review process progressed, it became clear that there was a need to add further clarity to the steps we were undertaking. To facilitate the team reviewing process, the first three stages were recast as six steps: a) identifying the research questions and purposes; b) organising the reviewing process; c) identifying key search terms and search strategies; d) deciding on the inclusion and exclusion criteria; e) screening titles and abstracts for study inclusion; and (f) reviewing full texts.

A significant challenge emerged in the early steps of the reviewing process: *how to strike a suitable balance between breadth, depth and comprehensiveness to account for the use of adaptive expertise in the literature?* Our collective and developing understanding of this vast and evolving body of literature in different education fields led us to regularly revisit our broad research question, to clarify our purposes and further refine our question. In reviewing and discussing selected articles that often represent a variety of approaches undertaken in the field, the research team became clearer about the focus of the review and its purpose. These discussions resulted in further refining of our search strategies and revising our inclusion and exclusion criteria, which are essential steps in undertaking a scoping review. As such, even though the review was well planned and systematic, it was not predetermined or linear, but rather an iterative and ongoing refinement process.

Considering that we were new to scoping reviews, we were constantly revising our understanding and enactment of this approach, similar to those who are experienced in undertaking scoping reviews (Peters et al., 2020). As such, we consulted the PRISMA-ScR (Tricco et al., 2018), which is built on the foundations of Arksey and O'Malley's (2005) guidelines and the elaboration of this foundational work by Levac and colleagues (2010). As the scoping review process unfolded, it became clear to us that scoping review as a methodology itself is undergoing constant revision in an iterative process; it is a methodology in flux. As a result, there is a need for a broad discussion of methodological rigour mixed with flexibility when undertaking research syntheses and in particular scoping reviews.

Methodological issue 3: Applying purposeful sampling strategies

Rather than identifying an exhaustive and well-defined body of evidence that addresses a precise question, scoping review involves purposeful sampling of 'information-rich' cases to inform the mapping of an emerging research area (Suri, 2011), such as teacher adaptive expertise, in response to a broad question. This search for information-rich cases was reflected in *our framing of the purposes for a scoping review of adaptive expertise*: to examine the types of available evidence; clarify key concepts and characteristics associated with this construct; and identify research gaps in the existing literature. The focus on informative cases was also reflected in the three subsequent research questions that evolved through our review process and that guided our review: 1) How is adaptive expertise conceptualised, developed and measured in the context of education? 2) Which aspect of the adaptive expertise is domain specific and which aspect is domain general? 3) What implications can we draw from this review to inform our understanding of teacher adaptive expertise in interdisciplinary science and mathematics?

This purposeful sampling strategy allowed us to be selective in our use of search terms and strategies. Our selection was based on our initial review of seminal papers such as the one by Hatano and Inagaki (1984) and a few other key authors in teacher education studies, such as Carbonell et al. (2014) and Yoon (2015, 2019). The selection of search terms was further refined as we screened the titles and abstracts resulting from our initial searches. Key terms, such as adaptive teaching, adaptability, adaptive performance and reflective expertise, emerged during the initial review and screening process. This is because authors tended to associate teacher adaptive expertise with these terms in the conceptualisation of their research, which appeared in the title or the keyword list of the articles.

Our search strategy (Figure 16.1) consisted of systematically scoping three databases; Scopus, ERIC and Web of Science. This resulted in 484 studies which we imported into Covidence (https://www.covidence.org/), an online systematic review management system (Babineau, 2014), for possible screening, with 159 duplicates removed and 324 studies then selected for the first stage of screening (using title and abstract).

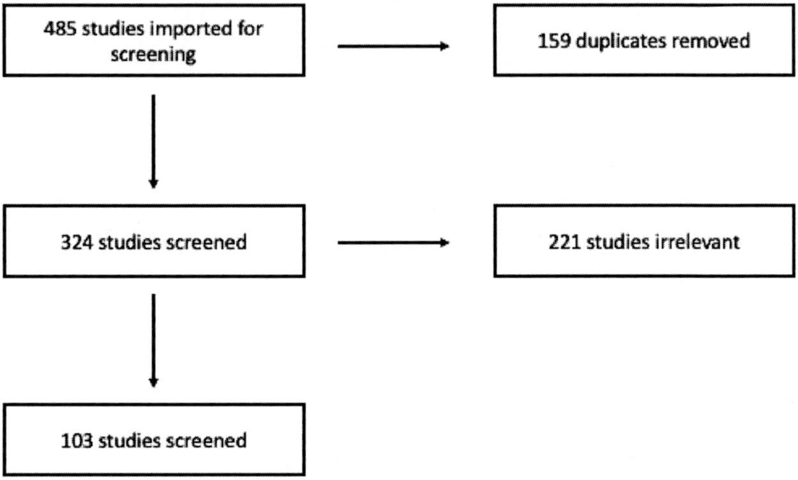

#

Figure 16.1: The different selection steps in the first stages of the scoping review

This purposeful sampling strategy was also employed in *deciding on the inclusion and exclusion criteria.* We determined the following six criteria (Table 16.2) through an iterative process as we applied the criteria to the literature and then met as a group to discuss the suitability of particular criteria and how to apply these in a consistent way. The first iteration is shown in regular font, the second iteration additions are shown in <u>underlined</u> font, and the third iteration additions are shown in **bold** font.

Table 16.2: The inclusion and exclusion criteria for the scoping review

Criterion	Inclusion	Exclusion
1. Source	Peer-reviewed journals	Conference papers/proceedings, book chapters, non-peer-reviewed journals
2. Language	English	Non-English
3. Type of publication	Empirical research	Literature reviews, editorials, position papers or commentaries

4. Research focus	Adaptive expertise is a substantial focus **Adaptive teaching is clearly conceptualised and operationalised**	Adaptive expertise is only mentioned in passing Adaptive systems **Adaptive teaching is mentioned but without clear conceptualisation**
5. Subject domain	All	All
6. Participants	Pre-service and in-service teachers Teachers of all levels (e.g., schools, university) and fields (e.g., medical, engineering)	School and university students

As an example of how we justified these inclusion and exclusion criteria, we determined that only empirical studies were appropriate for our scoping review due to the ambiguity in defining and using this term in non-academic literature (e.g., policy documents). In addition, our interest in adaptive expertise as it plays out in actual classrooms and education settings meant that literature reviews and conceptual papers were not relevant beyond constituting background for our research synthesis, and therefore these papers were excluded from this review.

Through the title and abstract screening process, we continually discussed the definition and terminology associated with adaptive expertise. Although adaptive expertise appears in the titles and abstracts of searched studies, other studies use the term 'adaptive teaching', with a similar meaning to adaptive expertise. In order to manage this situation, we selected several studies on adaptive teaching, and two or three researchers screened each article and discussed how the term was explained and applied in the paper. Consequently, inclusion and exclusion criteria regarding 'adaptive teaching' were added to the search approach as part of our incremental process of purposeful sampling and reviewing.

Methodological issue 4: Working with multiple, overlapping constructs

Through our search process, we found that adaptive expertise studies span a broad range of education contexts, from schools to higher education and in-service learning of professionals. Also, we discovered that different terms other than adaptive expertise were used to refer to the same or similar ideas. In some cases, the use of different terminologies is just a matter of preference of lexicon use. For example, 'adaptive teaching competency' (Vogt & Rogalla, 2009) and 'adaptive teaching expertise' (Soslau, 2012) are defined similarly, as a teachers' ability to manage and adjust planning and teaching to individual learning processes and activities. Yet, in other cases, the use of different terms reflect different meanings and epistemological orientations. For example, the term 'adaptive teaching' (Corno, 2008) tends to be widely used in literacy and mathematics education research, that focuses on the social dynamics of the classroom, and refers to how practising teachers identify and address student differences related to learning (Corno, 2008; Parsons et al., 2018). This contrasts with how adaptive teaching is referred to in mathematics education as "in-the-moment (Wager, 2014) or on-the-fly decision making (Sherin et al., 2008) or even as a teachable moment (Muir et al., 2010)" (Gallagher et al., 2020, p. 6). The term adaptive expertise is more frequently used in medical and health education, although emphases differ in that medical education tends to focus on clinical reasoning abilities of physicians and in health education adaptive expertise is associated with a philosophy of care that supports a patient-centred approach.

These different lenses and perspectives on adaptive expertise may derive from their differing epistemic alignments, for example a more cognitive perspective (e.g., medical education) in contrast with a socio-cultural perspective (e.g., school education and some health education). The plurality of perspectives, emerging from different foundational fields and epistemological positions, raised a dilemma for us as researchers: To what extent should we aim to converge the different definitions and terms to resolve the differences between the variations we discovered? Is it possible or even desirable to bring them together? What is lost and what is gained in doing so? How might we respect the diversity of perspectives, yet advance scholarship of the field? At the point of writing this chapter, we are still working through these issues.

Conclusion

In this chapter, we have chronicled our journey of applying a research synthesis methodology of scoping review to explore research literature on adaptive expertise in diverse education fields. In searching for a suitable methodology for this research synthesis, we have documented our discussion and decision making in relation to four methodological issues: selecting a review approach, identifying reviewing processes, applying sampling strategies and working with overlapping constructs. Several issues warrant further discussion.

First, during this process, it occurred to us that scoping reviews are underpinned by different methodological principles compared to systematic reviews. Scoping review methodology lends itself to an inductive approach to research and aligns better with purposeful sampling, instead of a comprehensive synthesis of evidence that is used in systematic reviews. In our case, we are interested in learning more about the construct of adaptive expertise: how is it used, in what literature, by whom and for what purposes? The question then arises as to what is the basis of our purposeful sampling? Since research into adaptive expertise is still emerging and constantly evolving in different fields, researchers tend to use different terms related to adaptive expertise and for different purposes. An ongoing challenge facing the research team is how we can work with and resolve the tensions in the varied conceptualisations of adaptive expertise in this diverse body of literature.

Second, it is important to recognise the significant challenge of striking a balance between breadth, depth and comprehensiveness in undertaking a scoping review, such as accounting for the use of adaptive expertise in the literature in our case. In our review, we have been constantly refining our research questions and clarifying purposes in order to reflect our collective and evolving understanding of the literature. While we started with Arksey and O'Malley's (2005) steps for a scoping review and carried these out, as the process unfolded, we found ourselves engaging in an iterative process of finding additional literature, refining search strategies when encountering new ideas or uncertainties, reviewing articles, and then revising the inclusion and exclusion criteria. Our methodology for undertaking the scoping review evolved in line with standard practice, extending from Arksey and O'Malley (2005) to Levac and colleagues (2010) and PRISMA-ScR (Tricco et al., 2018) as we progressed with the review. This iterative process helped the research team to clarify the scope of the review and develop a shared understanding of the literature under review and how to review it.

Moreover, important challenges were also raised in regard to the role of the reviewers. Among a team of five reviewers, an issue emerged as to how we could work efficiently to scope a large number of studies identified in the initial search (324 studies) and the significant number of studies selected for full-text review (103 studies). We needed to divide labour in an equitable and productive way to ensure consistent and shared understandings amongst the team (converging on a protocol) that had to emerge from the different opinions of the reviewers. It was decided early on that two reviewers would be (randomly) allocated to each article for both abstract and full text screening processes, enabled by the use of Covidence. When encountering conflicts in review decisions, these articles were re-evaluated and resolved together in a follow-up team discussion involving all reviewers. This process enabled the resolving of conflicts in opinions when they arose through a communal process. While the issues associated with working in a sizeable team of reviewers is not exclusive to scoping reviews, the iterative and more flexible process of the scoping review meant that there was a great deal of coordination and negotiation needed amongst team members.

Finally, we agree with Peters et al. (2020) that researchers undertaking scoping reviews need to continue to work to improve the clarity and usefulness of the standards for undertaking such reviews. In undertaking a scoping review, we must accept there is an element of methodological uneasiness that we have to endure as part of the review process. However, this also comes with the flexibility to adjust our review practises as we proceed with the synthesis process, which ultimately makes for research syntheses that are more responsive to changes in the literature and in the practises of the researchers themselves.

Implications for future research in STEM education

Our research project focuses on teacher adaptive expertise in the field of STEM Education. The STEM education community needs to acknowledge the rapidly evolving nature of the field as reflected in the literature. Our scoping reviews ought to lead to systematic reviews, as a body of relevant literature is accumulated, that can inform teaching and learning practice as well as our own research practices. Currently, it is medical researchers and medical educationalists who are driving the scoping review methodology agenda, much has occurred with the adoption of the ‘what works’ agenda by education researchers from the medical field (Pampaka et al., 2016). As education researchers and more specifically STEM education researchers, we need to actively participate in ongoing discussions concerning the definition of, and protocols for undertaking, research syntheses in general, and scoping reviews, in particular. This participation could be in the form of formal as well as informal discussions with others who are also engaging in research syntheses, and/or contributing to scoping review working groups (such as the scoping review working group at JBI). This type of participation could contribute to shaping the structure and implementation of PRISMA-ScR, and in ways that benefit education research. We anticipate that through sharing our methodological experiences of working with a scoping review in this chapter, that we can stimulate discussion on the processes and outcomes of this particular qualitative review process, and provide insights on some aspects of the process that can improve the methodological practices and experience for others.

References

Arksey, H. & O'Malley, L. (2005). Scoping studies: Towards a methodological framework. *International Journal of Social Research Methodology, 8*(1), 19-32.

Babineau J. (2014). Product review: Covidence (systematic review software). *Journal of the Canadian Health Libraries Association/ Journal de l'Association des bibliothèques de la sante´ du Canada, 35*(2), 68-71.

Burkhardt, H. & Schoenfeld, A. H. (2003). Improving educational research: Toward a more useful, more influential, and better-funded enterprise. *Educational researcher, 32*(9), 3-14.

Corno, L.Y.N. (2008). On teaching adaptively. *Educational Psychologist, 43*(3), 161-173.

Carbonell, K. B., Stalmeijer, R. E., Könings, K., Segers, M. R., & van Merrienboer, J. J. G. (2014). How experts deal with novel situations: A review of adaptive expertise. *Educational Research Review, 12*, 14-29.

Gallagher, M. A., Parsons, S. A., & Vaughn, M. (2020). Adaptive teaching in mathematics: A review of the literature, *Educational Review*. DOI: 10.1080/00131911.2020.1722065.

Grimshaw J. (2020). A guide to knowledge synthesis: A knowledge synthesis chapter. Retrieved from https://cihr-irsc.gc.ca/e/41382.html

Hammerness, K., Darling-Hammond, L., & Bransford, J. (2005). How teachers learn and develop. In L. Darling-Hammond & J. Bransford

(Eds.), *Preparing teachers for a changing world: What teachers should learn and be able to do* (pp. 358-389). Jossey-Bass.

Hanneke, R., Asada, Y., Lieberman, L., Neubauer, L., & Fagen, M. (2017). The scoping review method: Mapping the literature in 'structural change' public health interventions. In *SAGE Research Methods Cases*. https://dx.doi.org/10.4135/9781473999008.

Hatano, G. & Inagaki, K. (1984). Two courses of expertise. *Research & Clinical Center for Child Development*, *82-83* (Ann Rpt), 27-36.

Kao, S. S., Peters, M. D., & Ooi, E. H. (2017). Pediatric tonsillectomy quality of life assessment instruments: A scoping review protocol. *JBI Evidence Synthesis*, *15*(5), 1222-1227.

Levac, D., Colquhoun, H., & O'Brien, K.K. (2010). Scoping studies: Advancing the methodology. *Implementation Science, 5*(69), 1-9.

Marshall-Webb, M., Peters, M. D., Bright, T., & Watson, D. I. (2018). Effectiveness of Nissen fundoplication versus anterior and posterior partial fundoplication for treatment of gastro-esophageal reflux disease: A systematic review protocol. *JBI Evidence Synthesis, 16*(5), 1095-1102.

Muir, T., Beswick, K., & Williamson, J. (2010). Up, close and personal: Teachers' responses to an individualised professional learning opportunity. *Asia-Pacific Journal of Teacher Education, 38*(2), 129-146.

Pampaka, M., Williams, J., & Homer, M. (2016). Is the educational 'what works' agenda working? Critical methodological developments. *International Journal of Research & Method in Education, 39*(3), 231-236.

Parsons S. A., Vaughn M., Scales, R. Q., Gallagher, M. A., Parsons, A. W., Davis, S. G., Pierczynski, M., & Allen, M. (2018). Teachers' instructional adaptations: A research synthesis. *Review of Educational Research, 88*(2), 205-242.

Peters, M. D. J., Marnie, C., Tricco, A. C., Pollock, D., Munn, Z., Alexander, L., McInerney, P., Godfrey, C. M., & Khalil, H. (2020). Updated methodological guidance for the conduct of scoping reviews. *JBI Evidence Synthesis,18(*10), 2119-2126.

Sherin, M. G., Russ, R. S., Sherin, B. L., & Colestock, A. (2008). Professional vision in action: An exploratory study. *Issues in Teacher Education, 17*(2), 27.

Soslau, E. (2012). Opportunities to develop adaptive teaching expertise during supervisory conferences. *Teaching and Teacher Education, 28*(5), 768-779.

Suri, H. (2011). Purposeful sampling in qualitative research synthesis. *Qualitative Research Journal, 11*(2), 63-75.

Suri, H. (2020). Ethical considerations of conducting systematic reviews in educational research. In O. Zawacki-Richter, M. Kerres, S. Bedenlier, M. Bond, & K. Buntins (Eds.), *Systematic reviews in educational research* (pp. 41-54). Springer VS.

Suri, H. & Clarke, D. (2009). Advancements in research synthesis methods: From a methodologically inclusive perspective. *Review of Educational Research*, *79*(1), 395-430.

Timms, M. J., Moyle, K., Weldon, P. R., & Mitchell, P. (2018). *Challenges in STEM learning in Australian schools.* Australian Council for Educational Research.

Tricco, A. C., Lillie, E., Zarin, W., O'Brien, K., Colquhoun, H., Kastner, M., Levac, D., Ng, C., Sharpe, J. C., Wilson, K., Kenny, M., Warren, R., Wilson., C., Stelfox, H. T., & Straus, S. E. (2016). A scoping review on the conduct and reporting of scoping reviews. *BMC Medical Research Methodology*, *16*(1), 1-10.

Tricco, A. C., Lillie, E., Zarin, W., O'Brien, K. K., Colquhoun, H., Levac, D., Moher, D., Peters, M. D. J., Horsley, T., Weeks, L., Hempel, S., Akl, E. A., Chang, C., McGowan, J., Stewart, L., Hartling, S., Aldcroft, L., Wilson, M. G., Garritty, C., Lewis, S., Godfrey, C. M., Macdonald, M. T., Langlois, E. V., Soares-Weiser, K., Moriarty, J., Clifford, T., Tunçalp, O., & Straus, S. E. (2018). PRISMA extension for scoping reviews (PRISMA-ScR): Checklist and explanation. *Annals of Internal Medicine*, *169*(7), 467-473.

Von Esch, K. S. & Kavanagh, S. S. (2018). Preparing mainstream classroom teachers of English learner students: Grounding practice-based designs for teacher learning in theories of adaptive expertise development. *Journal of Teacher Education*, *69*(3), 239-251.

Vogt, F. & Rogalla, M. (2009). Developing adaptive teaching competency through coaching. *Teaching and Teacher Education, 25*(8), 1051-1060.

Wager, A. A. (2014). Noticing children's participation: Insights into teacher positionality toward equitable mathematics pedagogy. *Journal for Research in Mathematics Education, 45*(3), 312-350.

Yoon, S. A., Koehler-Yom, J., Anderson, E., Lin, J., & Klopfer, E. (2015). Using an adaptive expertise lens to understand the quality of teachers' classroom implementation of computer-supported complex systems curricula in high school science. *Research in Science & Technological Innovation, 33*(2), 237-251.

Yoon, S. A., Evans, C., Miller, K., Anderson, E., & Koehler, J. (2019). Validating A Model for Assessing Science Teacher's Adaptive

Expertise with Computer-Supported Complex Systems Curricula and Its Relationship to Student Learning Outcomes. *Journal of Science Teacher Education*, *30*(8), 890-905.

Lihua Xu (Bed, MEd, PhD) - Deakin University, School of Education, 75 Pigdons Rd, Waurn Ponds, VIC 3220, Australia. lihua.xu@deakin.edu.au
Lihua Xu is a Senior Lecturer in Education at Deakin University, Melbourne. Her research aims to impact on and improve teaching and learning practices in science and mathematics. She has researched and written extensively on the complexity of science and mathematics teaching practices in classrooms around the world, the conditions for innovative and sustained curriculum practices in schools, and student reasoning within and across the disciplinary boundaries of science and mathematics.
https://orcid.org/0000-0003-3292-1296

Kennedy Kam Ho Chan (BSc, Mphil, PhD) - The University of Hong Kong, Faculty of Education, Pok Fu Lam Road, Hong Kong Island, Hong Kong. kennedyckh@hku.hk
Kennedy Chan is a science education researcher and teacher educator with a strong commitment to understanding and developing science teachers' expertise. His research interests include science teacher professional knowledge and the use of video in teacher education. He is interested in how science teachers develop their professional knowledge for high-quality science teaching and how they develop innovative strategies to support teacher professional learning from their teaching experience.
https://orcid.org/0000-0003-1063-2813

Amanda Berry (PhD) - Faculty of Education, Monash University, 29 Ancora Imparo Way, Clayton, VIC 3800, Australia.
amanda.berry@monash.edu
Amanda Berry is a Professor in STEM Education in the School of Curriculum, Teaching and Inclusion in the Faculty of Education, Monash University. Her research focuses on the development of science teachers' knowledge and the ways in which that knowledge is shaped and articulated through teacher preparation, beginning teaching and in-service learning. Amanda holds current positions as Editor of the journal *Studying Teacher Education*, Associate Editor of *Research in Science Education* and Convenor of the STEM Special Interest Group of the Australian Association for Research in Education.
https://orcid.org/0000-0002-9902-7582

Joseph Paul Ferguson (BSc, BA(Hons), MTeach, PhD) - Deakin University, School of Education, Faculty of Arts & Education, 221 Burwood Highway, Burwood, VIC 3125, Australia.
joe.ferguson@deakin.edu.au
Joseph is an educational researcher and teacher educator interested in investigating the various manifestations of reasoning in the science classroom, in particular creative reasoning. Through an exploration of the philosophies and theories that underpin inquiry practices and the associated methodologies that enable this research, Joseph seeks to add to efforts to clarify what it means for teachers and students to enact inquiry in the classroom. Joseph is particularly interested in the affordances of video-based methodologies, informed by film theory and philosophy, to maximise the potential of such research to support teachers and students in their teaching and learning of science.
https://orcid.org/0000-0002-0971-3256

Gahyoung Kim (BScE, MPhys, PhD) - Monash University, School of Curriculum Teaching and Inclusive Education, Wellington Rd, Clayton, VIC 3800, Australia. Jinny.kim@monash.edu.au
Gahyoung is educational researcher interested in designing and implementing community-based education in formal and informal setting on Socioscientific Issues. Especially, engaging students in learning and applying the knowledge to everyday life as an activist in their community. Gahyoung has the diverse teaching experience for pre-service teachers and informal educators. Also, her research is included quantities and qualitative methodologies.
https://orcid.org/0000-0002-3583-2101

Jan van Driel (PhD) - The University of Melbourne, Melbourne Graduate School of Education, 100 Leicester Street, Carlton, VIC 3010, Australia. j.vandriel@unimelb.edu.au
Jan van Driel is a Professor of Science Education and leader of the Mathematics, Science & Technology Education Group in the Melbourne Graduate School of Education at the University of Melbourne. His research interests include science teacher knowledge, teacher education and professional learning, science and gender, and interdisciplinary science and STEM education. Currently, he is co-editor-in-Chief of the *International Journal of Science Education*, a member of the Education Committee of Council of the Australian Academy of Science and of the co-design team of

the Primary Mathematics and Science Program of the Dept. of Education and Training in Victoria.
https://orcid.org/0000-0002-8185-124X

Colleen Vale (PhD) - Faculty of Education, Monash University, 29 Ancora Imparo Way, Clayton, VIC 3800, Australia. Colleen.Vale@monash.edu
Colleen Vale is Professor of Mathematics Education in the School of Curriculum, Teaching and Inclusion in the Faculty of Education at Monash University. Equity and social justice in mathematics education is a focus of her research and scholarly activity. Her research interests and expertise include gender equity and digital technologies in mathematics education, out-of-field teaching in STEM, learning and teaching mathematical reasoning, interdisciplinary teaching in STEM and school mathematics leadership.
https://orcid.org/0000-0002-9070-7071

Wanty Widjaja (BEd, MEd, PhD) - Deakin University, School of Education, 221 Burwood Highway, VIC 3125, Australia.
w.widjaja@deakin.edu.au
Wanty Widjaja is an Associate Professor in Mathematics Education at Deakin University. Her research activity aims to transform pedagogical approaches in mathematics classrooms as evident in the major foci: Lesson Study, mathematical reasoning and problem solving, and interdisciplinary approaches in teaching mathematics and science in primary and secondary schools.
https://orcid.org/0000-0002-7288-6088

CHAPTER 17

DIGITAL BIBLIOMETRIC VISUALISATION TOOLS: CONSTRUCTING MAPS TO NAVIGATE THE LITERATURE

MELINDA KIRK AND
JOSEPH PAUL FERGUSON

Abstract

Digital bibliometric visualisation tools (DBVT) are dynamic, continually evolving and provide a productive approach to mapping and navigating research literature. Although an increasingly applied method and tool for academics, there is limited research focused on DBVT and their application in conducting scoping reviews. In this chapter, we explore the affordances and limitations of DBVT by considering their application within a PRISMA for Scoping Review Research (PRISMA-ScR) framework. We explore three DBVT–*ConnectedPapers*, *Litmaps* and *Scite*–as applied to the research literature on creativity in science education, to determine how these tools can be used to complete an effective scoping review. We suggest possible methodological implications for both experienced and neophyte researchers in STEM education in using DBVT to discover new (research) territory and get a lay of the (literature) land, with the aim to find research direction with the result of identifying research opportunities to ultimately enable new discoveries.

Keywords: Scoping review, bibliometric mapping, visualisation, navigation

> ...In that Empire, the Art of Cartography attained such Perfection that the map of a single Province occupied the entirety of a City, and the map of the Empire, the entirety of a Province. In time, those Unconscionable Maps no longer satisfied, and the Cartographers Guilds struck a Map of the Empire

> whose size was that of the Empire, and which coincided point for point with it. The following Generations, who were not so fond of the Study of Cartography as their Forebears had been, saw that that vast Map was Useless, and not without some Pitilessness was it, that they delivered it up to the Inclemencies of Sun and Winters. In the Deserts of the West, still today, there are Tattered Ruins of that Map, inhabited by Animals and Beggars; in all the Land there is no other Relic of the Disciplines of Geography. (Borges, 1946/1999, p. 325)

Literature as a field to be mapped

As science education researchers, we spend our time and energy immersed in various fields. To be clear, this is not just 'field' as a bounded body of knowledge and practises, but more importantly field in an ethnographic sense; places full of people, non-human organisms and abiotic components, ideas, and possibilities (Amit, 2000). These sociocultural, physical and research fields never simply pre-exist in totality for the researcher to discover, but rather are constructed as the researcher undertakes fieldwork (i.e., encountering people, objects) (Goffman, 1989). For us, this includes the context of our research; school classrooms (both physical and virtual, as we have become accustomed to during the ongoing COVID pandemic), science education centres, and myriad other spatio-temporal localities. If we take a moment to reflect on our experiences, we soon realise that the research literature that we engage with as an essential part of our research practises also constitutes a distinct field, within which we conduct fieldwork as we move within and between texts. In suggesting so, we draw on Kristmundsdottir's (2006) notion of "biography as field" (p. 163) and her proposal that there are "voices in the field" (p. 169) of the texts we read and write. Therefore, literature is a particular territory that we can, and indeed must, traverse to do research as we listen to such voices.

However, in order to meaningfully and productively travel across and through these literature landscapes as a field, it is necessary to map these textual territories. Just as we use field in an ethnographic sense, we use 'mapping' in an ethnographic sense; "maps represent the ethnographer's interpretation of people, places, and events" (Martin, 2019, paragraph 1). And they primarily do so in a visual way; maps are a powerful visual representational form. Without maps, we get lost and become unable to 'hear' the voices of teachers, students, and other science education researchers in the field of literature. It is important to point out that not only does such mapping of literature enable navigation of existing territories–the locating of things as well as the locating of things relative to other things, i.e., connections and links–but these maps can assist with finding research direction and the construction of new literature landscapes. However, as we write this chapter and you read our text, we invoke the spirit of Borges (1946/1999) to always remember that 'the map is not the territory'. As such, our mapping of literature ought to always be selective and purposeful in generating new understandings and practises; not constructing a map so large and detailed that it replicates the field of science education point-for-point (in which case it ceases to be a map). It is therefore necessary that we use (and in some cases develop) appropriate cartographic tools and practises to undertake this mapping–this is first and foremost a methodological challenge.

Digital bibliometric visualisation tools

Scoping reviews have commonly been described as a means to explore and 'map the research' (e.g., Khalil et al., 2021). Particularly in an emerging and dynamic field of research, such as our research focused on creativity in primary science, the mapping of literature is beneficial to gain insight into the direction research is taking and to gain insight into what ground has been covered. The literature on creativity in science education is growing so rapidly that those new to this field often struggle to know 'what's out there' and those established in the field struggle to notice 'what's new.' This is a landscape defined by myriad concepts, theories and approaches that researchers have deployed and continue to implement to understand the what, how, and why of creativity in the specific context of science (Tytler et al., 2020). It is a highly rewarding yet very challenging field to explore–so much literature that is linked in increasingly complex ways–which means that those of us interested in inhabiting and potentially expanding this landscape need to map it. But how do we do so in a visually compelling way such that our maps are able to purposefully inform our research? These need to be maps which enable us to 'see' what is there and from this infer what might yet develop which includes our own potential contributions to the field.

Within the context of our scoping review–creativity in primary science education, which is a newly emerging field of research–digital bibliometric visualisation tools (DBVT) make dynamic mapping possible as they constitute algorithmically driven systems that not only aid the navigation of this rapidly changing field through showing originary publications and the latest trends, but also exposing the apparent links between these publications and trends. We seek to demonstrate how the visual representation of the literature, which DBVT afford, make apparent not only the connectivity of the research but also the identification of possibilities for novel and purposeful research.

Bibliographic maps within our research are employed to aid the identification of relevant literature to be included in our scoping review. We argue that employing large-scale literature databases, bibliographic mapping, and bibliometric visualisation tools (also referred to as science mapping tools) are increasingly adopted tools as they can provide a more reliable and replicable method (Chen & Song, 2019) than manual selection and identification of literature.

Despite the rapid emergence of DBVT, it remains common practice for systematic reviews and scoping reviews to be conducted via databases that present lists of articles (Bales et al., 2019). Databases are commonly explored with lexical searches and the mnemonic for scoping reviews PCC (population, context, concept) used to determine research question considerations and subsequently search parameters (Tricco et al., 2018). Although scoping reviews are increasingly adopted, to date there have been limited examples or research focused on the utilisation of DBVT when conducting a scoping review (e.g., Gessler et al., 2021), with their application often limited to only the charting of literature at the stage of disseminating results. The way in which DBVT can support the identification, correlation, and representation of relevant scholarly publications (Chen & Song, 2019) as a key part of good methodological practice is discussed in this chapter. Methods of literature inclusion employed by these tools include author keyword coupling analysis (AKCA) (Yang et al., 2016); impact factor, with the applicator of the Journal Citation Ranking (JCR); citation performance; and connectivity. Different DBVT identify and arrange literature using different identification strategies. The intent of this chapter is not to explain the bibliometric functionality and programming, or our scoping review results, but rather to outline how three tools–*Connected Papers*, *Litmaps* and *Scite*–can be employed to conduct a scoping review in a methodologically rigorous and generative way. We describe affordances and challenges involved in using DBVT to identify significant literature and get a lay of the literature landscape in the field, in our case creativity in science education.

Scoping reviews

Scoping reviews serve a different purpose from a systematic review; they do not seek to answer a specific question (Oliveira et al., 2019; Tricco et al., 2018), rather they are aimed at generating a landscape of not only significant research or seminal research, but also identifying current and emerging areas of research in the field. They are employed to: discern relevant research and the body of knowledge in a specific field; make generalisations; and identify research gaps rather than develop conclusive synthesised results (Munn et al., 2018). A scoping review is sometimes completed prior to a systematic review to gain insight into specificities that could be further pursued in a systematic review; in this way scoping reviews and systematic reviews form a continuum (Tricco et al., 2018). Munn and colleagues (2018) built on Arskey's and O'Malley's (2005) and Levac's and colleague's (2010) purposes for a scoping review to include: identification of available evidence in a field; clarification of concepts and definitions in the literature; examination of research processes in a particular topic or field; as well as determining research gaps and opportunities (Oliveria et al., 2019). The key difference between a traditional literature review and a scoping review is the latter requires a transparent, replicable method to surmise an unbiased, rather than subjective, overview of the literature (Munn, et al., 2018).

Guidelines for reporting scoping reviews have emerged as a response to their increasing use (Tricco et al., 2018) and an identified need for clear guidelines (Tricco et al., 2016). The Joanna Briggs institute (JBI) published scoping review methodological guidelines in the form of PRISMA-ScR (Preferred Reporting Items for Systematic review and Meta-Analysis extension for Scoping Review) in 2015 (updated in 2017 and again in 2020) (Tricco et al., 2018). PRISMA, developed through various iterations since the 1980s, is an evidence-based minimum set of items or criteria, aimed at helping authors to transparently report a wide array of items in systematic reviews and meta-analyses. Built on Arskey's and O'Malley's (2005) earlier scoping review framework and Levac et al.'s (2010) elaborations, the PRISMA-ScR was developed as an extension of PRISMA to specifically improve the quality and consistency of methods for conducting scoping reviews and to provide a rigorous, transparent methodology for consistent reporting (Tricco et al., 2018). It was originally targeted at medical research; however, it is increasingly adopted in other fields including education. The PRISMA-ScR contains a checklist of 20 essential items and two optional items for inclusion, with specific recognition of the possible inclusion of evidence maps in a scoping review and with 'charting methods' replacing 'synthesis methods' within the PRISMA-ScR published guidelines (Peters et al., 2020).

The inclusion of charting methods and the creation of literature maps or charts in a scoping review is a key feature of this research synthesis approach that can be enabled by DBVT. Purposes for employing DBVT to conduct a scoping review, may differ among researchers. For the novice researcher, the purpose of employing DBVT to complete a scoping review is to identify relevant and key literature, determine the range and nature of the literature to develop a summation of the body of knowledge, and recognise key authors and plausible research foci. For the more experienced researcher, who is constantly striving to home in on what is most novel, the goal of forging connections between different research approaches/ traditions and connecting with peers internationally can be realised through using DBVT (Levac et al., 2010). Experienced researchers apply DBVT to clearly map connections and identify disconnections as well; subsequently research possibilities and opportunities for innovative research can emerge, as researchers seek to fill identified research gaps (Oliveria et al., 2019). Regardless of scoping review intention and the reasons for employing DBVT, we seek in this chapter to illustrate how their application is strongly aligned with the requirements of undertaking a reliable scoping review, through a rigorous and collaborative methodology that not only identifies literature but also enables the generative mapping of the literature landscape, an emerging requirement of, and increasing prerequisite of, a quality scoping review (Peters et al., 2020).

In what follows, we aim to cxplorc thc affordances of DBVT to conduct and present a scoping review in a transparent, reliable, and time-efficient way. We aim to exemplify how DBVT can complement traditional scoping reviews (Arksey & O'Malley, 2005; Levac et al., 2010), bibliometric literature mapping (Fellnhofer et al., 2019), hybrid literature mapping taxonomies (Chen & Song, 2019), and PRISMA-ScR (Tricco et al., 2018).

Bibliographic mapping

Conducting a scoping review has been explicitly likened to a "mapping review" of the literature (Peters et al., 2015, p. 141). Bibliometric mapping of literature is significant to research processes as it seeks to visually represent literature, make apparent research and information connections within a dynamic and continually emerging field of knowledge (Cobo et al., 2011). Bibliographic mapping, evidence mapping, science mapping, or literature mapping are terms employed for the visual representation of science or research in a field (Boyack & Klavans, 2010). We build on Belter's (2012) identification of three forms of bibliographic maps: those that visually represent the network of research based on collaboration networks; those that visually represent publication networks and semantic networks; and those that include contextual citation as part of the visualisation process.

Bibliographic citation is commonly the basis of literature maps, clearly depicting and aiding in the determination of significant, or highly cited, literature (Belter, 2012). The visual representation of the literature based on citation analysis is the result of four possible processes; direct citation, indirect citation, co-citation, and bibliographic coupling. The data and metrics required to conduct these complex citation analyses is made possible through DBVT as employed in our scoping review. DBVT clusters research, employing direct citation and in-direct citation analysis. Multiple generation citations, literature frequently cited together (co-citation) and literature with common citations (bibliographic coupling) identification, are enabled with DBVT searches. Within the DBVT 'maps', such as *Connected Papers* and *Litmaps* employed in our scoping review, literature is represented visually by nodes or seeds (we have adopted the 'seed' terminology for the purpose of our research and chapter). Both direct citation and indirect citation are visually represented, with the size based on the number of citations. The visual co-citation connection within the map, the edge or line drawn between publications that are both cited by the same publications, illustrates the direction or a lack of recognition and direction within the literature, through apparent scholarly connection and literature impact. Relationships and interconnectivity are also illustrated spatially, with the distance between literature depicting the 'distance' or 'closeness,' possible similarity and relatedness of research. Conversely, large distances between the seeds can illustrate disparity of research within the literature map. Smaller distances, and densely populated areas within a map, demonstrate a stronger relationship and interconnectivity of the research (Eitan et al., 2020).

Collaboration maps are based on authorship (i.e., citation) and assist with the identification of key institutions or dominant countries contributing to research. This could be considered a more global map (Bales et al., 2020). Chen and Song (2019) describe the mapping of scientific literature as being either global, local, or a hybrid approach. Global mapping comprises a broad stroke, comprehensive coverage of a discipline. Whereas localised mapping and more granular and hybrid mapping constitutes the overlay of global literature maps. Global level mapping of collaboration networks is beneficial for identifying areas or communities of research within the field, and significant research yet to be connected to a broader community (Bales et al., 2020). The visualisation of relationships between literature and research enables easier identification of research lacking collaboration or affirmation (i.e., limited citation). Blank spaces–'tundra, treeless (seedless) plains'–or less densely populated regions within the map, or individual unconnected seeds, can denote potential openings for the 'seeding' of new research collaborations, research community development, and possibly even new research directions. This is particularly relevant for a scoping review, as it does not seek to appraise the research itself, but rather aims to discover the breadth of relevant research and identify research opportunities (Tricco et al., 2018). Furthermore, authorship-based networks evident through citation analysis spaces and clustering of literature can uncover collaboration opportunities to be considered in relation to one's own research.

DBVT do not singularly rely on citations to establish collaboration networks or to identify relevant literature. Semantic, or key word analysis, is also embedded in their functionality or search options. In using DBVT, you can search via lexical searches, as per traditional Peer Reviewed Electronic Search Strategies (PRESS) (McGowan et al., 2015). Semantic network and publication network bibliographic maps have different methods and benefits. Lexical or semantically based bibliographic map searches are derived from the occurrence of words in networks of literature, based on the title, keyword lists, and abstracts. Like manual keyword searches conducted within traditional literature searches of databases, findings are dependent on, and restricted by, the identification of synonymity, homonymity (Belter, 2012), and truncation (different word endings, plural, and tense variations). Searching via lexical searches is recommended in the PRISMA-ScR (Tricco et al., 2018) and PRESS (McGowan et al., 2015). We argue, however, that this lexical means of searching within traditional research databases (e.g., ERIC) is not as efficient as utilising DBVT. The embedded functionality of keyword similarity searches with DBVT, adds reliability of search and provides a more efficient search tool. The duality of the functionality to create a visual map (e.g., *Litmaps*) of confirmed literature, is also in accordance and compliance with the increasingly recognised PRISMA-ScR and the fundamental intent of a scoping review to map the literature (Khalil et al., 2021).

A new and emerging approach to scoping reviews

In recent years, guidelines and protocols have been developed to support and ensure quality reviews of literature have been established (Colquhoun et al, 2014; Tricco et al., 2018). To navigate our way through the literature, it is necessary to establish a clear and robust process for systematic literature reviews, including a scoping review protocol (i.e., methodology). Pham and colleagues (2014) acknowledge scoping reviews as "systematic maps" (p. 374); we recognise that this denotes the significance of establishing systematic, rigorous, and clear processes and protocols to 'chart' the evidence and 'map the literature.' The need to visually map the literature that is identified for inclusion in a scoping review is an attribute of the increasingly adopted PRISMA-ScR (Page & Moher, 2017).

As with a systematic review, clear parameters (e.g., inclusion/exclusion protocols) and search methods need to be established before commencing a scoping review. Clear protocols are especially important when adopting DBVT to not only ensure replicable methods and reliable literature selection, but also to provide clear guidance–to avoid 'getting lost' in the vast literature landscape that these tools (DBVT) can avail. Furthermore, DBVT connectively enables visual mapping not only in the process of identification, but the presentation of the selected literature in a visual form, which is a requirement of PRISMA-ScR (see Table 17.1).

Scoping review and bibliographic mapping

The PRISMA-ScR (Page & Moher, 2017) stages as well as Arskey's and O'Malley's (2005) scoping review framework and Levac et al.'s (2010) elaborations are widely considered essential in the undertaking of scoping reviews in a methodologically rigorous and generative way. We have extended this framing to consider Chen and Song's (2019) systematic taxonomy of literature mapping, delineated within Fellnohofer's (2019) identified four systematic taxonomy stages, to clearly explicate and establish the positioning and role of DBVT in our scoping review so we can identify, collaboratively review, and visually map, the literature (Table 17.1). It was necessary to analyse these frameworks and stages in conjunction to ensure the plausible application of DBVT to our scoping review. Together these frameworks provide a clear orientation for conducting the scoping review such that it foregrounds and values a literature map, as the interrelationship of the scoping review framing (Arskey & O'Malley, 2005; Levac, et al., 2010), bibliographic mapping, and taxonomy stages (Tricco et al., 2018; Fellnhofer, 2019; Chen & Song, 2019) are interactively considered (Table 17.1). *ConnectedPapers, Litmaps* and *Scite* as DBVT will be explored in the following section.

Table 17.1: Taxonomy interrelationships

PRISMA (ScR) Protocol Structure Summary *(Adapted from Tricco, et.al., 2018)*	**Scoping Review Framework** **Adapted from Arskey's & O'Malleys scoping review framework** with Levac et al.'s enhancements *(Arskey & O'Malley, 2005 p 22-23 and Levac, Colquhorn & O'Brien, 2010, p.4-6)*	**Hybrid literature mapping: Systematic taxonomy stages** *(Adapted from Chen & Song, 2019)*	**Bibliographic mapping and literature visualisation analysis stages in application** *(Adapted from Fellnhofer, 2019 p.26)*
Background **Objectives**	**1.Identifying the research questions** Clarifying and linking the purpose and research questions		
Eligibility criteria **Sources of evidence**	**2. Identify relevant literature** Balancing feasibility with breadth and comprehensiveness **3. Study selection** Using an iterative team approach to select studies and extract data	**1.Peer reviewed publications are visualised**	a) **Data collection and identification of documents** b) **Dataset quality checks:** *Checking Scopus Journal Rating as well as citation index*
Charting methods	**4. Extract map and chart the data** Incorporating a numerical summary and qualitative thematic analysis	**2. Research clusters are identified with visualisation software**	c) **Creation of networking maps and clustered identification based on bibliographic coupling** *using Connected Papers, then repeated with Litmaps.*
Results	**5. Collate summarise and report the results** Identify the implications of the study findings for policy, practice, or research	**3. Clusters are discussed in relationship to other publications, prominent authors, prominent journals, citations**	d) **Creation and analysis of further network maps** *using Litmaps to insert all identified literature, creating a new literature map* e) **Topic cluster interpretation** *using Litmaps create map overlays to represent different themes or focus (key themes, key literature, key research, key terminology).* *Using Scite to analyse and interpret the different types of citations and literature connectivity* *Citavi - to record, collaborate and confirm selected literature (linked to LIT platform directly)*
Conclusions relating to the ScR questions and objectives	**6. Optional consultation and review** (Arskey & O'Malley, 2005) Or adoption of consultation (Levac et al., 2010)	**4.Recommendations for future research**	**Writing of scoping review:** *Dissemination of findings and application to own research*

Establishing our search protocol

Published guidance for conducting systematic scoping reviews, such as Peters and colleagues (2015), advise that the rationale for inclusion criteria should be established and explained in the scoping review background section. However, they also advise that the background section should recognise the existing knowledge in this field. We argue that within a newly emerging field, such as our research exploring creativity in science, this is often the purpose of the scoping review to establish what is already 'out there'. The broadly accepted Participants, Concepts and Context (PCC) approach to framing a scoping review, however, does provide 'orientation,' for not only the research questions but also a foundation for our lexical search parameters. Three different search strategies have been employed to ensure we adequately identify the breadth of literature in this newly emerging research field. We have adopted Chen & Song's (2019) three retrieval strategies: 1) a query-based lexical search, as per traditional data searches and to maximise the citation analysis capabilities of DBVT; 2) a forward expansion article search with an identified seminal paper; and 3) a backward expansion strategy with an identified recently published relevant authoritative paper (which we adopted and extended to a fourth search through the iterative process of key author identification and consideration) (Figure 17.1).

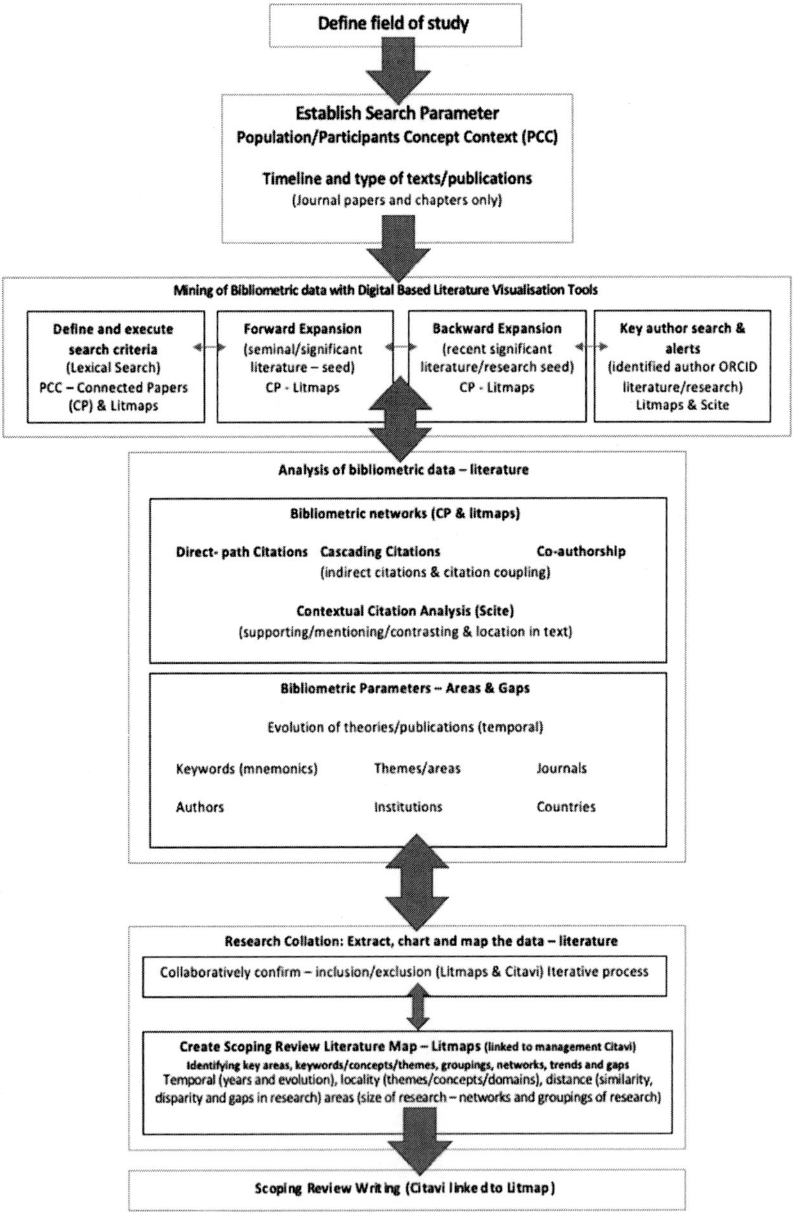

Figure 17.1: DBVT scoping review protocol (CP = *ConnectedPapers*)

ConnectedPapers was employed first in our scoping review to identify relevant literature. *ConnectedPapers* utilises a similarity-based metric of both co-citation and bibliographic coupling, also known as co-references, to identify literature relatedness. This recognises overlapping citations and references to establish research interconnectivity (Eitan et al., 2020). The facilitation of bibliographic coupling that is used by *ConnectedPapers*, enables the identification of not only relevant literature, but also new and emerging research as recent and uncited works can be identified in the search (Boyack & Klavans, 2010). Without bibliographic coupling, relevant recent but uncited research would not be exposed. The *ConnectedPapers* database is Semantic Scholar Paper Corpus, with the search engine primarily focused on title searches, making it suitable for a forward and backward expansion citation search strategy (Chen & Song, 2019) through the insertion of a significant or seminal 'seed' paper. We identified 'Creative little scientists: Exploring pedagogical synergies between inquiry-based and creative approaches in early years science' (Cremin et al., 2015) as a significant and influential paper for our research. This paper aligned with our research inquiry (i.e., PCC), which is student creativity in science in the primary school context. The forward citation of this as a preliminary seed resulted in the identification of relevant literature, as the interconnectivity to other papers was visually displayed or mapped (Figure 17.2).

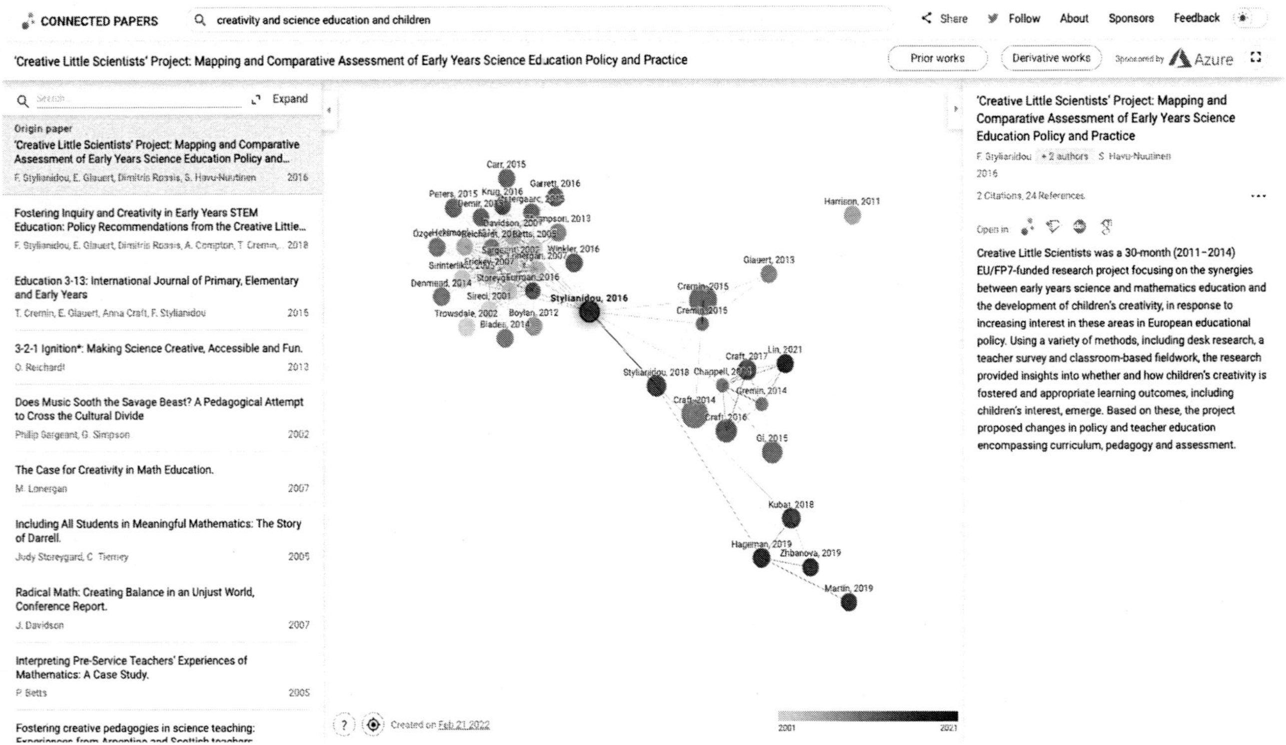

Figure 17.2: Backward expansion citation search shown in *ConnectedPapers*

This step unveiled significant related research, as citations, proximity to the seed article (primary education focused), and the easily seen abstracts and linking to full papers, which made this a time-efficient process to scope the literature as primary education focused STEM literature clustered in proximity, for ease of contextually relevant literature identification. Furthermore, the prior and derivative works functionality of *ConnectedPapers* also provided an additional means to determine seminal, and significant works, as commonly cited publications of papers within the automated relational map are identified. The 'derivative works' function of *ConnectedPapers*, for novice and accomplished researchers alike, enables the researcher to stay abreast of emerging literature, as papers that are cited by recognised research usually survey and report the field or recent relevant publications (Eitan et al., 2020).

Through this forward and backward search method using *ConnectedPapers*, prolific researchers and key authors within the field became apparent (e.g., Anna Craft, Bill Lucas). The significance of literature is immediately apparent through the visual depiction of citations (i.e., size of nodes and citation rate) and lines of connectivity (i.e., literature relatedness–co-citation). Through the connectivity of *ConnectedPapers* to Semantic Scholar, it was also possible to follow identified key authors in the field and receive alerts. Notifications of related new publications can also be enabled within *Litmaps* in connection with the accepted literature. This is aligned with the premise of a scoping review; to be abreast of all relevant research, including emerging and recent literature. We used the functionality within *Litmaps* to search entire key author works, through directly linking and uploading papers from their ORCID profile, for inclusion consideration. As a scoping review seeks to scope all the relevant literature, the capacity to enter the ORCID of influential and significant researchers in the field, aids the efficacy and connectivity of the literature search. The functionality to import already saved publications in EndNote, Mendeley, and Zotero libraries (Litmaps, 2021) also reduces time and enables researcher collaboration, as experienced researchers on the scoping review team may already have identified key literature for consideration.

A flexible and iterative forward expansion and backward expansion (Chen & Song, 2019) approach was completed with *Litmaps*. Duplicating and supplementing the literature search in *Litmaps*, following *ConnectedPapers,* served to triangulate the search, providing a confirmatory process to ensure validity through a repeated process. The ability to add seeds (i.e., literature) to the map within *Litmaps*, unlike other DBVT, enabled ease of not only the identification of literature, but also enabled an iterative process to collaboratively confirm relevance and connectivity of identified literature.

Research collation and reporting stage: Mapping our literature

As *Litmaps* is a collaborative platform, multiple researchers can work simultaneously on this platform to review and confirm literature inclusion and exclusion. Hence, *Litmaps* was employed in our scoping review to both confirm literature and create our own scoping review literature map (Figure 17.3). The best way to map the literature within a scoping review is an identified challenge for many researchers that often requires creative solutions (Khalil et al., 2021). *Litmaps* afforded a generative approach to progressing our scoping review and specifically our literature map. Confirmed literature was inserted to construct our own model of the literature or scoping review map. The *Litmaps* platform allows flexibility in the layout of literature maps, with 288 different possibilities of displaying the literature (Litmaps, 2021). *Litmaps* supports multiple views and visualisation foci. For instance, the x-axis and y-axis can be set to 'title similarity' to focus on the similarity of titles and research. In order to understand citation versus time, the publication date is set to the x-axis and the citations to the y-axis, which aids hierarchical and core literature determination, in conjunction with temporality. The temporal display of the literature, with the most recent literature on the right, enables easy identification of emerging research, concepts and authors within the field, as well as temporal research directionality to determine emerging shifts in focus and knowledge.

Temporal arrangement of literaturc and citation indexes alone, however, do not ensure quality analysis of literature, the realisation of the significance of literature, or 'where' or 'how' the literature fits into the larger 'literature landscape'. Citation numbers alone are a crude measure of literature importance (Rife et al., 2021) and positioning. The consideration of citation counts, co-citation, and citation coupling (*ConnectedPapers* and *Litmaps*), provides surface level or 'datum' of connectivity and meaning. Analysing the 'terrain' or context of the literature is essential to gain insight into the true scope or 'latitude' of the literature in the field. *Scite* is a growing DBVT that can be employed to contextualise citations and literature within the broader field and landscape (Nicholson et al., 2021) to enable the navigation of the literature terrain. Beyond utilising citations metrics (e.g., Impact Factor and H-Index), the *Scite* platform algorithmically performs a nuanced contextual citation analysis, which involves the identification of the location of citations (e.g., introduction, method, results, discussion) and the type of citation (supporting, mentioning, or contrasting) (Nicholson et al., 2021). This is significant as these two features determine literature actual-connectivity and relatedness. The adoption of *Scite* to review and confirm inclusion of literature within our scoping review provided an efficient and comprehensive means of selecting relevant literature and connectivity, supporting the identification of influential research.

Analysis of literature within our scoping review, revealed emerging 'fields' or 'areas' within our search, or themes (e.g., domain specific creativity, pedagogy focused literature, assessment focused literature, culture linked, collaboration linked, critical thinking linked) within our field of research. The adoption of the overlaying capability with *Litmaps*, enabled the creation of a 'choropleth mapping'[1] as thematic areas are coloured differently and the density of distributions of themes/subjects apparent through spatial distancing of colours is made possible (Figure 17.3).

DBVT-enabled scoping reviews: Methodological affordances, challenges & implications

As with traditional scoping review approaches, keyword searches can be performed in DBVT, and this is advantageous in this platform as key-word co-occurrence functionality is embedded and automated within these DBVT, which directly correlate literature in searches to establish literature relatedness (thus increasing search efficiency). We suggest that this embedded feature provides a triangulation of results beyond manual selection of literature. Traditional keyword-based searches alone, have been recognised within scoping methodology critique to be problematic as similar research can have a mismatch of vocabulary.

Beyond literature identification through lexicon and key word searches, literature identification through citation is advantageous. With DBVT, the 'mining' of bibliometric data is possible and automated; direct citations, research citing another paper, relevancy and connections are revealed, even if devoid of like-keywords (Chen, 2018). While direct citation and 'main-path analysis' of literature citations is important and often reflects the direct impact of a work, the inclusion of indirect citations, also enabled by DBVT, reveals citation influence and impact of literature in a broader sense (Boyack

---

[1] We consider it valuable to expand the cartography metaphors used in discussing literature and literature reviews to geology and geomorphology metaphors as they help in making apparent the methodological challenges and affordances of approaches such as DBVT-enabled scoping reviews. A choropleth map is a statistical thematic map that uses intensity of colour to correspond with an aggregate summary of a geographic characteristic within spatial enumeration units.

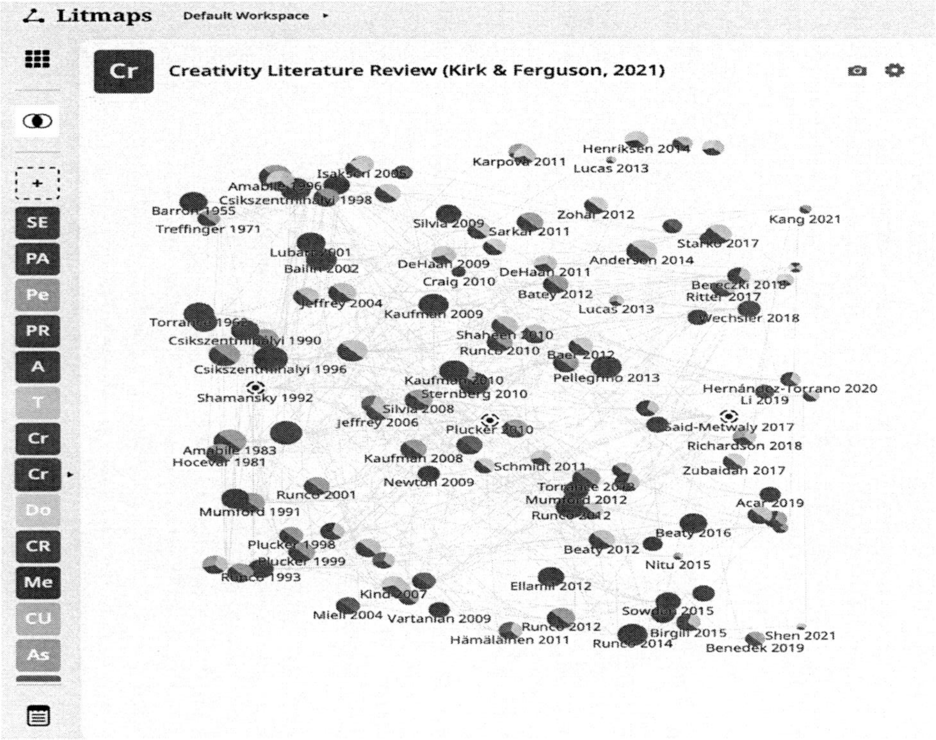

Figure 17.3: Choropleth literature map in *Litmaps*: research themes identified

& Klavans, 2010). Multiple generation citations, or indirect citation identification, as well as bibliographic coupling, are all afforded by DBLT (including *Connected Papers, Litmaps* and *Scite*). This particular means of generating "cascading citations" (Chen, 2018, p. 6) illustrates literature's direct impact, cumulative citations, literature relationships (Wren et al., 2004) as well as the distance, temporality and similarity, of literature.

A flexible forward and backward citation method is particularly beneficial to the beginning researcher who may have a limited knowledge of key authors. For the more experienced researcher, DBVT avails the opportunity to establish research connectivity with key research or one's own research. The ability to add multiple papers to *Litmaps* in the research identification stage, reduces time spent pursuing possibly unrelated literature in other search engines. The display of identified texts, title, abstract, and number of citations, aids the determination of relatedness, applicability, and warrant of full text exploration in a timely manner, which aids the inclusion, exclusion, and selection processes.

The collaborative functionality of *Litmaps*, with the ability for multiple researchers to interact with the tool in simultaneous ways, was advantageous at multiple stages of the scoping review. Collaboration, especially at the inclusion and exclusion stage but also the consultative review stage, is essential to conducting a reliable and compliant scoping review (Peters et al., 2020). *Litmaps* enabled us to undertake: literature searches; inclusion/exclusion criteria determination; map creation; clustering to establish research interconnectivity; overlay to display themes and fields within our literature search; and the visualisation of literature relationships and relevance to our own research (Litmaps, 2021).

The adoption of *Scite* within our methodology elevated our conception of literature mapping to include the consideration of 'topography' within our literature map. An in-depth analysis of the literature landscape as 'bathymetry'–the mapping terminology derived from the Greek words for deep and measure–was enabled, as beyond 'surface level' analysis of citation was made possible. Disputed literature could be likened to 'continental rifting' that can result in 'fault lines' being made apparent, resulting in the possible 'declination' or 'erosion' of knowledge and understanding in a given field, and the identification of a need for research to 'bridge the divide' between theories or between theory and practice. The consideration of 'hypsometry'–the measurement of the height of the land, or the elevation of the literature–is an affordance of *Scite*. Confirmatory citations can elevate literature, 'like mountains'–the 'accumulation,' or 'sedimentation,' building up of citations–and the formation of connected literature or 'mountain ranges'. However this often requires time to emerge, develop, and stabilise.

Although we were able to create our own literature map in *Litmaps*, utilising DBVT alone was insufficient to preclude 'getting lost' in the literature. To aid the 'navigation' and 'direction' of our research, it was necessary to efficiently manage the literature, to ultimately be able to 'plot' our own scoping review. Subsequently, *Citavi*, a cloud-based software, was used to record the synthesis, supporting the screening, coding, and organisation of the literature. *Citavi* was selected as it can directly import references and papers from *Litmaps* via the *Citavi* plugin function. Unlike conventional organisational software (e.g., Excel spreadsheets), *Citavi* doubles as a reference management system. However, distinct from other referencing systems (e.g., Endnote), *Citavi* is collaborative, making it ideal for conducting a scoping review. Similar to platforms such as Mendeley, which is also a widely adopted research collaboration platform, *Citavi* papers can be annotated and shared. *Citavi* enabled not only the recording of our literature but also excerpts, quotes, images, and notes identified by us were extracted to create an automatically referenced draft paper, in our case a scoping review. Map keys or legends are essential to enable the unearthing of information within a map. Within our scoping review, *Citavi* proved to be 'key' to effectively charting our literature map and preparing and writing our DBVT-enabled scoping review.

There's gold in them thar hills![2]

Within our scoping review, DBVT enabled us to map the current lay of the literature landscape that is the field of creativity in science education. It must be noted that our literature map depicts the scope of the current creativity in primary science literature only; the literature 'terrain' and 'landscape' is continually evolving. New territory is continually being created and discovered, and needs to be analysed, with worthy discoveries or literature ascertained and 'charted' to be recognised, in order to heighten the research impact and to provide orientation to ultimately 'bridge the divide' between theory and practice (which is a primary goal of our research).

To effectively chart current literature, we recommend DBVT. Our scoping-literature-hybrid-map process utilised both a global and local approach

[2] This is of course a catch cry of Yosemite Sam in the Looney Tunes cartoons.

(Chen & Song, 2019) to depict and grapple with the structure of creativity literature in STEM, with a focus on primary education. Significant research areas, and their connections, were made apparent through the clear visualisation that the DBVT maps afforded. This included: the size of literature seed (clearly depicting the displays, the uptake and impact of research); the density and distances between seeds (illustrating the closeness and disparity of research); the direction research is taking that is temporally analysed (through the visible timeline of identified literature), and the interconnectivity of research (through the co-citation lines). Through the iterative process of conducting our scoping review, it became apparent, however, that our charting of the literature needed to include 'horizontal datum' (consideration of 'position' or type of literature and the 'distances' or likenesses of research) as well as the 'vertical datum' of the literature (mapping of the 'surface level' heights). However, beyond recognising the 'latitude', or 'elevation'–citation numbers–of the literature, we needed to 'dig deeper' to below the surface level datum, to consider the context of literature citations and research. The adoption of *ConnectedPapers*, *Litmaps,* and *Scit*e together with the management software *Citavi*, enabled us to undertake this 'geological' work. We found 'gold' when we looked beneath the surface datum of the literature in our scoping review. Through citation, contextual, and collaborative analysis, we realised opportunities for ideas to be expanded, adopted, adapted, and even challenged. It is noted that gold nuggets often appear in 'faults,' or for us the consideration of alternative perspectives. We liken the process of conducting our scoping review–searching the literature–to circulating through gold bearing rocks (i.e., rich literature), to reveal and pick up gold and concentrate it in new localities; to forge and find direction towards our own gold, that is innovative and influential research. All of which is not possible without useful maps.

References

Amit, V. (2000). Introduction: Constricting the field. In V. Amit (Ed.), *Constructing the field: Ethnographic fieldwork in the contemporary world* (pp. 1-18). Routledge.

Arskey, H. & O'Malley, L. (2005). Scoping studies: Towards a methodological framework. *International Journal of Social Research Methodology, 8*(1), 19-32.

Bales, M.E., Wright, D., Oxley, P., & Wheeler, T.R. (2019). Bibliometric visualization and analysis software: State of the art, workflows, and best practices. Samuel J. Wood Library, Weill Cornell Medicine.

Belter, C. (2012). Visualizing networks of scientific research. *Online*, *36*(3), 14.

Borges, J. L. (1946/1999). On exactitude in science. In *Jorge Luis Borges, Collected Fictions* (p. 325) (Trans. Hurley, H.). Penguin Books.

Boyack, K. W. & Klavans, R. (2010). Co-citation analysis, bibliographic coupling, and direct citation: Which citation approach represents the research front most accurately? *Journal of the American Society for Information Science & Technology, 61*(12), 2389-2404.

Chen, C. (2018). Cascading citation expansion. *Journal of Information Science Theory and Practice, 6*(2), 6-23.

Chen, C. & Song, M. (2019). Visualizing a field of research: A methodology of systematic scientometric reviews. *PLoS ONE 14*(10).

Cobo, M. J., López-Herrera, A. G., Herrera-Viedma, E., & Herrera, F. (2011). Science mapping software tools: Review, analysis, and cooperative study among tools. *Journal of the American Society for Information Science & Technology, 62*(7), 1382-1402.

Colquhoun, H. L., Levac, D. E., O'Brien, K. K., Straus, S. E., Tricco, A. C., Perrier, L., Kastner, M., & Moher, D. (2014). Scoping reviews: Time for clarity in definition, methods, and reporting. *Journal of Clinical Epidemiology, 67*(12), 1291-1294.

Cremin, T., Glauert, E. B., Craft, A., Compton, A., & Stylianidou, F. (2015). Creative little scientists: Exploring pedagogical synergies between inquiry-based and creative approaches in Early Years science. *Education 3-13, 43*(4), 404-419.

Eitan, A., Smolyansky, E., & Harpaz, I. (2020). Connected Papers. Retrieved from https://www.connectedpapers.com/

Fellnhofer, K. (2019). Toward a taxonomy of entrepreneurship education research literature: A bibliometric mapping and visualization. *Educational Research Review*, *27*, 28-55.

Gessler, M., Nägele, C., & Stalder, B.E. (2021). Scoping review on research at the boundary between learning and working: A bibliometric mapping analysis of the last decade. *International Journal for Research in Vocational Education and Training*, *8*(4), 170-206.

Goffman, E. (1989). On fieldwork. *Journal of Contemporary Ethnography, 18*(2), 123-132.

Khalil, H., Peters, M. D., Tricco, A. C., Pollock, D., Alexander, L., McInerney, P., & Munn, Z. (2021). Conducting high quality scoping reviews: Challenges and solutions. *Journal of Clinical Epidemiology,* 130, 156-160.

Kristmundsdottir, S. D. (2006). Far from the Trobriands? Biography as field. In. S. Coleman & P. Collins (Eds.), *Locating the field: Space, place and context in anthropology* (pp. 163-178). Berg.

Levac, D., Colquhoun, H., & O'Brien, K. K. (2010). Scoping studies: Advancing the methodology. *Implementation Science, 5*(1), 1-9.

Litmaps [Computer software]. (2021). Retrieved from https://app.litmaps.co

McGowan, J., Sampson, M., Salzwedel, D. M., Cogo, E., Foerster, V., & Lefebvre, C. (2016). PRESS Peer Review of Electronic Search Strategies: 2015 Guideline statement. *Journal of Clinical Epidemiology*, *75*, 40–46. https://doi.org/10.1016/j.jclinepi.2016.01.021

Martin, T. (2019). *Ethnographic mapping.* Retrieved from https://cuny.manifoldapp.org/projects/ethnographies-of-work.

Munn, Z., Peters, M. D. J., Stern, C. Tufanaru, C., McArthur, A., & Aromataris, E. (2018). Systematic review or scoping review? Guidance for authors when choosing between a systematic or scoping review approach. *BMC Medical Research Methodology,* 18, 143. https://doi.org/10.1186/s12874-018-0611-x

Nicholson, J. M., Mordaunt, M., Lopez, P., Uppala, A., Rosati, D., Rodrigues, N. P., ... & Rife, S. C. (2021). Scite: A smart citation index that displays the context of citations and classifies their intent using deep learning. *Quantitative Science Studies*, *2*(3), 882-898.

Oliveira, O. J. d., Silva, F. F. d., Juliani, F., Barbosa, L. C. F. M., & Nunhes, T. V. (2019). Bibliometric method for mapping the state-of-the-art and identifying research gaps and trends in literature: An essential instrument to support the development of scientific projects. In S. Kunosic, & E. Zerem (Eds.), *Scientometrics recent advances*. IntechOpen. https://doi.org/10.5772/intechopen.85856.

Page, M. J. & Moher, D. (2017). Evaluations of the uptake and impact of the Preferred Reporting Items for Systematic reviews and Meta-Analyses (PRISMA) statement and extensions: A scoping review. *Systematic Review, 6*(1), 263.

Peters, M. D. J., Godfrey, C. M., Khalil, H., McInerney, P., Parker, D., & Soares, C. B. (2015). Guidance for conducting systematic scoping reviews: *International Journal of Evidence-Based Healthcare*, *13*(3), 141-146.

Peters, M. D. J., Godfrey, C., McInerney, P., Munn, Z., Tricco, A. C., & Khalil, H. (2020). Chapter 11: Scoping Reviews (2020 version). In E. Aromataris & Z. Munn (Eds.), *JBI Manual for Evidence Synthesis*. Available from https://synthesismanual.jbi.global

Pham, M. T., Rajić, A., Sargeant, J. M., Papadopoulos, A., Mcewen, S. A., & Greig, J. D. (2014). A scoping review of scoping reviews: Advancing the approach and enhancing the consistency. *Research Synthesis Methods*, *5*(4), 371–385. https://doi.org/10.1002/jrsm.1123

Rife, S. C., Rosati, D., & Nicholson, J. M. (2021). Scite: The next generation of citations. *Learned Publishing*, *34*, 454-456. https://doi.org/10.1002/leap.1379

Tricco, A. C., Lillie, E., Zarin, W., O'Brien, K., Colquhoun, H., Kastner, M., Levac, D., Ng, C., Sharpe, J. C., Wilson, K., Kenny, M., Warren, R., Wilson, C., Stelfox, H. T., & Straus, S. E. (2016). A scoping review on the conduct and reporting of scoping reviews. *BMC Medical Research Methodology*, *16*(1), 1-10.

Tricco, A. C., Lillie, E., Zarin, W., O'Brien, K. K., Colquhoun, H. L., Levac, D. E., Moher, D., Peters, M. D., Horsley, T., Weeks, L., Hempel, S., Akl, E. A., Chang, C. S., McGowan, J., Stewart, L. A., Hartling, L., Aldcroft, A., Wilson, M. G., Garritty, C. M., Lewin, S., Godfrey, C. M., Macdonald, M., Langlois, E. V., Soares-Weiser, K., Moriarty, J., Clifford, T., Tunalp, Z., & Straus, S. E. (2018). PRISMA Extension for Scoping Reviews (PRISMA-ScR): Checklist and Explanation. *Annals of Internal Medicine, 169*, 467-473.

Tytler, R., Prain, V., Ferguson, J. P., & Cripps Clark, J. (2020). Methodological challenges in researching, teaching and learning student capabilities: The "creativity" case. In P. White, R. Tytler, J. P. Ferguson, & J. Cripps Clark (Eds.), *Methodological approaches to STEM education research – Volume 1* (pp. 1-20). Cambridge Scholars Publishing.

Wren, J. D, Bekeredjian, R., Stewart, J. A., Shohet, R. V., & Garner, H. R. (2004). Knowledge discovery by automated identification and ranking of implicit relationships. *Bioinformatics, 20*(3), 389-398.

Yang, S. L., Han, R. Z., Wolfram, D., & Zhao, Y. H. (2016). Visualizing the intellectual structure of information science (2006–2015): Introducing author keyword coupling analysis. *Journal of Informetrics, 10*(1), 132-150.

Melinda Kirk (BEd, MEd) - Deakin University, School of Education, Faculty of Arts & Education, 221 Burwood Highway, Burwood, VIC 3125, Australia. melinda.kirk@deakin.edu.au
Melinda is a PhD candidate, educational researcher, and sessional lecturer. An experienced educator she has a special interest in primary interdisciplinary and creative education. She is currently working as a Research Assistant on the international, Australian Research Council (ARC) Project, the Interdisciplinary Maths and Science (IMS) Learning Project. How collaborative learning can be cultivated to support critical and creative thinking in an interdisciplinary Primary classroom is the focus of her PhD research.
https://orcid.org/0000-0002-0058-6289

Joseph Paul Ferguson (BSc, BA(Hons), MTeach, PhD) - Deakin University, School of Education, Faculty of Arts & Education, 221 Burwood Highway, Burwood, VIC 3125, Australia.
joe.ferguson@deakin.edu.au
Joseph is an educational researcher and teacher educator interested in investigating the various manifestations of reasoning in the science classroom, in particular creative reasoning. Through an exploration of the philosophies and theories that underpin inquiry practices and the associated methodologies that enable this research, Joseph seeks to add to efforts to clarify what it means for teachers and students to enact inquiry in the classroom. Joseph is particularly interested in the affordances of video-based methodologies, informed by film theory and philosophy, to maximise the potential of such research to support teachers and students in their teaching and learning of science.
https://orcid.org/0000-0002-0971-3256